Lecture Notes in Physics

Springer
Berlin
Heidelberg
New York
Hong Kong
London
Milan
Paris
Tokyo

The Editorial Policy for Edited Volumes

The series *Lecture Notes in Physics* (LNP), founded in 1969, reports new developments in physics research and teaching - quickly, informally but with a high degree of quality. Manuscripts to be considered for publication are topical volumes consisting of a limited number of contributions, carefully edited and closely related to each other. Each contribution should contain at least partly original and previously unpublished material, be written in a clear, pedagogical style and aimed at a broader readership, especially graduate students and nonspecialist researchers wishing to familiarize themselves with the topic concerned. For this reason, traditional proceedings cannot be considered for this series though volumes to appear in this series are often based on material presented at conferences, workshops and schools.

Acceptance

A project can only be accepted tentatively for publication, by both the editorial board and the publisher, following thorough examination of the material submitted. The book proposal sent to the publisher should consist at least of a preliminary table of contents outlining the structure of the book together with abstracts of all contributions to be included. Final acceptance is issued by the series editor in charge, in consultation with the publisher, only after receiving the complete manuscript. Final acceptance, possibly requiring minor corrections, usually follows the tentative acceptance unless the final manuscript differs significantly from expectations (project outline). In particular, the series editors are entitled to reject individual contributions if they do not meet the high quality standards of this series. The final manuscript must be ready to print, and should include both an informative introduction and a sufficiently detailed subject index.

Contractual Aspects

Publication in LNP is free of charge. There is no formal contract, no royalties are paid, and no bulk orders are required, although special discounts are offered in this case. The volume editors receive jointly 30 free copies for their personal use and are entitled, as are the contributing authors, to purchase Springer books at a reduced rate. The publisher secures the copyright for each volume. As a rule, no reprints of individual contributions can be supplied.

Manuscript Submission

The manuscript in its final and approved version must be submitted in ready to print form. The corresponding electronic source files are also required for the production process, in particular the online version. Technical assistance in compiling the final manuscript can be provided by the publisher's production editor(s), especially with regard to the publisher's own LaTeX macro package which has been specially designed for this series.

LNP Homepage (springerlink.com/series/lnp/)

On the LNP homepage you will find:

–The LNP online archive. It contains the full texts (PDF) of all volumes published since 2000. Abstracts, table of contents and prefaces are accessible free of charge to everyone. Information about the availability of printed volumes can be obtained.

–The subscription information. The online archive is free of charge to all subscribers of the printed volumes.

–The editorial contacts, with respect to both scientific and technical matters.

–The author's / editor's instructions.

R. Conte F. Magri M. Musette
J. Satsuma P. Winternitz

Direct and Inverse Methods in Nonlinear Evolution Equations

Lectures Given at the C.I.M.E. Summer School
Held in Cetraro, Italy, September 5-12, 1999

Editor: Antonio M. Greco

Fondazione
C.I.M.E.

Springer

Authors

Robert Conte
CEA, Saclay, Service de Physique
de l'Etat Condensé (SPEC)
91191 Gif-sur-Yvette CX, France

Franco Magri
Università degli Studi Bicocca
Dipartimento di Matematica
Via Bicocca degli Arcimbold 8
20126 Milano, Italy

Micheline Musette
Vrije Universiteit Brussel
Fak. Wetenschappen DNTK
Pleinlaan 2, 1050 Brussels, Belgium

Junkichi Satsuma
University Tokyo, Graduate School
of Mathematical Sciences
Komaba 3-8-1, 153 Tokyo, Japan

Pavel Winternitz
C.R.D.E., Université de Montreal
H3C 3J7 Montreal, Quebec, Canada

Editor

Antonio M. Greco
Università di Palermo
Dipartimento di Matematica
Via Archirafi 34, 90123 Palermo, Italy

C.I.M.E. activity is supported by:
Ministero dell'Università Ricerca Scientifica e Tecnologica, Consiglio Nazionale delle Ricerche and E.U. under the Training and Mobility of Researchers Programme.

Cataloging-in-Publication Data applied for
Bibliographic information published by Die Deutsche Bibliothek Die Deutsche Bibliothek lists this publication in the Deutsche Nationalbibliografie; detailed bibliographic data is available in the Internet at <http://dnb.ddb.de>.

ISSN 0075-8450
ISBN 3-540-20087-8 Springer-Verlag Berlin Heidelberg New York

Springer-Verlag Berlin Heidelberg New York
a member of BertelsmannSpringer Science+Business Media GmbH

springeronline.com

Typesetting by the authors/editors
Cover design: *design & production*, Heidelberg

Printed on acid-free paper
55/3141/du - 5 4 3 2 1 0

Preface

This book contains the lectures given at the *Centro Internazionale Matematico Estivo* (CIME), during the session **Direct and Inverse Method in Non Linear Evolution Equations**, held at Cetraro in September 1999.

The lecturers were **R. Conte** of the Service de physique de l'état condensé, CEA Saclay, **F. Magri** of the University of Milan, **M. Musette** of Dienst Theoretical Naturalness, Verite Universities Brussels, **J. Satsuma** of the Graduate School of Mathematical Sciences, University of Tokyo and **P. Winternitz** of the Centre de recherches mathématiques, Université de Montréal.

The courses face from different point of view the theory of the exact solutions and of the complete integrability of non linear evolution equations.

The Magri's lectures develop the geometrical approach and cover a large amount of topics concerning both the finite and infinite dimensional manifolds, Conte and Musette explain as Painlevé analysis and its various extensions can be extensively applied to a wide range of non linear equations. In particular Conte deals with the ODEs case, while Musette deals with the PDEs case. The Lie's method is the main subject of Winternitz's course where is shown as any kind of possible symmetry can be used for reducing the considered problem, and eventually for constructing exact solutions.
Finally Satsuma explains the bilinear method, introduced by Hirota, and, after considering in depth the algebraic structure of the completely integrable systems, presents modification of the method which permits to treat, among others, the ultra-discrete systems.

All lectures are enriched by several examples and applications to concrete problems arising from different contexts. In this way, from one hand the effectiveness of the used methods is pointed out, from the other hand the interested reader can experience directly the different geometrical, algebraical and analytical machineries involved.

I wish to express my appreciation to the authors for these notes, updated to the summer 2002, and to thank all the participants of this CIME session.

Padua, March 2003 *Antonio M. Greco*

Contents

Exact solutions of nonlinear partial differential equations by singularity analysis
Robert Conte ... 1
1 Introduction ... 1
2 Various levels of integrability for PDEs, definitions ... 2
3 Importance of the singularities: a brief survey of the theory of Painlevé ... 9
4 The Painlevé test for PDEs in its invariant version ... 11
4.1 Singular manifold variable φ and expansion variable χ ... 11
4.2 The WTC part of the Painlevé test for PDEs ... 14
4.3 The various ways to pass or fail the Painlevé test for PDEs ... 17
5 Ingredients of the "singular manifold method" ... 18
5.1 The ODE situation ... 19
5.2 Transposition of the ODE situation to PDEs ... 19
5.3 The singular manifold method as a singular part transformation ... 20
5.4 The degenerate case of linearizable equations ... 21
5.5 Choices of Lax pairs and equivalent Riccati pseudopotentials ... 21
Second-order Lax pairs and their privilege ... 21
Third-order Lax pairs ... 23
5.6 The admissible relations between τ and ψ ... 24
6 The algorithm of the singular manifold method ... 24
6.1 Where to truncate, and with which variable? ... 27
7 The singular manifold method applied to one-family PDEs ... 29
7.1 Integrable equations with a second order Lax pair ... 29
The Liouville equation ... 30
The AKNS equation ... 32
The KdV equation ... 33
7.2 Integrable equations with a third order Lax pair ... 35
The Boussinesq equation ... 35
The Hirota-Satsuma equation ... 37
The Tzitzéica equation ... 38
The Sawada-Kotera and Kaup-Kupershmidt equations ... 43
The Sawada-Kotera equation ... 44

The Kaup-Kupershmidt equation ... 45
7.3 Nonintegrable equations, second scattering order ... 49
The Kuramoto-Sivashinsky equation ... 49
7.4 Nonintegrable equations, third scattering order ... 52
8 Two common errors in the one-family truncation ... 53
8.1 The constant level term does not define a BT ... 53
8.2 The WTC truncation is suitable iff the Lax order is two ... 54
9 The singular manifold method applied to two-family PDEs ... 54
9.1 Integrable equations with a second order Lax pair ... 55
The sine-Gordon equation ... 55
The modified Korteweg-de Vries equation ... 57
The nonlinear Schrödinger equation ... 59
9.2 Integrable equations with a third order Lax pair ... 59
9.3 Nonintegrable equations, second and third scattering order ... 60
The KPP equation ... 60
The cubic complex Ginzburg-Landau equation ... 65
The nonintegrable Kundu-Eckhaus equation ... 68
10 Singular manifold method *versus* reduction methods ... 69
11 Truncation of the unknown, not of the equation ... 72
12 Birational transformations of the Painlevé equations ... 74
13 Conclusion, open problems ... 76
References ... 77

The method of Poisson pairs in the theory of nonlinear PDEs
Franco Magri, Gregorio Falqui, Marco Pedroni ... 85
1 Introduction: The tensorial approach and the birth of the method of Poisson pairs ... 85
1.1 The Miura map and the KdV equation ... 86
1.2 Poisson pairs and the KdV hierarchy ... 88
1.3 Invariant submanifolds and reduced equations ... 90
1.4 The modified KdV hierarchy ... 94
2 The method of Poisson pairs ... 96
3 A first class of examples and the reduction technique ... 101
3.1 Lie–Poisson manifolds ... 101
3.2 Polynomial extensions ... 102
3.3 Geometric reduction ... 103
3.4 An explicit example ... 104
3.5 A more general example ... 108
4 The KdV theory revisited ... 109
4.1 Poisson pairs on a loop algebra ... 109
4.2 Poisson reduction ... 110
4.3 The GZ hierarchy ... 112
4.4 The central system ... 113
4.5 The linearization process ... 115
4.6 The relation with the Sato approach ... 117

5 Lax representation of the reduced KdV flows 120
5.1 Lax representation 120
5.2 First example 122
5.3 The generic stationary submanifold 124
5.4 What more? 125
6 Darboux–Nijenhuis coordinates and separability 125
6.1 The Poisson pair 126
6.2 Passing to a symplectic leaf 128
6.3 Darboux–Nijenhuis coordinates 130
6.4 Separation of variables 131
References 134

Nonlinear superposition formulae of integrable partial differential equations by the singular manifold method
Micheline Musette 137
1 Introduction 137
2 Integrability by the singularity approach 138
3 Bäcklund transformation: definition and example 139
4 Singularity analysis of nonlinear differential equations 139
4.1 Nonlinear ordinary differential equations 139
4.2 Nonlinear partial differential equations 142
5 Lax Pair and Darboux transformation 143
5.1 Second order scalar scattering problem 144
5.2 Third order scalar scattering problem 145
5.3 A third order matrix scattering problem 146
6 Different truncations in Painlevé analysis 147
7 Method for a one-family equation 149
8 Nonlinear superposition formula 151
9 Results for PDEs possessing a second order Lax pair 151
9.1 First example: KdV equation 151
9.2 Second example: MKdV and sine-Gordon equations 153
10 PDEs possessing a third order Lax pair 156
10.1 Sawada-Kotera, KdV_5, Kaup-Kupershmidt equations ... 156
10.2 Painlevé test 157
10.3 Truncation with a second order Lax pair 158
10.4 Truncation with a third order Lax pair 158
10.5 Bäcklund transformation 159
10.6 Nonlinear superposition formula for Sawada-Kotera 160
10.7 Nonlinear superposition formula for Kaup-Kupershmidt . 161
10.8 Tzitzéica equation 165
References 166

Hirota bilinear method for nonlinear evolution equations

Junkichi Satsuma 171
1 Introduction 171
2 Soliton solutions 172
2.1 The Burgers equation 172
2.2 The Korteweg-de Vries equation 173
2.3 The nonlinear Schrödinger equation 174
2.4 The Toda equation 175
2.5 Painlevé equations 176
2.6 Difference vs differential 177
3 Multidimensional equations 180
3.1 The Kadomtsev-Petviashvili equation 180
3.2 The two-dimensional Toda lattice equation 181
3.3 Two-dimensional Toda molecule equation 184
3.4 The Hirota-Miwa equation 185
4 Sato theory 187
4.1 Micro-differential operators 187
4.2 Introduction of an infinite number of time variables 189
4.3 The Sato equation 192
4.4 Generalized Lax equation 194
4.5 Structure of tau functions 195
4.6 Algebraic identities for tau functions 200
4.7 Vertex operators and the KP bilinear identity 204
4.8 Fermion analysis based on an infinite dimensional Lie algebra 207
5 Extensions of the bilinear method 210
5.1 q-discrete equations 210
5.2 Special function solution for soliton equations 212
5.3 Ultra discrete soliton system 215
5.4 Trilinear equations 218
References 221

Lie groups, singularities and solutions of nonlinear partial differential equations

Pavel Winternitz 223
1 Introduction 223
2 The symmetry group of a system of differential equations 225
2.1 Formulation of the problem 225
Prolongation 226
Symmetry group: Global approach, use the chain rule 227
Symmetry group: Infinitesimal approach 227
Reformulation 227
2.2 Prolongation of vector fields and the symmetry algorithm . . . 228
2.3 First example: Variable coefficient KdV equation 230
2.4 Symmetry reduction for the KdV 232
2.5 Second example: Modified Kadomtsev-Petviashvili equation . . 235

3 Classification of the subalgebras of a finite dimensional Lie algebra 238
3.1 Formulation of the problem 238
3.2 Subalgebras of a simple Lie algebra 239
3.3 Example: Maximal subalgebras of $o(4,2)$ 240
3.4 Subalgebras of semidirect sums 244
3.5 Example: All subalgebras of $sl(3,\mathcal{R})$ classified under the group $SL(3,\mathcal{R})$ 248
3.6 Generalizations 252
4 The Clarkson-Kruskal direct reduction method and conditional symmetries 252
4.1 Formulation of the problem 252
4.2 Symmetry reduction for Boussinesq equation 253
4.3 The direct method 254
4.4 Conditional symmetries 256
4.5 General comments 261
5 Concluding comments 263
5.1 References on nonlinear superposition formulas 263
5.2 References on continuous symmetries of difference equations . 264
References 264

List of Participants 274

Index 277

Exact solutions of nonlinear partial differential equations by singularity analysis

Robert Conte

Service de physique de l'état condensé, CEA Saclay, F–91191 Gif-sur-Yvette Cedex, France; *conte@drecam.saclay.cea.fr*

Summary. Whether integrable, partially integrable or nonintegrable, nonlinear partial differential equations (PDEs) can be handled from scratch with essentially the same toolbox, when one looks for analytic solutions in closed form. The basic tool is the appropriate use of the singularities of the solutions, and this can be done without knowing these solutions in advance. Since the elaboration of the *singular manifold method* by Weiss *et al.*, many improvements have been made. After some basic recalls, we give an interpretation of the method allowing us to understand why and how it works. Next, we present the state of the art of this powerful technique, trying as much as possible to make it a (computerizable) algorithm. Finally, we apply it to various PDEs in $1+1$ dimensions, mostly taken from physics, some of them chaotic: sine-Gordon, Boussinesq, Sawada-Kotera, Kaup-Kupershmidt, complex Ginzburg-Landau, Kuramoto-Sivashinsky, etc.

1 Introduction

Our interest is to find explicitly the "macroscopic" quantities which materialize the integrability of a given *nonlinear* differential equation, such as particular solutions or first integrals. We mainly handle partial differential equations (PDEs), although some examples are taken from ordinary differential equations (ODEs). Indeed, the methods described in these lectures apply equally to both cases.

These methods are based on the *a priori* study of the singularities of the solutions. The reader is assumed to possess a basic knowledge of the singularities of nonlinear *ordinary* differential equations, the Painlevé property for ODEs and the Painlevé test. All this prerequisite material is well presented in a book by Hille [63] while Cargèse lecture notes [26] contain a detailed exposition of the methods, including the Painlevé test for ODEs. Many applications are given in a review [110].

As a general bibliography on the subject of these lectures, we recommend Cargèse lecture notes [37] and a shorter subset of these with emphasis on the various so-called truncations [24].

Throughout the text, we exclude linear equations, unless explicitly stated.

R. Conte, Exact solutions of nonlinear partial differential equations by singularity analysis, Lect. Notes Phys. **632**, 1–83 (2003)
http://www.springerlink.com/

2 Various levels of integrability for PDEs, definitions

In this section, we review the required definitions (exact solution, Bäcklund transformation, Lax pair, singular part transformation, etc).

The most important point is the global nature of the information which is looked for. The existence theorem of Cauchy (for ODEs) or Cauchy-Kowalevski (for PDEs) is of no help for this purpose. Indeed, it only states a local property and says nothing on what happens outside the disk of definition of the Taylor series. Therefore it cannot distinguish between chaotic equations and integrable ones.

Still from this point of view, Laurent series are not better than Taylor series. For instance, the Bianchi IX cosmological model is a six-dimensional dynamical system

$$\sigma^2(\operatorname{Log} A)'' = A^2 - (B-C)^2, \text{ and cyclically, } \sigma^4 = 1, \tag{1}$$

which is undoubtedly chaotic [115]. Despite the existence of the Laurent series [43]

$$\begin{aligned} A/\sigma &= \chi^{-1} + a_2\chi + O(\chi^3),\ \chi = \tau - \tau_1, \\ B/\sigma &= b_0\chi + b_1\chi^2 + O(\chi^3), \\ C/\sigma &= c_0\chi + c_1\chi^2 + O(\chi^3), \end{aligned} \tag{2}$$

which depends on six independent arbitrary coefficients, $(\tau_1, b_0, c_0, b_1, c_1, a_2)$, a wrong statement would be to conclude to the absence of chaos.

This leads us to the definition of the first one of several needed global mathematical objects.

Definition 2.1. *One calls* ***exact solution*** *of a nonlinear PDE any solution defined in the whole domain of definition of the PDE and which is given in closed form, i.e. as a finite expression.*

The opposite of an exact solution is of course not a wrong solution, but what Painlevé calls "une solution illusoire", such as the above mentioned series.

Note that a multivalued expression is not excluded. From this definition, an exact solution is *global*, as opposed to *local*. This generically excludes series or infinite products, unless their domain of validity can be made the full domain of definition, like for linear ODEs.

Example 2.1. The Kuramoto-Sivashinsky (KS) equation

$$u_t + uu_x + \mu u_{xx} + \nu u_{xxxx} = 0,\ \nu \neq 0, \tag{3}$$

describes, for instance, the fluctuation of the position of a flame front, or the motion of a fluid going down a vertical wall, or a spatially uniform oscillating chemical reaction in a homogeneous medium (see Ref. [84] for a review), and it is well known for its chaotic behaviour. An exact solution is the solitary wave of Kuramoto and Tsuzuki [75] in which the wavevector k is fixed

$$u = 120\nu \left(\frac{k}{2}\tanh\frac{k}{2}\xi\right)^3 + \left(\frac{60}{19}\mu - 30\nu k^2\right)\frac{k}{2}\tanh\frac{k}{2}\xi + c,$$
$$\xi = x - ct - x_0,\ k^2 = \frac{11\mu}{19\nu} \text{ or } -\frac{\mu}{19\nu}, \tag{4}$$

which depends on two arbitrary constants (c, x_0). On the contrary, the Laurent series

$$u = 120\nu\xi^{-3} + \frac{60}{19}\mu\xi^{-1} + c - \frac{120\times 11}{19^2}\mu^2\xi + u_6\xi^3 + O(\xi^4), \tag{5}$$

which depends on three arbitrary constants (c, x_0, u_6), is not an exact solution, since no closed form expression is yet known for the sum of this series.

There exists a powerful tool to build exact solutions, this is the Bäcklund transformation. For simplicity, but this is not a restriction, we give the basic definitions for a PDE defined as a single scalar equation for one dependent variable u and two independent variables (x, t).

Definition 2.2. *(Refs. [7] vol. III chap. XII, [34]) A* ***Bäcklund transformation*** *(BT) between two given PDEs*

$$E_1(u, x, t) = 0,\ E_2(U, X, T) = 0 \tag{6}$$

is a pair of relations

$$F_j(u, x, t, U, X, T) = 0,\ j = 1, 2 \tag{7}$$

with some transformation between (x, t) and (X, T), in which F_j depends on the derivatives of $u(x, t)$ and $U(X, T)$, such that the elimination of u (resp. U) between (F_1, F_2) implies $E_2(U, X, T) = 0$ (resp. $E_1(u, x, t) = 0$). The BT is called the ***auto-BT*** *or the* ***hetero-BT*** *according as the two PDEs are the same or not.*

Example 2.2. The sine-Gordon equation (we identify sine-Gordon and sinh-Gordon since an affine transformation on u does not change the integrability nor the singularity structure)

$$\text{sine-Gordon} \ : E(u) \equiv u_{xt} + 2a\sinh u = 0 \tag{8}$$

admits the auto-BT

$$(u + U)_x + 4\lambda \sinh\frac{u - U}{2} = 0, \tag{9}$$
$$(u - U)_t - \frac{2a}{\lambda}\sinh\frac{u + U}{2} = 0, \tag{10}$$

in which λ is an arbitrary complex constant, called the *Bäcklund parameter.*

Given the obvious solution $U = 0$ (called *vacuum*), the two equations (7)–(8) are Riccati ODEs with constant coefficients for the unknown $e^{u/2}$,

$$(e^{u/2})_x = \lambda(1 - (e^{u/2})^2), \tag{11}$$
$$(e^{u/2})_t = -a(1 - (e^{u/2})^2)/(2\lambda), \tag{12}$$

therefore their general solution is known in closed form

$$e^{u/2} = \tanh\theta,\ \theta = \left(\lambda x - \frac{a}{2\lambda}t - z_0\right), \tag{13}$$

with (λ, z_0) arbitrary. This solution is called the *one-soliton solution*, it is also written as

$$\tanh(u/4) = -e^{-2\theta},\ u_x = 4\lambda \operatorname{sech} 2\theta,\ u_t = -2a\lambda^{-1} \operatorname{sech} 2\theta. \tag{14}$$

By iteration, this procedure gives rise to the N-soliton solution [76, 1], an exact solution depending on $2N$ arbitrary complex constants (N values of the Bäcklund parameter λ, N values of the shift z_0), with N an arbitrary positive integer. A remarkable feature of the SG-equation, due to the fact that at least one of the two ODEs (7)–(8) is of order one, is that this N-soliton can be obtained from N different copies of the one-soliton by a simple algebraic operation, i.e. without integration (see Musette's lecture [91]).

Example 2.3. The Liouville equation

$$\text{Liouville: } E(u) \equiv u_{xt} + \alpha e^u = 0 \tag{15}$$

admits two BTs. The first one

$$(u - v)_x = \alpha\lambda e^{(u+v)/2}, \tag{16}$$
$$(u + v)_t = -2\lambda^{-1} e^{(u-v)/2}, \tag{17}$$

is a BT to a linearizable equation called the d'Alembert equation

$$\text{d'Alembert: } E(v) \equiv v_{xt} = 0. \tag{18}$$

The second one is an auto-BT

$$(u + U)_x = -4\lambda \sinh\frac{u - U}{2}, \tag{19}$$
$$(u - U)_t = \lambda^{-1}\alpha e^{(u+U)/2}. \tag{20}$$

The first of these two BTs allows one to obtain the general solution of the nonlinear Liouville equation, see Sect. 7.

This ideal situation (generation of the general solution) is exceptional and the generic case is the generation of particular solutions only, as in the sine-Gordon example.

The importance of the BT is such that it is often taken as a definition of *integrability*.

Definition 2.3. *A PDE in N independent variables is* ***integrable*** *if at least one of the following properties holds.*

1. *Its general solution is an explicit closed form expression, possibly presenting movable critical singularities.*
2. *It is linearizable.*
3. *For $N > 1$, it possesses an auto-BT which, if $N = 2$, depends on an arbitrary complex constant, the Bäcklund parameter.*
4. *It possesses a hetero-BT to another integrable PDE.*

Although partially integrable and nonintegrable equations, i.e. the majority of physical equations, admit no BT, they retain part of the properties of (fully) integrable PDEs, and this is why the methods presented in these lectures apply to both cases as well. For instance, the KS equation admits the *vacuum* solution $u = 0$ and, in Sect. 2, an iteration will be built leading from $u = 0$ to the solitary wave (4); the nonintegrability manifests itself in the finite number of times this iteration provides a new result ($N = 1$ for the KS equation, and one cannot go beyond (4) [30]).

For various applications of the BT, see Ref. [51].

When a PDE has some good reasons to possess such features, such as the reasons developed in Sect. 4, we want to find the BT if it exists, since this is a generator of exact solutions, or a degenerate form of the BT if the BT does not exist, and we want to do it by singularity analysis *only*.

Before proceeding, we need to define some other elements of integrability.

Definition 2.4. *Given a PDE, a* ***Lax pair*** *is a system of two linear differential operators*

$$\textit{Lax pair}:\ L_1(U,\lambda),\ L_2(U,\lambda), \tag{21}$$

depending on a solution U of the PDE and, in the 1+1-dimensional case, on an arbitrary constant λ, called the ***spectral parameter****, with the property that the vanishing of the commutator $[L_1, L_2]$ is equivalent to the vanishing of the PDE $E(U) = 0$.*

A Lax pair can be represented in several, equivalent ways.

The *Lax representation* [30] is a pair of linear operators (L, P) (scalar or matrix) defined by

$$L_1 = L - \lambda,\ L_2 = \partial_t - P,\ L_1\psi = 0,\ L_2\psi = 0,\ \lambda_t = 0, \tag{22}$$

in which the elimination of the scalar λ yields

$$L_t = [P, L], \tag{23}$$

i.e. , thanks to the *isospectral* condition $\lambda_t = 0$, a time evolution analogous to the one in Hamiltonian dynamics.

The *zero-curvature representation* is a pair (L, M) of linear operators independent of (∂_x, ∂_t)

$$\begin{aligned}&L_1 = \partial_x - L,\ L_2 = \partial_t - M,\ L_1\boldsymbol{\psi} = 0,\ L_2\boldsymbol{\psi} = 0,\\ &[\partial_x - L, \partial_t - M] = L_t - M_x + LM - ML = 0.\end{aligned} \tag{24}$$

The common order N of the matrices is called the *order* of the Lax pair.

The *projective Riccati representation* is a first order system of $2N - 2$ Riccati equations in the unknowns $\psi_j/\psi_1, j = 2, \dots, N$, equivalent to the zero-curvature representation (24).

The *scalar representation* is a pair of scalar linear PDEs, one of them of order higher than one,

$$\begin{aligned}&L_1\psi = 0,\ L_2\psi = 0,\\ &X \equiv [L_1, L_2] = 0.\end{aligned} \tag{25}$$

In $1 + 1$-dimensions, one of the PDEs can be made an ODE (i.e. involving only x- or t-derivatives), in which case the order of this ODE is called the order of the Lax pair.

The *string representation* or *Sato representation* [70]

$$[P, Q] = 1. \tag{26}$$

This very elegant representation, reminiscent of Hamiltonian dynamics, uses the Sato definition of a *microdifferential operator* (a differential operator with positive and negative powers of the differential operator ∂) and of its *differential part* denoted $()_+$ (the subset of its nonnegative powers), e.g.

$$Q = \partial_x^2 - u, \tag{27}$$

$$L = Q^{1/2}, \tag{28}$$

$$\left(L^3\right)_+ = \partial_x^3 - (3/4)\{u, \partial_x\}, \tag{29}$$

$$\left(L^5\right)_+ = \partial_x^5 - (5/4)\{u, \partial_x^3\} + (5/16)\{3u^2 + u_{xx}, \partial_x\}, \tag{30}$$

in which $\{a, b\}$ denotes the anticommutator $ab + ba$. See Ref. [46] for a tutorial presentation.

Example 2.4. The sine-Gordon equation (8) admits the zero-curvature representation

$$(\partial_x - L)\begin{pmatrix}\psi_1\\ \psi_2\end{pmatrix} = 0\ L = \begin{pmatrix}U_x/2 & \lambda\\ \lambda & -U_x/2\end{pmatrix}, \tag{31}$$

$$(\partial_t - M)\begin{pmatrix}\psi_1\\ \psi_2\end{pmatrix} = 0,\ M = -(a/2)\lambda^{-1}\begin{pmatrix}0 & e^U\\ e^{-U} & 0\end{pmatrix}, \tag{32}$$

equivalent to the Riccati representation, with $y = \psi_1/\psi_2$,

$$y_x = \lambda + U_x y - \lambda y^2, \tag{33}$$

$$y_t = -\frac{a}{2}\lambda^{-1}e^U + \frac{a}{2}\lambda^{-1}e^{-U}y^2. \tag{34}$$

Example 2.5. The matrix nonlinear Schrödinger equation

$$iQ_t - (b/a)Q_{xx} - 2abQRQ = 0,\ -iR_t - (b/a)R_{xx} - 2abRQR = 0, \quad (35)$$

in which (Q, R) are rectangular matrices of respective orders (m, n) and (n, m), and (i, a, b) constants, admits the zero-curvature representation ([83] Eq. (5))

$$(\partial_x - L)\psi = 0,\ (\partial_t - M)\psi = 0, \quad (36)$$
$$L = aP + \lambda G,\ M = (-aGP^2 + GP_x + 2\lambda P + (2/a)\lambda^2 G)b/i, \quad (37)$$

in which λ is the spectral parameter, P and G matrices of order $m+n$ defined as

$$P = \begin{pmatrix} 0 & Q \\ -R & 0 \end{pmatrix},\ G = \begin{pmatrix} 1_m & 0 \\ 0 & -1_n \end{pmatrix}. \quad (38)$$

The matrix G characterizes the internal symmetry group $\mathrm{GL}(m, \mathcal{C}) \otimes \mathrm{GL}(n, \mathcal{C})$. The lowest values

$$m = 1, n = 1,\ Q = (u),\ R = (U), \quad (39)$$

define the AKNS system (Sect. 9.1), whose reduction $U = \bar{u}$ is the usual scalar nonlinear Schrödinger equation.

Example 2.6. The 2 + 1-dimensional Ito equation [68]

$$E(u) \equiv \left(u_{xxxt} + 6\alpha^{-1}u_{xt}u_{xx} + a_1 u_{tt} + a_2 u_{xt} + a_3 u_{xx} + a_4 u_{ty}\right)_x = 0 \quad (40)$$

has a Lax pair whose scalar representation is

$$L_1 \equiv \partial_x^3 + a_1\partial_t + (a_2 + 6\alpha^{-1}U_{xx})\partial_x + a_4\partial_y - \lambda \quad (41)$$
$$L_2 \equiv \partial_x\partial_t - \mu\partial_x + (\frac{a_3}{3} + 2\alpha^{-1}U_{xt}) \quad (42)$$
$$\alpha[L_1, L_2] = 2E(U) + 6U_{xxx}L_2. \quad (43)$$

In the 2 + 1-dimensional case $a_4 \neq 0$, the parameter λ can be set to 0 by the change $\psi \mapsto \psi e^{\lambda y}$. This is the reason of the precision at the end of item 2 in definition 2.4. This pair has the order four in the generic case $a_1 \neq 0$, although neither L_1 nor L_2 has such an order.

Example 2.7. The string representation of the Lax pair of the derivative of the first Painlevé equation is

$$[P, Q] = [((\partial_x^2 - u)^3)_+, \partial_x^2 - u] = -(1/4)u_{xxx} + (3/4)uu_x = 1. \quad (44)$$

Example 2.8. The Sato representation of the Lax pair for the whole Korteweg-de Vries hierarchy is

$$\partial_{t_m} L = [(L^{2m-1})_+, L],\ L = Q^{1/2},\ Q = \partial_x^2 - u,\ m = 1, 3, 5, \ldots \quad (45)$$

From the singularity point of view, the Riccati representation is the most suitable, as will be seen.

The last main definition we need is the singular part transformation, which we used to call (improperly) Darboux transformation (for the definition of a Darboux transformation [13], see Musette's lecture [91] in this volume).

Definition 2.5. *Given a PDE, a* ***singular part transformation*** *is a transformation between two solutions (u, U) of the PDE*

$$\textit{singular part transformation}: \ u = \sum_f \mathcal{D}_f \operatorname{Log} \tau_f + U \tag{46}$$

linking their difference to a finite number of linear differential operators $\mathcal{D}_f$ (f like family) acting on the logarithm of functions τ_f.

In the definition (46), it is important to note that, despite the notation, each function τ_f is in fact the ratio of the "tau-function" of u by that of U.

Lax pairs, Bäcklund and singular part transformations are not independent. In order to exhibit their interrelation, one needs an additional information, namely the link

$$\forall f \ : \ \mathcal{D}_f \operatorname{Log} \tau_f = F_f(\psi), \tag{47}$$

which most often is the identity $\tau = \psi$, between the functions τ_f and the function ψ in the definition of a scalar Lax pair.

Example 2.9. The (integrable) sine-Gordon equation (8) admits the singular part transformation

$$u = U - 2(\operatorname{Log} \tau_1 - \operatorname{Log} \tau_2), \tag{48}$$

in which (τ_1, τ_2) is a solution (ψ_1, ψ_2) of the system (31)–(32).

Then its BT (7)–(8) is the result of the elimination [5] of τ_1/τ_2 between the singular part transformation (48) and the Riccati form of the Lax pair (33)–(34), with the correspondence $\tau_f = \psi_f, f = 1, 2$. This elimination reduces to the substitution $y = e^{-(u-U)/2}$ in the Riccati system (33)–(34), and this is one of the advantages of the Riccati representation. Therefore the Bäcklund parameter and the spectral parameter are identical notions.

Example 2.10. The (nonintegrable) Kuramoto-Sivashinsky equation admits the degenerate singular part transformation

$$u = U + (60\nu\partial_x^3 + (60/19)\mu\partial_x) \operatorname{Log} \tau, \tag{49}$$

in which $U = c$ (vacuum) and τ is the general solution ψ of the linear system (a degenerate second order scalar Lax pair)

$$L_1\psi \equiv (\partial_x^2 - k^2/4)\psi = 0, \tag{50}$$

$$L_2\psi \equiv (\partial_t + c\partial_x)\psi = 0, \tag{51}$$

$$[L_1, L_2] \equiv 0. \tag{52}$$

The solution u defined by (49) is then the solitary wave (4), and this is a much simpler way to write it, because the logarithmic derivatives in (49) take account of the whole nonlinearity.

Since, roughly speaking, the BT is equivalent to the couple (singular part transformation, Lax pair), one can rephrase as follows the iteration to generate new solutions. Let us symbolically denote

$E(u) = 0$ the PDE,
$\mathrm{Lax}(\psi, \lambda, U) = 0$ a scalar Lax pair,
F the link (47) $\mathcal{D}\,\mathrm{Log}\,\tau = F(\psi)$ from ψ to τ,
$u =$ singular part transformation(U, τ) the singular part transformation.

The iteration is the following, see e.g. [60].

1. (initialization) Choose $u_0 =$ a particular solution of $E(u) = 0$; set $n = 1$; perform the following loop until some maximal value of n;
2. (start of loop) Choose $\lambda_n =$ a particular complex constant;
3. Compute, by integration, a particular solution ψ_n of the linear system $\mathrm{Lax}(\psi, \lambda_n, u_{n-1}) = 0$;
4. Compute, without integration, $\mathcal{D}\,\mathrm{Log}\,\tau_n = F(\psi_n)$;
5. Compute, without integration,
$u_n =$ singular part transformation(u_{n-1}, τ_n);
6. (end of loop) Set $n = n + 1$.

Depending on the choice of λ_n at step 2, and of ψ_n at step 3, one can generate either the N-soliton solution, or solutions rational in (x, t), or a mixture of such solutions.

3 Importance of the singularities: a brief survey of the theory of Painlevé

A classical theorem states that a function of one complex variable without any singularity in the analytic plane (i.e. the complex plane compactified by addition of the unique point at infinity) is a constant. Therefore a function with singularities is characterized, as shown by Mittag-Leffler, by the knowledge of its singularities in the analytic plane. Similarly, if u satisfies an ODE or a PDE, the structure of singularities of the general solution characterizes the level of integrability of the equation. This is the basis of the theory of the (explicit) integration of nonlinear ODEs built by Painlevé, which we only briefly introduce here [for a detailed introduction, see Cargèse lecture notes: Ref. [26] for ODEs, Ref. [37] for PDEs].

To integrate an ODE is to acquire a global knowledge of its general solution, not only the local knowledge ensured by the existence theorem of Cauchy. So, the most demanding possible definition for the "integrability" of an ODE is the single valuedness of its general solution, so as to adapt this solution to any kind of initial conditions. Since even linear equations may fail to have this property, e.g. $2xu' + u = 0,\ u = cx^{-1/2}$, a more reasonable definition is the following one.

Definition 3.1. *The Painlevé property (PP) of an ODE is the uniformizability of its general solution.*

In the above example, the uniformization is achieved by the change of independent variable $x = X^2$. This definition is equivalent to the more familiar one.

Definition 3.2. *The Painlevé property (PP) of an ODE is the absence of movable critical singularities in its general solution.*

Definition 3.3. *The Painlevé property (PP) of a PDE is its integrability (Definition 2.3) and the absence of movable critical singularities in its general solution.*

Let us recall that a singularity is said *movable* (as opposed to *fixed*) if its location depends on the initial conditions, and *critical* if multivaluedness takes place around it. Indeed, out of the four configurations of singularities (critical or noncritical) and (fixed or movable), only the configuration (critical and movable) prevents uniformizability: one does not know where to put the cut since the point is movable.

Wrong definitions of the PP, alas repeatedly published, consist in replacing in the definition "movable critical singularities" by "movable singularities other than poles", or "its general solution" by "all its solutions". Even worse definitions only refer to Laurent series. See Ref. [26], Sect. 2.6, for the arguments of Painlevé himself.

The mathematicians like Painlevé want to integrate whole classes of ODEs (e.g. second order algebraic ODEs). We will only use their methods for a given ODE or PDE, with the aim of deriving the elements of integrability described in Sect. 2 (exact solutions, ...). This *Painlevé analysis* is twofold ("double méthode", says Painlevé).

1. Build necessary conditions for an ODE or a PDE to have the PP (this is called the *Painlevé test*).
2. When all these conditions are satisfied, or at least some of them, find the global elements of integrability. In the integrable case this is achieved either (ODE case) by explicitly integrating or (PDE case) by finding an auto-BT (like equations (7)–(8) for sine–Gordon) or a BT towards another PDE with the PP (like (16)–(17) between the d'Alembert and Liouville equations). In the partially integrable case, only degenerate forms of the above can be expected, as described in Sect. 2.

4 The Painlevé test for PDEs in its invariant version

When the PDE reduces to an ODE, the Painlevé test (for shortness we will simply say the test) reduces by construction to the test for ODEs, presented in detail elsewhere [26] and assumed known here.

We will skip those steps of the test which are the same for ODEs and for PDEs (e.g., diophantine conditions that all the leading powers and all the Fuchs indices be integer), and we will concentrate on the features which are specific to PDEs, namely the description of the movable singularities, the optimal choice of the expansion variable for the Laurent series, the advantage of the homographic invariance.

4.1 Singular manifold variable φ and expansion variable χ

Consider a nonlinear PDE

$$E(u,x,t,\dots)=0. \tag{53}$$

To test movable singularities for multivaluedness without integrating, which is the essence of the test, one must first describe them, then, among other steps, check the existence near each movable singularity of a Laurent series which represents the general solution.

For PDEs, the singularities are not isolated in the space of the independent variables $(x,t,\dots)$, but they lay on a codimension one manifold

$$\varphi(x,t,\dots)-\varphi_0=0, \tag{54}$$

in which the *singular manifold variable* φ is an arbitrary function of the independent variables and φ_0 an arbitrary movable constant. Even in the ODE case, the movable singularity can be defined as $\varphi(x)-\varphi_0=0$, since the implicit functions theorem allows this to be locally inverted to $x-x_0=0$; the arbitrary function φ thus introduced may then be used to construct exact solutions which would be impossible to find with the restriction $\varphi(x)=x$ [122, 98].

One must then define from $\varphi-\varphi_0$ an *expansion variable* χ for the Laurent series, for there is no reason to confuse the roles of the singular manifold variable and the expansion variable. Two requirements must be respected: firstly, χ must vanish as $\varphi-\varphi_0$ when $\varphi\to\varphi_0$; secondly, the structure of singularities in the φ complex plane must be in a one-to-one correspondence with that in the χ complex plane, so χ must be a homographic transform of $\varphi-\varphi_0$ (with coefficients depending on the derivatives of φ).

The Laurent series for u and E involved in the Kowalevski-Gambier part of the test are defined as

$$u=\sum_{j=0}^{+\infty}u_j\chi^{j+p},\ -p\in\mathcal{N},\ E=\sum_{j=0}^{+\infty}E_j\chi^{j+q},\ -q\in\mathcal{N}^* \tag{55}$$

with coefficients u_j, E_j independent of χ and only depending on the derivatives of φ.

To illustrate our point, let us take as an example the Korteweg-de Vries equation

$$E \equiv bu_t + u_{xxx} - (6/a)uu_x = 0 \tag{56}$$

(this is one of the very rare locations where this equation can be taken as an example; indeed, usually, things work so nicely for KdV that it is hazardous to draw general conclusions from its single study).

The choice $\chi = \varphi - \varphi_0$ originally made by Weiss *et al.* [65] makes the coefficients u_j, E_j invariant under the two-parameter group of translations $\varphi \mapsto \varphi + b'$, with b' an arbitrary complex constant and therefore they only depend on the differential invariant grad φ of this group and its derivatives:

$$u = 2a\varphi_x^2\chi^{-2} - 2a\varphi_{xx}\chi^{-1} + ab\frac{\varphi_t}{6\varphi_x} + \frac{2a}{3}\frac{\varphi_{xxx}}{\varphi_x} - \frac{a}{2}\left[\frac{\varphi_{xx}}{\varphi_x}\right]^2 + O(\chi), \chi = \varphi - \varphi_0. \tag{57}$$

There exists a choice of χ for which the coefficients exhibit the highest invariance and therefore are the shortest possible (all details are in Sect. 6.4 of Ref. [26]), this best choice is [6]

$$\chi = \frac{\varphi - \varphi_0}{\varphi_x - \dfrac{\varphi_{xx}}{2\varphi_x}(\varphi - \varphi_0)} = \left[\frac{\varphi_x}{\varphi - \varphi_0} - \frac{\varphi_{xx}}{2\varphi_x}\right]^{-1}, \ \varphi_x \neq 0, \tag{58}$$

in which x denotes one of the independent variables whose component of grad φ does not vanish. The expansion coefficients u_j, E_j are then invariant under the six-parameter group of homographic transformations

$$\varphi \mapsto \frac{a'\varphi + b'}{c'\varphi + d'}, \ a'd' - b'c' \neq 0, \tag{59}$$

in which a', b', c', d' are arbitrary complex constants. Accordingly, these coefficients only depend on the following elementary differential invariants and their derivatives: the *Schwarzian*

$$S = \{\varphi; x\} = \frac{\varphi_{xxx}}{\varphi_x} - \frac{3}{2}\left(\frac{\varphi_{xx}}{\varphi_x}\right)^2, \tag{60}$$

and one other invariant per independent variable $t, y, \ldots$

$$C = -\varphi_t/\varphi_x, \ K = -\varphi_y/\varphi_x, \ \ldots \tag{61}$$

The reason for the minus sign in the definition of C is that, under the travelling wave reduction $\xi = x - ct$, the variable C becomes the constant c. These two invariants are linked by the cross-derivative condition

$$X \equiv ((\varphi_{xxx})_t - (\varphi_t)_{xxx})/\varphi_x = S_t + C_{xxx} + 2C_x S + CS_x = 0, \quad (62)$$

identically satisfied in terms of φ.

For our KdV example, the final Laurent series, as compared with the initial one (57), is remarkably simple:

$$u = 2a\chi^{-2} - ab\frac{C}{6} + \frac{2aS}{3} - 2a(bC - S)_x\chi + O(\chi^2),\ \chi = (58). \quad (63)$$

For the practical computation of (u_j, E_j) as functions of (S, C) only, i.e. what is called the invariant Painlevé analysis, the above explicit expressions of (S, C, χ) in terms of φ are *not* required, the variable φ completely disappears, and the only necessary information is the gradient of the expansion variable χ defined by Eq. (58). This gradient is a polynomial of degree two in χ (this is a property of homographic transformations), whose coefficients only depend on S, C:

$$\chi_x = 1 + \frac{S}{2}\chi^2, \quad (64)$$

$$\chi_t = -C + C_x\chi - \frac{1}{2}(CS + C_{xx})\chi^2. \quad (65)$$

The above choice (58) of χ which generates homographically invariant coefficients is the simplest one, but it is only particular. The general solution to the above two requirements which also generates homographically invariant coefficients is defined by an affine transformation on the inverse of χ [38]

$$Y^{-1} = B(\chi^{-1} + A),\ B \neq 0. \quad (66)$$

Since a homography conserves the Riccati nature of an ODE, the variable Y satisfies a Riccati system, easily deduced from the canonical one (64)–(65) satisfied by χ, see (115)–(116).

A frequent worry is: is there any restriction (or advantage, or inconvenient) to perform the test with χ or Y rather than with $\varphi - \varphi_0$? The precise answer is: the three Laurent series are equivalent (their set of coefficients are in a one-to-one correspondence, only their radii of convergence are different). As a consequence, the Painlevé test, which involves the *infinite* series, is insensitive to the choice, and the costless choice (the one which minimizes the computations) is undoubtedly χ defined by its gradient (64)–(65) (to perform the test, one can even set, following Kruskal [69], $S = 0, C_x = 0$). If the same question were asked not about the test but about the second stage of Painlevé analysis as formulated at the end of Sect. 3, the answer would be quite different, and it is given in Sect. 6.1.

Finally, let us mention a useful technical simplification. From its definition (58), the variable χ^{-1} is a logarithmic derivative, so the system (64)–(65) can be integrated once

$$\begin{aligned}
\Psi &= (\varphi - \varphi_0)\varphi_x^{-1/2}, && (67)\\
(\operatorname{Log}\Psi)_x &= \chi^{-1}, && (68)\\
(\operatorname{Log}\Psi)_t &= -C\chi^{-1} + \frac{1}{2}C_x. && (69)
\end{aligned}$$

This feature helps to process PDEs which can be defined in either conservative or potential form when the conservative field u has a simple pole, such as the Burgers equation

$$\begin{aligned}
&E(u) \equiv bu_t + (u^2/a + u_x)_x = 0, && (70)\\
&F(v) \equiv bv_t + v_x^2/a + v_{xx} + G(t) = 0,\ u = v_x,\ E = F_x. && (71)
\end{aligned}$$

Despite its (unique) logarithmic term, the ψ-series for v

$$\begin{aligned}
v = {} & a \operatorname{Log}\Psi + v_0 + (2v_{0,x} - abC)\chi\\
& +(F(v_0) - aS/2 + abC_x/2)\chi^2 + O(\chi^3), \qquad (72)
\end{aligned}$$

in which v_0 is arbitrary, is "shorter" than the Laurent series for u

$$\begin{aligned}
u = {} & a\chi^{-1} + (ab/2)C + u_2\chi\\
& + \left[(a/4)(b^2(C_t + CC_x) + 2bC_{xx} - S_x - u_{2,x})\right]\chi^2 + O(\chi^3), \qquad (73)
\end{aligned}$$

in which u_2 is arbitrary, and the resulting series for $F(v)$, which is not a ψ-series but a Laurent series, is *much* shorter than the Laurent series for $E(u)$. See Sect. 7.3 for an application.

4.2 The WTC part of the Painlevé test for PDEs

As mentioned at the beginning of Sect. 4, we do not give here all the detailed steps of the test nor all the necessary conditions which it generates (this is done in Sect. 6.6 of Ref. [26]). We mainly state the notation to be extensively used throughout next sections.

The WTC part [65] of the full test, when rephrased in the equivalent invariant formalism [23], consists in checking the existence of all Laurent series (55) able to represent the general solution, maybe after suitable perturbations [29, 95] not describe here.

The gradient of the expansion variable χ is given by (64)–(65), with the cross-derivative condition (78). This condition may be used to eliminate, depending on the PDE, either derivatives $S_{mx,nt}$, with $n \geq 1$, or derivatives $C_{mx,nt}$, with $m \geq 3$, and all equations later written are already simplified in either way.

The *first step* is to find all the admissible values (p, u_0) which define the leading term of the series for u. Such an admissible couple is called a *family of movable singularities* (the term *branch* should be avoided for the confusion which it induces with branching, i.e. multivaluedness).

The recurrence relation for the next coefficients u_j, after replacement of (p, u_0),

$$\forall j \geq 1: \; E_j \equiv P(j)u_j + Q_j(\{u_l \mid l < j\}) = 0 \tag{74}$$

depends linearly on u_j and nonlinearly on the previously computed coefficients u_l.

The *second step* is to compute the *indicial equation*

$$P(i) = 0 \tag{75}$$

(a determinant in the multidimensional case of a system of PDEs). Its roots are called the *Fuchs indices* of the family because they are indeed the characteristic indices of a linear differential equation near a Fuchsian singularity (the name *resonances* sometimes given to these indices refers to no resonance phenomenon and should also be avoided). One then requires that all indices be integer and obey a rank condition which, for a single PDE, reduces to the condition that all indices be distinct. The value $i = -1$ is always a Fuchs index.

The *third and last step* is to require that, for any admissible family and any Fuchs index i (a signed integer), the *no-logarithm condition*

$$\forall i \in \mathcal{Z}, \; P(i) = 0: \; Q_i = 0 \tag{76}$$

holds true, so that the coefficient u_i is an arbitrary function of the independent variables. In the multidimensional case, this is the condition of orthogonality between the vector $\mathbf{Q}_i$ and the adjoint of the linear operator $\mathbf{P}(i)$. Whenever there exist negative integers in addition to the ever present value -1 counted with multiplicity one, the condition (76) can only be tested by a perturbation [29].

This ends this subset of the test which, let us insist on the terminology, is only aimed at building *necessary* conditions for the PP.

The Laurent series for u built in this way depends on at most N arbitrary functions (if N denotes the differential order), namely the coefficients u_i introduced at the N Fuchs indices, including φ for the index -1.

Any item u_j, E_j, Q_j depends, through the elementary invariants (S, C), on the derivatives of φ up to the order $j+1$, so the dependences are as follows: $u_0 = f(C)$, $u_1 = f(C, C_x, C_t)$, $u_2 = f(C, C_x, C_t, C_{xx}, C_{xt}, C_{tt}, S)$, ...

Let us take an example.

Example 4.1. The Kolmogorov-Petrovskii-Piskunov (KPP) equation [73, 99]

$$E(u) \equiv bu_t - u_{xx} + 2d^{-2}(u - e_1)(u - e_2)(u - e_3) = 0, \; e_j \text{ distinct}, \tag{77}$$

encountered in reaction-diffusion systems (an additional convection term uu_x is quite important in physical applications to prey-predator models [113]) possesses the two families (d denotes any square root of d^2)

$$p = -1, \; u_0 = d, \tag{78}$$

each family has the same two indices $(-1, 4)$, and the Laurent series for each family reads

$$u = d\chi^{-1} + (s_1/3 - (bd/6)C) \tag{79}$$
$$- (d/6)\left((b^2/6)C^2 - 6a_2 - S - bC_x\right)\chi + O(\chi^2), \tag{80}$$

with the notation

$$s_1 = e_1 + e_2 + e_3,\ a_2 = \left((e_2 - e_3)^2 + (e_3 - e_1)^2 + (e_1 - e_2)^2\right)/(18d^2). \tag{81}$$

At index $i = 4$, the two no-log conditions, one for each sign of d [18],

$$Q_4 \equiv C[(bdC + s_1 - 3e_1)(bdC + s_1 - 3e_2)(bdC + s_1 - 3e_3)$$
$$- 3b^2d^3(C_t + CC_x)] = 0 \tag{82}$$

are not identically satisfied, so the PDE fails the test.

It is time to define a quantity which, although useless for the test itself, is of first importance at the second stage of Painlevé analysis, which will be developed in Sects. 5 to 9. This quantity is defined from the finite subset of nonpositive powers of the Laurent series for u.

Definition 4.1. *Given a family (p, u_0), the singular part operator $\mathcal{D}$ is defined as*

$$\operatorname{Log}\varphi \mapsto \mathcal{D}\operatorname{Log}\varphi = u_T(0) - u_T(\infty), \tag{83}$$

in which the notation $u_T(\varphi_0)$, which emphasizes the dependence on the movable constant φ_0, stands for the principal part (T like truncation) of the Laurent series (55), i.e. the finite subset of its nonpositive powers

$$u_T(\varphi_0) = \sum_{j=0}^{-p} u_j\chi^{j+p}. \tag{84}$$

For most PDEs, this operator is linear.
For the Laurent series already considered (63), (72), (73), (80), the operator is, respectively, $\mathcal{D} = -2a\partial_x^2, a, a\partial_x, d\partial_x$. For the Kuramoto-Sivashinsky equation (3), there exists a unique Laurent series (55) with $p = -3$ (given by (5) for a particular value of χ, and by the derivative of (347) for any χ), with a singular part operator equal to

$$\mathcal{D} = 60\nu\partial_x^3 + (60/19)\mu\partial_x. \tag{85}$$

This is precisely the third order linear operator on the rhs of (49).

4.3 The various ways to pass or fail the Painlevé test for PDEs

If one processes a multidimensional PDE the coefficients of which depend on some parameters μ,

$$\mathbf{E}(\mathbf{u}, \mathbf{x}; \mu) = 0, \tag{86}$$

(boldface means multicomponent), the Painlevé test generates the following output:

1. leading order (p, u_0), Fuchs indices i and singular part operator $\mathcal{D}$ for each admissible family,
2. diophantine conditions that all singularity orders p and all Fuchs indices i be integer, conditions whose solution creates constraints of the type
$$F(\mu, C) = 0, \tag{87}$$
3. no-log conditions
$$\forall i \; \forall n \; \forall u_j \; : \; \mathbf{Q}_i^n(\mu, S, C, u_j) = 0, \tag{88}$$
arising from any integer Fuchs index i, in which n is the Fuchsian perturbation order [29] if necessary, u_j are the arbitrary coefficients u_{arb} introduced at earlier Fuchs indices j.

In particular, the Laurent series (55) are of no use and should not be computed beyond the highest Fuchs integer. All this output (items 1 and 3) is easily produced with a computer algebra program and, in all further examples, we will simply list these results without any more detail.

Strictly speaking, the answer provided by the test to the question "Has the PDE the PP?" is either *no* (at least one of the necessary conditions fails) or *maybe* (all necessary conditions are satisfied, and the PDE may possess the PP but this still has to be proven). It is never *yes*, as shown by the famous counterexample of Picard (the second order ODE with the general solution $\wp(\lambda \operatorname{Log}(c_1 x + c_2), g_2, g_3)$, which therefore has the PP iff $2\pi i \lambda$ is a period of the Weierstrass elliptic function $\wp$, a transcendental condition impossible to generate by a finite algebraic procedure).

Now that the necessary part (i.e. the Painlevé test) of Painlevé analysis is finished, let us turn to the question of sufficiency.

To reach our goal which is to obtain as many analytic results as possible, we do not adopt such a drastic point of view, but the opposite one. Instead of the logical *and* performed by the mathematician on all the necessary conditions generated by the test, we perform a logical *or* operation on these conditions. Therefore the above Painlevé test must be performed to its end, i.e. without stopping even in case of failure of some condition, so as to collect all the necessary conditions. Turning to sufficiency, these conditions have to be examined independently in the hope of finding some global element of integrability. An application of this point of view to the Lorenz model, a third order ODE, can be found in Sect. 6.7 of Ref. [26].

If the PDE under study possesses a singlevalued exact solution, there must exist a Laurent series (55) which represents it locally. Therefore the practical criterium to be implemented deals with the existence of *particular* Laurent series, and the result of the test belongs to one of the following mutually exclusive situations.

1. (The best situation) Success of the test, at least for some values of μ selected by the test. The PDE may have the PP, and one must look for its BT;
2. There exists at least one value of $(\mu, \varphi, u_{\text{arb}})$ which ensures the existence of a particular Laurent series. For these values, an exact solution may exist;
3. There exists at least one value of $(\mu, \varphi, u_{\text{arb}})$ enforcing some of, but not all, the no-log conditions of at least one particular Laurent series. Quite probably no exact solution exists, but there may exist a conservation law (a first integral for an ODE);
4. (The worst situation) There is no value of $(\mu, \varphi, u_{\text{arb}})$ enforcing at least one of the no-log conditions of the various series. Quite probably the PDE is chaotic and possesses no exact solution at all.

Examples of these various situations are, respectively:

1. All the PDEs which have the PP (sine-Gordon, Korteweg-de Vries, ...), but also the counterexample of Picard quoted above;
2. The equation of Kuramoto and Sivashinsky (3), with the particular Laurent series (5);
3. The Lorenz model for $b = 2\sigma$, for which the no-log condition at $i = 4$ is violated and there exists a first integral;
4. The Rössler dynamical system for which the unique family has the never satisfied condition $Q_2 \equiv 16 = 0$.

5 Ingredients of the "singular manifold method"

The methods to handle the integrable and nonintegrable situations are the same, simply a more or less important result is obtained.

The goal is to find a (possibly degenerate) couple (singular part transformation, Lax pair) in order to deduce the Bäcklund transformation or, if a BT does not exist, to generate some exact solutions.

The full Laurent series is of no help, for this is not an exact solution according to the definition in Sect. 2. Since this is the only available piece of information and since a finite (closed form) expression is required to represent an exact solution, let us represent, following the idea of Weiss, Tabor, and Carnevale [65], an unknown exact solution u as the sum of a singular part, built from the finite principal part of the Laurent series (i.e. the finite number of terms with negative powers), and of a regular part made of one term denoted U. This assumption is identical to that of a singular part transformation (46), in which nothing would be specified on U.

This method is widely known as the *singular manifold method* or *truncation method* because it selects the beginning of the Laurent series and discards ("truncates") the remaining infinite part.

Since its introduction by WTC [65], it has been improved in many directions [38, 47, 58, 40, 35, 49, 41], and we present below the current status of the method.

5.1 The ODE situation

For the six ordinary differential equations (ODE) (P1)–(P6) which bear his name, Painlevé proved the PP by showing [105, 106] the existence of one (case of P1) or two (P2–P6) function(s) $\tau = \tau_1, \tau_2$, called tau-functions, linked to the general solution u by logarithmic derivatives

$$\text{P1} \;:\; u = \mathcal{D}_1 \operatorname{Log} \tau \tag{89}$$

$$\text{Pn},\; n = 2, \dots, 6 :\; u = \mathcal{D}_n (\operatorname{Log} \tau_1 - \operatorname{Log} \tau_2) \tag{90}$$

where the operators $\mathcal{D}_n$ are linear:

$$\mathcal{D}_1 = -\partial_x^2,\; \mathcal{D}_2 = \mathcal{D}_4 = \pm\partial_x,\; \mathcal{D}_3 = \pm e^{-x}\partial_x, \tag{91}$$

$$\mathcal{D}_5 = \pm x e^{-x} (2\alpha)^{-1/2} \partial_x,\; \mathcal{D}_6 = \pm x(x-1) e^{-x} (2\alpha)^{-1/2} \partial_x. \tag{92}$$

These functions τ_1, τ_2 satisfy third order nonlinear ODEs and they have the same kind of singularities than solutions of *linear* ODEs, namely they have no movable singularities at all; they are entire functions for P1–P5, and their only singularities for P6 are the three fixed critical points $(\infty, 0, 1)$.

ODEs cannot possess an auto-BT, since the number of independent arbitrary coefficients in a solution cannot exceed the order of the ODE. They can however possess a Schlesinger transformation (see definition Sect. 11).

5.2 Transposition of the ODE situation to PDEs

For PDEs, similar ideas prevail. The analogue of (89)–(90), with an additional rhs U, is now the singular part transformation (46), and the scalar(s) ψ to which the scalar(s) τ are linked by (47) are assumed to satisfy a linear system, the Lax pair.

Another interesting observation must be made. There seems to exist two and only two classes of integrable 1 + 1-dimensional PDEs, at least at the level of the base member of a hierarchy: those which have only one family of movable singularities, and those which have only pairs of families with opposite principal parts, similarly to the distinction between P1 on one side and P2–P6 on the other side. Among the $1+1$-dimensional integrable equations, those with one family include KdV, the AKNS, Hirota-Satsuma and Boussinesq equations; they also include the Sawada-Kotera, Kaup-Kupershmidt and Tzitzéica equations because only one of their two families is relevant [41, 9]. Equations with pairs of opposite families include sine-Gordon, mKdV and Broer-Kaup (two families each), NLS (four families).

5.3 The singular manifold method as a singular part transformation

As qualitatively described in Sect. 5, the singular manifold method looks very much like a resummation of the Laurent series, just like the geometric series

$$\sum_{j=0}^{+\infty} x^j, \ x \to 0, \tag{93}$$

becomes a finite sum in the resummation variable $X = x/(1-x)$

$$\sum_{J=0}^{1} X^J, \ X \to 0. \tag{94}$$

The principle of the method is the following [65]. One first notices that the (infinite) Laurent series (55) in the variable $\varphi - \varphi_0$ can be rewritten as the sum of two terms

$$u = \mathcal{D} \operatorname{Log} \tau + \text{ regular part}. \tag{95}$$

The first term $\mathcal{D} \operatorname{Log} \tau$, built from the singular part operator defined in Sect. 4.2, is a finite Laurent series and, if τ is any variable fulfilling the two requirements for an expansion variable enunciated in Sect. 4.1, it captures all the singularities of u when $\varphi \to \varphi_0$. The second term, temporarily called "regular part" for this reason, is yet unspecified. The sum of these two terms is therefore a finite Laurent series (hence the name *truncated series*), and the variable τ is a *resummation variable* which has made the former infinite series in $\varphi - \varphi_0$ a finite one. One then tries to identify this resummation (95) with the definition of a singular part transformation (46). This involves two features. The first feature is to uncover a link (47) between τ and a scalar component ψ of a Lax pair. The second feature is to prove that the left over "regular part" is indeed a second solution to the PDE under study.

5.4 The degenerate case of linearizable equations

The Burgers equation (71), under the transformation of Forsyth (Ref. [52] p. 106),

$$v = a \operatorname{Log} \tau, \ \tau = \psi, \tag{96}$$

is linearized into the heat equation

$$b\psi_t + \psi_{xx} + G(t)\psi = 0. \tag{97}$$

This can be considered as a degenerate singular part transformation (46), in which U is identically zero and ψ satisfies a single linear equation, not a pair of linear equations, so this fits the general scheme.

Another classical example is the second order particular Monge-Ampère equation $s + pq = 0$, linearized into the d'Alembert equation $s = 0$:

$$s + pq \equiv u_{xt} + u_x u_t = 0, \tag{98}$$

$$u = \operatorname{Log} \tau, \ \tau = \psi, \ \psi_{xt} = 0. \tag{99}$$

5.5 Choices of Lax pairs and equivalent Riccati pseudopotentials

To fix the ideas, we list here a few usual second order and third order Lax pairs depending on undetermined coefficients, together with the constraints imposed on these coefficients by the commutativity condition.

It is sometimes appropriate to represent an n-th order Lax pair by the $2(n-1)$ equations satisfied by an equivalent $(n-1)$-component pseudopotential $\mathbf{Y}$ of Riccati type, the first component of which is chosen as

$$Y_1 = \psi_x/\psi, \tag{100}$$

in which ψ is a scalar component of the Lax pair.

Second-order Lax pairs and their privilege

The general second-order scalar Lax pair reads, in the case of two independent variables (x, t),

$$L_1\psi \equiv \psi_{xx} - d\psi_x - a\psi = 0, \tag{101}$$

$$L_2\psi \equiv \psi_t - b\psi_x - c\psi = 0, \tag{102}$$

$$[L_1, L_2] \equiv X_0 + X_1\partial_x, \tag{103}$$

$$\tag{104}$$

$$X_0 \equiv -a_t + a_x b + 2ab_x + c_{xx} - c_x d = 0, \tag{105}$$

$$X_1 \equiv -d_t + (b_x + 2c - bd)_x = 0. \tag{106}$$

For the inverse scattering method to apply, the coefficients (d, a) of the x-part (101) are required to depend linearly on the field U of the PDE.

The Lax pair (101)–(102) is identical to a linearized version of the Riccati system satisfied by the most general expansion variable Y defined by (66), under the correspondence

$$Y = B^{-1}\frac{\psi}{\psi_x}, \ B \ne 0, \tag{107}$$

$$d = 2A, \ a = A_x - A^2 - S/2, \ b = -C, \ c = C_x/2 + AC + \int A_t \mathrm{d}x, \tag{108}$$

and the commutator of the Lax pair is (78).

In particular, when the coefficient d is zero or when, by a linear change $\psi \mapsto e^{\int d \mathrm{d}x/2} \psi$, it can be set to zero without altering the linearity of a on U, the correspondence is [38]

$$\chi = \frac{\psi}{\psi_x}, \ B = 1, \ A = 0, \tag{109}$$

$$d = 0, \ a = -S/2, \ b = -C, \ c = C_x/2, \tag{110}$$

$$L_1\psi \equiv \psi_{xx} + \frac{S}{2}\psi = 0, \tag{111}$$

$$L_2\psi \equiv \psi_t + C\psi_x - \frac{C_x}{2}\psi = 0, \tag{112}$$

$$2[L_1, L_2] \equiv X = S_t + C_{xxx} + CS_x + 2C_x S = 0. \tag{113}$$

Therefore second order Lax pairs are privileged in the singularity approach, in the sense that their coefficients can be identified with the elementary homographic invariants S, C of the invariant Painlevé analysis and, if appropriate, A, B. Conversely, and this has historically been the reason of some errors described in Sect. 8.2, at the stage of searching for the BT, these homographic invariants S, C are useless when the Lax order is higher than two and they should not be considered.

As explained in Sect. 5.3, given a Lax pair, one should define from it either one or two scalars ψ_f. Consider the second order Lax pair defined by the gradient of Y. Then, for one-family PDEs, this unique scalar ψ is defined by (107). For two-family PDEs, the two scalars ψ_f are defined by

$$Y = \frac{\psi_1}{\psi_2}, \tag{114}$$

which leads to the zero-curvature representation of the Lax pair

$$(\partial_x - L)\begin{pmatrix}\psi_1\\ \psi_2\end{pmatrix} = 0,\ L = \begin{pmatrix} -A - B^{-1}B_x/2 & B^{-1} \\ B(A_x - A^2 - S/2) & A + B^{-1}B_x/2\end{pmatrix}, \tag{115}$$

$$(\partial_t - M)\begin{pmatrix}\psi_1\\ \psi_2\end{pmatrix} = 0, \tag{116}$$

$$M = \begin{pmatrix} AC + C_x/2 - B^{-1}B_t/2 & -CB^{-1} \\ B((CS + C_{xx})/2 + A_t + CA^2 + C_xA) & -AC - C_x/2 + B^{-1}B_t/2\end{pmatrix}.$$

The reason why the Riccati form is the most suitable characterization of the Lax pair is that it allows two linearizations [40, 49], namely (107) and (114), depending on whether the PDE has one family or two opposite families.

Third-order Lax pairs

The general third-order scalar Lax pair is defined as

$$\begin{aligned}
L_1\psi &\equiv \psi_{xxx} - f\psi_{xx} - a\psi_x - b\psi = 0, && (117)\\
L_2\psi &\equiv \psi_t - c\psi_{xx} - d\psi_x - e\psi = 0, && (118)\\
[L_1, L_2] &\equiv X_0 + X_1\partial_x + X_2\partial_x^2, && (119)\\
X_0 &\equiv -b_t - ae_x + e_{xxx} + b_{xx}c + 2bcf_x + bc_xf - e_{xx}f \\
&\quad +3bc_{xx} + 3b_xc_x + 3bd_x + b_xd = 0, && (120)\\
X_1 &\equiv -a_t + 3e_{xx} + 2b_xc + a_{xx}c + d_{xxx} + 3ac_{xx} + 2ad_x \\
&\quad +3a_xc_x + 3bc_x + a_xd + 2acf_x + ac_xf - 2e_xf - d_{xx}f = 0, && (121)\\
X_2 &\equiv -f_t + (cf^2 + cf_x + 2c_xf + df + 2ac + c_{xx} + 3d_x + 3e)_x = 0. && (122)
\end{aligned}$$

An equivalent two-component pseudopotential is the projective Riccati one $\mathbf{Y} = (Y_1, Y_2)$ [38, 39] (written below, for simplicity, in the case $f = 0$)

$$\begin{aligned}
Y_1 &= \frac{\psi_x}{\psi},\ Y_2 = \frac{\psi_{xx}}{\psi}, && (123)\\
Y_{1,x} &= -Y_1^2 + Y_2, && (124)\\
Y_{2,x} &= -Y_1Y_2 + aY_1 + b, && (125)\\
Y_{1,t} &= -(dY_1 + cY_2)Y_1 + (ac + d_x)Y_1 + (c_x + d)Y_2 + e_x + bc && (126)\\
&= (cY_2 + dY_1 + e)_x, && (127)\\
Y_{2,t} &= -(dY_1 + cY_2)Y_2 + (2ac_x + a_xc + bc + d_{xx} + ad + 2e_x)Y_1 \\
&\quad +(c_{xx} + 2d_x + ac)Y_2 + 2bc_x + b_xc + bd + e_{xx}, && (128)\\
Y_{1,tx} - Y_{1,xt} &= X_1 + X_2Y_1, && (129)\\
Y_{2,tx} - Y_{2,xt} &= -X_0 + X_2Y_2. && (130)
\end{aligned}$$

When there is no reason to distinguish between x and t, for instance because the PDE is invariant under the permutation (Lorentz transformation)

$$\mathcal{P}:\ (\partial_x, \partial_t) \to (\partial_t, \partial_x), \tag{131}$$

it is natural to consider the following third-order matrix Lax pair, invariant under (131), defined in the basis (ψ_x, ψ_t, ψ) [9],

$$(\partial_x - L)\begin{pmatrix}\psi_x\\ \psi_t\\ \psi\end{pmatrix} = 0,\ L = \begin{pmatrix}f_1 & f_2 & f_3\\ g_1 & g_2 & g_3\\ 1 & 0 & 0\end{pmatrix}, \tag{132}$$

$$(\partial_t - M)\begin{pmatrix}\psi_x\\ \psi_t\\ \psi\end{pmatrix} = 0,\ M = \begin{pmatrix}g_1 & g_2 & g_3\\ h_1 & h_2 & h_3\\ 0 & 1 & 0\end{pmatrix}. \tag{133}$$

In the equivalent projective Riccati components (Y_1, Y_2)

$$Y_1 = \frac{\psi_x}{\psi},\ Y_2 = \frac{\psi_t}{\psi}, \tag{134}$$

with the property $Y_{1,t} = Y_{2,x}$, it is defined as

$$Y_{1,x} = -Y_1^2 + f_1 Y_1 + f_2 Y_2 + f_3, \tag{135}$$
$$Y_{2,x} = -Y_1 Y_2 + g_1 Y_1 + g_2 Y_2 + g_3, \tag{136}$$
$$Y_{1,t} = -Y_1 Y_2 + g_1 Y_1 + g_2 Y_2 + g_3, \tag{137}$$
$$Y_{2,t} = -Y_2^2 + h_1 Y_1 + h_2 Y_2 + h_3. \tag{138}$$

The nine functions $f_j, g_j, h_j, j = 1,2,3$, must satisfy six cross-derivative conditions $X_j = 0$

$$(Y_{1,x})_t - (Y_{1,t})_x = X_0 + X_1 Y_1 + X_2 Y_2 = 0, \tag{139}$$
$$(Y_{2,x})_t - (Y_{2,t})_x = X_3 + X_4 Y_1 + X_5 Y_2 = 0, \tag{140}$$

easy to write explicitly. It is worth noticing that there exists no such invariant second-order matrix Lax pair.

5.6 The admissible relations between τ and ψ

By elimination of ∂_t, one of the two PDEs defining the BT to be found can be made an ODE, e.g. (64) or (152). This nonlinear ODE, with coefficients depending on U and, in the 1+1-dimensional case, on an arbitrary constant λ, has the property [41] of being linearizable. This very strong property restricts the admissible choices (47) to a finite number of possibilities, and full details can be found in Musette lecture [91].

6 The algorithm of the singular manifold method

We now have all the ingredients to give a general exposition of the method in the form of an algorithm. The present exposition largely follows the lines of Ref. [41]. The various situations thus implemented are:

one-family and two-opposite-family PDEs, second or higher order Lax pair, various allowed links between the two sets of functions (τ, ψ).

Consider a PDE (53) with only one family of movable singularities or exactly two families of movable singularities with opposite values of u_0, and denote $\mathcal{D}$ the *singular part operator* of either the unique family or anyone of the two opposite families.

First step. Assume a singular part transformation defined as

$$u = U + \mathcal{D}(\operatorname{Log} \tau_1 - \operatorname{Log} \tau_2), \ E(u) = 0, \tag{141}$$

with u a solution of the PDE under consideration, U an unspecified field which most of the time will be found to be a second solution of the PDE, τ_f the "entire" function (or more precisely ratio of entire functions) attached to each family f. For one-family PDEs, one denotes $\tau_1 = \tau, \tau_2 = 1$, so the singular part transformation assumption (3) becomes

$$u = U + \mathcal{D} \operatorname{Log} \tau, \ E(u) = 0. \tag{142}$$

A consequence of the assumption (3) is the existence of the involution

$$\forall f : \ (u, U, \tau_f) \mapsto (U, u, \tau_f^{-1}), \tag{143}$$

since the operator $\mathcal{D}$ is linear, and, for two-family PDEs, of the involution

$$\forall (u, U) : \ (\mathcal{D}, \tau_1, \tau_2) \mapsto (-\mathcal{D}, \tau_2, \tau_1). \tag{144}$$

Second step. Choose the order two, then three, then ..., for the unknown Lax pair, and define one or two (as many as the number of families) *scalars* ψ_f from the component(s) of its wave vector (e.g. the scalar wave vector if the PDE has one family and the pair is defined in scalar form). Such sample Lax pairs and scalars can be found in Sect. 5.5.

Third step. Choose an explicit link F

$$\forall f : \ \mathcal{D} \operatorname{Log} \tau_f = F(\psi_f), \tag{145}$$

the same for each family f, between the functions τ_f and the scalars ψ_f defined from the Lax pair. According to Sect. 5.6, at each scattering order, there exists only a finite number of choices (94), among them the most frequent one

$$\forall f : \ \tau_f = \psi_f. \tag{146}$$

Fourth step. Define the "truncation" and solve it, that is to say: with the assumptions (3) for a singular part transformation, (94) for a link between τ_f and ψ_f, (101)–(102) or (117)–(118) or other for the Lax pair in ψ, express $E(u)$ as a polynomial in the derivatives of ψ_f which is irreducible *modulo* the Lax pair. For the above pairs and a one-family PDE, this amounts to eliminate any derivative of ψ of order in (x,t) higher than or equal to $(2,0)$ or $(0,1)$ (second order case) or to $(3,0)$ or $(0,1)$ (third order), thus resulting in a polynomial of one variable ψ_x/ψ (second order) or two variables $\psi_x/\psi, \psi_{xx}/\psi$ (third order)

$$E(u) = \sum_{j=0}^{-q} E_j(S,C,U)(\psi/\psi_x)^{j+q} \text{ (one-family PDE, second order)}, \quad (147)$$

$$E(u) = \sum_{k\geq 0}\sum_{l\geq 0} E_{k,l}(a,b,c,d,e,U)(\psi_x/\psi)^k(\psi_{xx}/\psi)^l$$
$$\text{(one-family PDE, third order).} \quad (148)$$

Since one has no more information on this polynomial $E(u)$ except the fact that it must vanish, one requests that it identically vanishes, by solving the set of *determining equations*

$$\forall j \quad E_j(S,C,U) = 0 \text{ (one-family PDE, second order)} \quad (149)$$
$$\forall k\ \forall l \quad E_{k,l}(a,b,c,d,e,U) = 0 \text{ (one-family PDE, third order)} \quad (150)$$

for the unknown coefficients (S,C) or (a,b,c,d,e) as functions of U, and one establishes the constraint(s) on U by eliminating (S,C) or (a,b,c,d,e). The strategy of resolution is developed in Sect. 7.3.

The constraints on U reflect the integrability level of the PDE. If the only constraint on U is to satisfy some PDE, one is on the way to an auto-BT if the PDE for U is the same as the PDE for u, or to a remarkable correspondence (hetero-BT) between the two PDEs.

The second, third and fourth steps must be repeated until a success occurs. The process is successful if and only if all the following conditions are met

1. U comes out with one constraint exactly, namely: to be a solution of some PDE,
2. (if an auto-BT is desired) the PDE satisfied by U is identical to (53),
3. the vanishing of the commutator $[L_1, L_2]$ is equivalent to the vanishing of the PDE satisfied by U,
4. in the 1+1-dimensional case only and if the PDE satisfied by U is identical to (53), the coefficients depend on an arbitrary constant λ, the spectral or Bäcklund parameter.

At this stage, one has obtained the singular part transformation and the Lax pair.

Fifth step. Obtain the two equations for the BT by eliminating ψ_f [5] between the singular part transformation and the Lax pair. This sometimes uneasy operation when the order n of the Lax pair is too high may become elementary by considering the equivalent Riccati representation of the Lax pair and eliminating the appropriate components of $\mathbf{Y}$ rather than ψ. Assume for instance that $\tau = \psi$, $\mathcal{D} = \partial_x$, and the PDE has only one family. Then Eq. (3) reads

$$Y_1 = u - U \tag{151}$$

with Y_1 defined in (100), and the BT is computed as follows: eliminate all the components of $\mathbf{Y}$ but Y_1 between the equations for the gradient of $\mathbf{Y}$, then in the resulting equations substitute Y_1 as defined in (151).

If the computation of the BT requires the elimination of Y_2 between (124)–(128), this BT is

$$Y_{1,xx} + 3Y_1Y_{1,x} + Y_1^3 - aY_1 - b = 0, \tag{152}$$
$$Y_{1,t} - (cY_{1,x} + cY_1^2 + dY_1 + e)_x = 0, \tag{153}$$
$$(Y_{1,xx})_t - (Y_{1,t})_{xx} = X_0 + X_1Y_1 + X_2Y_1^2 = 0, \tag{154}$$

in which Y_1 is replaced by an expression of $u - U$, e.g. (151).

Although, let us repeat it, the method equally applies to integrable as well as nonintegrable PDEs, examples are split according to that distinction, to help the reader to choose his/her field of interest.

6.1 Where to truncate, and with which variable?

This section is self-contained, and mainly destinated to persons accustomed to perform the WTC truncation. Although some paragraphs might be redundant with Sect. 6, it may help the reader by presenting a complementary point of view.

Let us assume in this section that the unknown Lax pair is second order. Then the truncation defined in the fourth step of Sect. 6 is performed in the style of Weiss *et al.* [65], i.e. with a single variable. This WTC truncation consists in forcing the series (55) to terminate; let us denote p and q the singularity orders of u and $E(u)$, $-p'$ the rank at which the series for u stops, and $-q'$ the corresponding rank of the series for E

$$u = \sum_{j=0}^{-p'} u_j Z^{j+p}, \ u_0 u_{-p'} \neq 0, \ E = \sum_{j=0}^{-q'} E_j Z^{j+q}, \tag{155}$$

in which the *truncation variable* Z chosen by WTC is $Z = \varphi - \varphi_0$. Since one has no more information on Z, the method of WTC is to require the separate satisfaction of each of the *truncation equations*

$$\forall j = 0, \dots, -q' \ : E_j = 0. \tag{156}$$

In earlier presentations of the method, one had to prove by recurrence that, assuming that enough consecutive coefficients u_j vanish beyond $j = -p'$, then all further coefficients u_j would vanish. This painful task is useless if one defines the process as done above.

The first question to be solved is: what are the admissible values of p', i.e. those which respect the condition $u_{-p'} \neq 0$?

The answer depends on the choice of the truncation variable Z. In Sect. 4.1, three choices were presented, $Z =$ either $\varphi - \varphi_0$, χ or Y, respectively defined by equations (54), (58), (66), with the property that any two of their inverse are linearly dependent.

The advantage of χ or Y over $\varphi - \varphi_0$ is the following. The gradient of χ (resp. Y) is a polynomial of degree two in χ (resp. Y), so each derivation of a monomial aZ^k increases the degree by one, while the gradient of $\varphi - \varphi_0$ is a polynomial of degree zero in $\varphi - \varphi_0$, so each derivation decreases the degree by one. Consequently, one finds two solutions and only two to the condition $u_{-p'} \neq 0$ [108]

1. $p' = p, q' = q$, in which case the three truncations are identical, since the three sets of equations $E_j = 0$ are equivalent (the finite sum $\sum E_j Z^{j+q}$ is just the same polynomial of Z^{-1} written with three choices for its base variable),
2. for χ and Y only, $p' = 2p, q' = 2q$, in which case the two truncations are different since the two sets of equations $E_j = 0$ are inequivalent (they are equivalent only if $A = 0$).

To perform the first truncation $p' = p, q' = q$, one must then choose $Z = \chi$ since Y brings no more information and $\varphi - \varphi_0$ creates equivalent but lengthier expressions.

To perform the second truncation $p' = 2p, q' = 2q$, one must choose $Z = Y$, since χ would create the *a priori* constraint $A = 0$.

The second question to be solved is: given some PDE with such and such structure of singularities, and assuming that one of the above two truncations is relevant (which is a separate topic), which one should be selected?

The answer lies in the two elementary identities [32]

$$\tanh z - \frac{1}{\tanh z} = -2i \,\mathrm{sech} \left[2z + i\frac{\pi}{2}\right], \quad \tanh z + \frac{1}{\tanh z} = 2 \tanh \left[2z + i\frac{\pi}{2}\right]. \tag{157}$$

Let us explain why on two examples, the ODEs whose general solution is $\tanh(x - x_0)$ and $\mathrm{sech}(x - x_0)$, namely

$$E \equiv u' + u^2 - 1 = 0, \;\; u = \tanh(x - x_0), \tag{158}$$

$$E \equiv v'^2 + a^{-2} v^4 - v^2 = 0, \;\; v = a \,\mathrm{sech}(x - x_0), \tag{159}$$

(this is just for convenience that we do not set $a = 1$). Equation (158) has the single family

$$p = -1, q = -2, u_0 = 1, \text{Fuchs indices} = (-1), \tag{160}$$

and equation (159) has the two opposite families

$$p = -1, q = -4, v_0 = ia, \text{Fuchs indices} = (-1), \tag{161}$$

in which ia denotes any square root of $-a^2$. The first truncation

$$u = \sum_{j=0}^{-p} u_j \chi^{j+p}, \ E = \sum_{j=0}^{-q} E_j \chi^{j+q}, \ \forall j \ : E_j = 0, \tag{162}$$

generates the respective results

$$u = \chi^{-1}, \ S = -2, \tag{163}$$
$$v = ia\chi^{-1}, \ E_2 \equiv a^2(1 - S) = 0, \ E_3 \equiv 0, \ E_4 \equiv -a^2 S^2/4, \tag{164}$$

thus providing (after integration of the Riccati ODE (64)) the general solution of equation (158), and no solution at all for equation (159).

The second truncation

$$u = \sum_{j=0}^{-2p} u_j Y^{j+p}, \ E = \sum_{j=0}^{-2q} E_j Y^{j+q}, \ \forall j \ : E_j = 0, \tag{165}$$

generates the respective results

$$u = B^{-1}Y^{-1} + (1/4)BY, \ A = 0, \ S = -1/2, \ B \text{ arbitrary}, \tag{166}$$
$$v = iaB^{-1}Y^{-1} - (1/4)iaBY, \ A = 0, \ S = -1/2, \ B \text{ arbitrary}, \tag{167}$$

thus providing, thanks to the identities (157), the general solution for both equations.

The conclusions from this exercise which can be generalized are:

1. for PDEs with only one family, the second truncation brings no additional information as compared to the first one and is always useless;
2. for PDEs with two opposite families (two opposite values of u_0 for a same value of p), the first truncation can never provide the general solution and can only provide particular solutions, while the second one may provide the general solution.

This defines the guideline to be followed in the respective Sects. 7 and 9. The question of the relevance of the parameter B, which seems useless in the above two examples, is addressed in Sect. 9.

7 The singular manifold method applied to one-family PDEs

7.1 Integrable equations with a second order Lax pair

There is only one truncation variable, which must be chosen as χ.

Weiss introduced a nice notion, initially for one-family integrable equations with a second order Lax pair, later extended to two-family such equations by Pickering [49]. This is the following.

Definition 7.1. *([20]) Consider the set of $-q+p$ determining equations (149) $E_j = 0$, which depend on (S, C, U). One calls* ***singular manifold equation*** *(SME) the result of the elimination of U between them.*

In the two-family situation, these determining equations also depend on (A, B), see (165), and the extension of this definition [49] is to also require the elimination of (A, B).

Despite its name, originally restricted to integrable equations, the SME can be made of several equations in the nonintegrable case.

The SME has the following properties.

1. unicity, whatever be the integrability of the PDE,
2. invariance under homography by construction [6], i.e. dependence only on one Schwarzian S and as many C quantities as independent variables other than the one in the Schwarzian,
3. the SME set is made of one and only one equation if and only if the PDE is integrable.

Although one can define a SME whatever be the order of the Lax pair, it is inconsistent, as will be explained in Sect. 8.2, to do so whenever this order is higher than two.

The Liouville equation

It is convenient to consider, following Zhiber and Shabat [133], the equation

$$E(u) \equiv u_{xt} + \alpha e^{u} + a_1 e^{-u} + a_2 e^{-2u} = 0,\ \alpha \neq 0, \tag{168}$$

which has the advantage to include the Liouville equation $a_1 = a_2 = 0$, the sine-Gordon equation ($a_1 \neq 0, a_2 = 0$) and the Tzitzéica equation ($a_1 = 0, a_2 \neq 0$). As to the case $a_1 a_2 \neq 0$, it fails the test. Let us consider here the Liouville case. The results to be found are its auto-BT [87] and its hetero-BT to the d'Alembert equation. This will be achieved with two different truncations.

Although not algebraic in u, the PDE is algebraic in either e^{u} or e^{-u}.

Equation (168) always possesses the family

$$e^{u} \sim -(2/\alpha)\varphi_x \varphi_t (\varphi - \varphi_0)^{-2},\ \text{indices } (-1, 2),\ \mathcal{D} = (2/\alpha)\partial_x \partial_t. \tag{169}$$

For Liouville, this is the only family.

The special form of Liouville equation allows the assumption

$$e^{u} = \mathcal{D} \operatorname{Log} \tau + e^{U},\ E(u) = 0,\ \mathcal{D} = (2/\alpha)\partial_x \partial_t, \tag{170}$$

to be integrated twice to yield

$$u = -2 \operatorname{Log} \tau + V,\ E(u) = \sum_{j=0}^{2} E_j \tau^{j-2} = 0, \tag{171}$$

in which nothing is assumed on V.

The Liouville equation is nongeneric for the singular manifold method in the sense that it is linearizable into another equation (thus, it should even *not* be part of the Sect. 7.1).

Therefore we define the first truncation in an exceptional way, namely we do not assume any linear relations on $\tau \equiv \psi$ and just treat τ as the truncation variable. The three determining equations are then quite simple [9]

$$E_0 \equiv 2\tau_x\tau_t + \alpha e^V = 0, \tag{172}$$

$$E_1 \equiv \tau_{xt} = 0, \tag{173}$$

$$E_2 \equiv V_{xt} = 0, \tag{174}$$

and their general solution depends on two arbitrary functions of one variable

$$\tau = f(x) + g(t), \tag{175}$$

$$e^V = -\frac{2}{\alpha}\tau_x\tau_t = -\frac{2}{\alpha}f'(x)g'(t), \tag{176}$$

$$e^u = -\frac{2}{\alpha}\frac{\tau_x\tau_t}{\tau^2} = -\frac{2}{\alpha}\frac{f'(x)g'(t)}{(f(x)+g(t))^2}, \tag{177}$$

$$e^U = \tau^{-2}e^V + \frac{2}{\alpha}\frac{\tau_x\tau_t}{\tau^2} = 0. \tag{178}$$

Thus, the two fields u and V are the general solution of, respectively, the Liouville and d'Alembert equations. The hetero-BT between these two equations is provided by the elimination of f and g between (176), (177) and the $x-$ and $t-$derivatives of (171)

$$(u-v)_x = \alpha\lambda e^{(u+v)/2}, \tag{179}$$

$$(u+v)_t = -2\lambda^{-1}e^{(u-v)/2}, \tag{180}$$

in which v is another solution of d'Alembert equation defined by

$$e^v = (\lambda\tau_t)^{-2}e^V = -\frac{2}{\alpha}\lambda^{-2}\frac{f'(x)}{g'(t)}. \tag{181}$$

Remark. When performing the truncation (170), Tamizhmani and Lakshmanan [117] already found $e^U = 0, \tau_{xt} = 0$ as a *particular* solution, while the above truncation (171) proves it to be the *general* solution. Another difference between the two truncations is the presence of a field V in (171), which allows us to find in addition the hetero-BT between the Liouville and d'Alembert equations.

Let us now define the second truncation, by the assumption

$$u = -2\,\mathrm{Log}\,\tau + \tilde{W}, \tag{182}$$

and the link (146), with ψ solution of the Lax pair (101)–(102). Introducing the Riccati variable Y defined by (107), this second truncation is equivalent to [9]

$$u = -2 \operatorname{Log} Y + W,\ Y^{-1} = B(\chi^{-1} + A),$$
$$E(u) = \sum_{j=0}^{4} E_j(S, C, A, B, W) Y^{j-2},\ \forall j:\ E_j = 0, \tag{183}$$

and its result is recovered from the truncation of sine-Gordon in Sect. 9.1 by simply setting $a_1 = 0$.

The AKNS equation

The AKNS equation [1]

$$E(u) \equiv u_{xxxt} + 4\alpha^{-1}(2(u_x - \beta)u_{xt} + (u_t - \gamma)u_{xx}) = 0 \tag{184}$$

admits the single family

$$p = -1,\ q = -5,\ u_0 = \alpha,\ \text{indices } (-1, 1, 4, 6),\ \mathcal{D} = \alpha\partial_x, \tag{185}$$

so the assumption for the singular part transformation is (142). Let us choose at second step the scalar Lax pair (88)–(89) for ψ, at third step the link (146) between τ and ψ. Then there are only three non identically zero determining equations (149) [89]

$$E_2 \equiv 4\alpha SC + 8(U_t - \gamma) - 16C(U_x - \beta) = 0, \tag{186}$$
$$E_3 \equiv -\alpha(CS_x + 4SC_x) + 16C_x(U_x - \beta) - 8U_{xt} + 4CU_{xx} = 0, \tag{187}$$
$$E_5 \equiv E(U) + (\alpha/2)(2SS_t - CSS_x - S_{xxt} - S_xC_{xx}) - 2S_x(U_t - \gamma)$$
$$-4S_t(U_x - \beta) - 4SU_{xt} + 2(SC + C_{xx})U_{xx} = 0, \tag{188}$$

plus the ever present condition $X = 0$, Eqn. (113). Their detailed resolution for (U_x, U_t) is as follows. One eliminates U_t between E_2 and E_3

$$E_3 + E_{2,x} \equiv 3C(-4U_{xx} + \alpha S_x) = 0, \tag{189}$$

discards the nongeneric solution $C = 0, U_t = \gamma, S_t = 0$, introduces an arbitrary function of t after one integration, and solves for U_x

$$U_x - \beta = (\alpha/4)(S + 2\lambda(t)). \tag{190}$$

Then E_2 is solved for U_t

$$U_t - \gamma = \alpha\lambda(t)C. \tag{191}$$

The cross-derivative condition $U_{xt} = U_{tx}$ is solved for S_t

$$S_t = 4\lambda(t)C_x - 2\lambda'(t). \tag{192}$$

Substituting U_x, U_t, S_t and C_{xxx} taken from (78) in E_5, one obtains the condition

$$E_5 = 2\alpha\lambda(t)\lambda'(t) = 0, \tag{193}$$

which introduces the spectral parameter as the arbitrary constant λ.

The solution for (U_x, U_t) is

$$U_x - \beta = (\alpha/4)(S + 2\lambda),\ U_t - \gamma = \alpha\lambda C, \tag{194}$$

and the elimination of U defines the SME

$$\frac{S_t}{C_x} - 4\lambda = 0. \tag{195}$$

The solution for (S, C) is

$$S = (4/\alpha)(U_x - \beta) - 2\lambda,\ C = (U_t - \gamma)/(\alpha\lambda), \tag{196}$$

and its cross-derivative condition

$$X \equiv E(U)/(\alpha\lambda) = 0 \tag{197}$$

creates on the field U the only constraint that U satisfy the AKNS PDE.

The BT is the result of the substitution $\chi^{-1} = (u - U)/\alpha$ in (64)–(65).

The KdV equation

The Korteweg-de Vries equation for u (30) is defined in conservative form, so it is cheaper to process the potential form

$$E(v) \equiv bv_t + v_{xxx} - (3/a)v_x^2 + F(t) = 0,\ u = v_x. \tag{198}$$

Its unique family is

$$p = -1,\ q = -4,\ v_0 = -2a,\ \text{indices } (-1, 1, 6),\ \mathcal{D} = -2a\partial_x. \tag{199}$$

With the assumption

$$v = V + \mathcal{D}\,\mathrm{Log}\,\tau,\ E(v) = 0, \tag{200}$$

for the singular part transformation, the choice of the second-order scalar Lax pair (88)–(89) for ψ, the link (146) between τ and ψ, one generates the three determining equations

$$E_2 \equiv -2a(bC + 2S) - 12V_x = 0, \tag{201}$$
$$E_3 \equiv 2a(bC - S)_x = 0, \tag{202}$$
$$E_4 \equiv E(V) + \frac{S}{2}E_2 - \frac{1}{2}E_{3,x} = 0. \tag{203}$$

After one integration of E_3, the system (E_2, E_3) is solved for (S, C)

$$S = -2\lambda(t) - (2/a)V_x,\ bC = 4\lambda(t) - (2/a)V_x, \tag{204}$$

in which $\lambda(t)$ is an arbitrary integration function. Then E_4, as seen from its above written compacted expression, expresses that V satisfies the PDE. Last, the cross-derivative condition (78)

$$X \equiv -2\lambda'(t) - 2(E(V))_x/(ab) = 0 \tag{205}$$

introduces the spectral parameter as an arbitrary complex constant and proves that a Lax pair has been obtained for the conservative (not the potential) equation. This Lax pair can be written, at the reader's taste, either in the scalar representation (88)–(89), with $U = V_x$,

$$L_1 \equiv \partial_x^2 - U/a - \lambda, \tag{206}$$
$$L_2 \equiv b\partial_t + (4\lambda - 2U/a)\partial_x + U_x/a, \tag{207}$$
$$a[L_1, L_2] = bU_t + U_{xxx} - (6/a)UU_x, \tag{208}$$

or in the zero-curvature representation (115)–(116)

$$L = \begin{pmatrix} 0 & 1 \\ U/a + \lambda & 0 \end{pmatrix}, \tag{209}$$
$$M = b^{-1}\begin{pmatrix} -U_x/a & 2U/a - 4\lambda \\ -U_{xx}/a + 2(U/a + \lambda)(U/a - 2\lambda) & U_x/a \end{pmatrix}, \tag{210}$$

or in the Riccati representation for $\omega = \chi^{-1}$ (see (64)–(65) and (69))

$$\omega_x = -\frac{S}{2} - \omega^2 = \left(\frac{U}{a} + \lambda\right) - \omega^2, \tag{211}$$
$$\omega_t = (-C\omega + C_x/2)_x = b^{-1}((2U/a - 4\lambda)\omega - U_x/a)_x. \tag{212}$$

This last representation is by far the best one, for it allows one to deduce immediately two quite important informations, namely the auto-Bäcklund transformation of KdV and the hetero-Bäcklund transformation between KdV and mKdV. Firstly, the substitution of the inverse relation of (200)

$$\omega = (v - V)/(-2a) \tag{213}$$

in (211)–(212) provides the auto-BT for the conservative form of KdV

$$a(v + V)_x = -2a^2\lambda + (v - V)^2/2, \tag{214}$$
$$a(b(v + V)_t - 2F'(t)) = -(v - V)(v - V)_{xx} + 2(V_x^2 + v_x V_x + v_x^2), \tag{215}$$

after suitable differential consequences of the x-part have been added to the t-part in order to suppress λ and cubic terms in (215).

Secondly, the elimination of U between (211)–(212) leads to the mKdV equation (405) for w, with the identification $w = \alpha\omega, \nu = \lambda$; since conversely the elimination of ω leads to the KdV equation for U, the system (211)–(212) also represents the hetero-BT between KdV and mKdV ([29] Eq. (5.16), [120]).

As to the SME, it results from the elimination of V between (E_2, E_3, E_4)

$$bC - S - 6\lambda = 0. \tag{216}$$

Most of these results for KdV were found in the original paper of WTC [65].

Remark. The Bäcklund transformation (211)–(212) between $w = \alpha\omega$ and U is also called a *Miura transformation* from mKdV to KdV because the kdV field is an explicit algebraic transform of the mKdV field. The inverse of a Miura transformation is not a Miura transformation but a Bäcklund transformation.

7.2 Integrable equations with a third order Lax pair

Let us process a few PDEs which possess a third order Lax pair, and let us first perform their one-family truncation with the (wrong) assumption of a second order Lax pair, because this often provides interesting results.

The Boussinesq equation

The Boussinesq equation (Bq) is often defined in a two-component evolution form [132]

$$\mathrm{sBq}(u,r) \equiv \begin{cases} u_t - r_x = 0, \ (\alpha, \beta, \varepsilon) \text{ constant}, \\ r_t + \varepsilon^2((u+\alpha)^2 + (\beta^2/3)u_{xx})_x) = 0. \end{cases} \tag{217}$$

Let us consider its one-component "potential" form

$$\mathrm{pBq}(v) \equiv v_{tt} + \varepsilon^2 \left((v_x + \alpha)^2 + (\beta^2/3)v_{xxx}\right)_x = 0, \ u = v_x, \ r = v_t. \tag{218}$$

Equation (218) has only one family of movable singularities

$$p = -1, \ q = -5, \ \text{indices } (-1, 1, 4, 6), \ \mathcal{D} = 2\beta^2 \partial_x, \tag{219}$$

and it passes the Painlevé test [125]. Since (218) is a conservation law, the computations can be reduced by considering the "second potential Bq" equation

$$\mathrm{ppBq}(w) \equiv w_{tt} + \varepsilon^2 \left((w_{xx} + \alpha)^2 + (\beta^2/3)w_{xxxx}\right) = 0, \ u = v_x = w_{xx}, \tag{220}$$

whose single family is of the logarithmic type $w \sim 2\beta^2 \operatorname{Log} \chi$

$$p = 0^-, \ q = -4, \ \text{indices}(-1, 0, 1, 6), \ \mathcal{D} = 2\beta^2. \tag{221}$$

Let us assume for the would-be singular part transformation the relation

$$w = 2\beta^2 \operatorname{Log} \tau + W, \ \operatorname{ppBq}(w) = 0, \tag{222}$$

and for the link between τ and ψ the identity (146).

Let us first assume that ψ satisfies the second-order scalar Lax pair (88)–(89). This is equivalent to the usual WTC truncation in the invariant formalism [6]

$$\operatorname{ppBq}(w) \equiv \sum_{j=0}^{4} E_j \chi^{j-4} = 0, \tag{223}$$

and this generates the three determining equations

$$E_2 \equiv (4/3)\beta^2 \varepsilon^2 S - 2C^2 - 4\varepsilon^2 (W_{xx} + \alpha) = 0, \tag{224}$$

$$E_3 \equiv -2(C_t - CC_x - (\beta^2 \varepsilon^2/3) S_x) = 0, \tag{225}$$

$$E_4 \equiv (SE_2 - E_{3,x})/2 + C_x^2 + \beta^2 \operatorname{ppBq}(W) = 0. \tag{226}$$

From the last equation $E_4 = 0$, the desired solution $\operatorname{ppBq}(W) = 0$ cannot be generic, so this second-order assumption fails to provide the auto-BT. However, it does provide another information, namely a hetero-BT between the Boussinesq PDE and another PDE. Indeed, under the natural parametric representation of E_3 (which, by the way, would be the SME if the second order were the correct one),

$$S = 3z_t - 3(\beta\varepsilon)^2 z_x^2/2, \ C = (\beta\varepsilon)^2 z_x, \tag{227}$$

the field z, by the cross-derivative condition (78), satisfies the modified Boussinesq equation [67]

$$\operatorname{MBq}(z) \equiv z_{tt} + ((\beta\varepsilon)^2/3) z_{xxxx} + 2(\beta\varepsilon)^2 z_t z_{xx} - 2(\beta\varepsilon)^4 z_x^2 z_{xx} = 0. \tag{228}$$

Just like for the KdV equation (Sect. 7.1), this leads, after a short computation left to the reader, to the hetero-BT between the Boussinesq and the modified Boussinesq equations.

Going to third order, the assumption (222) and (146), with ψ solution of the scalar Lax pair (117)–(118), generates

$$\operatorname{ppBq}(w) \equiv \sum_{k=0}^{2} \sum_{l=0}^{2} E_{k,l} Y_1^k Y_2^l, \ k + l \le 2. \tag{229}$$

These six determining equations $E_{k,l} = 0$, plus the three cross-derivative conditions $X_j = 0, j = 0, 1, 2$, are solved as follows in the Gel'fand-Dikii case $f = 0$: [39, 41]

$$
\begin{array}{ll}
E_{02} \equiv (\beta\varepsilon)^2 - c^2 = 0 & \Rightarrow c = \beta\varepsilon, \\
E_{11} \equiv d = 0 & \Rightarrow d = 0, \\
E_{20} \equiv 3(V_x + \alpha) + 2\beta^2 a = 0 & \Rightarrow a = -3(V_x + \alpha)/(2\beta^2), \\
E_{10} \equiv \varepsilon V_{xx} - \beta e_x = 0 & \Rightarrow e_x = \beta^{-1}\varepsilon V_{xx}, \\
X_1 \equiv 3V_{xt} + 3\beta\varepsilon V_{xxx} + 4\beta^3\varepsilon b_x = 0 & \Rightarrow b = g(t) - 3(\beta^{-2}V_{xx} + \beta^{-3}\varepsilon^{-1}V_t)/4, \\
X_0 \equiv (3/(4\varepsilon\beta^2))\mathrm{pBq}(V) = 0 & \Rightarrow V \text{ satisfies the PDE (218)}, \\
E_{00} \equiv 2\beta^2 g'(t) = 0 & \Rightarrow g(t) = \lambda,
\end{array}
\tag{230}
$$

in which λ is an arbitrary constant. The coefficients a, b, c, d, e are

$$
\begin{aligned}
& a = -(3/2)\beta^{-2}(V_x + \alpha),\ b = \lambda - (3/4)\beta^{-2}V_{xx} - (3/4)\beta^{-3}\varepsilon^{-1}V_t, \\
& c = \beta\varepsilon,\ d = 0,\ e = \beta^{-1}\varepsilon(V_x + \alpha), && (231) \\
& X_0 = (3/(4\varepsilon\beta^2))\mathrm{pBq}(V), X_1 = 0, X_2 = 0, && (232)
\end{aligned}
$$

and they define a third-order Lax pair of the potential Boussinesq equation (218) [131, 132, 88].

The BT is just (152)–(153) or equivalently, after substitution of $Y_1 = (v - V)/(2\beta^2)$,

$$
\begin{aligned}
& (v - V)_{xx} + 3\beta^{-1}\varepsilon^{-1}(v + V)_t + 3\beta^{-2}(v - V)((v + V)_x + 2\alpha) \\
& +\beta^{-4}(v - V)^3 - 8\beta^2\lambda = 0, && (233) \\
& (v + V)_{xx} - \beta^{-1}\varepsilon^{-1}(v - V)_t + \beta^{-2}(v - V)(v - V)_x = 0. && (234)
\end{aligned}
$$

The Hirota-Satsuma equation

Defined as [24]

$$
\mathrm{HS}(w) \equiv [w_{xxt} + (6/a)w_x w_t]_x = 0,\ a \neq 0, \tag{235}
$$

it is better processed on its potential form

$$
\mathrm{pHS}(w) \equiv w_{xxt} + (6/a)w_x w_t + F(t) = 0,\ a \neq 0. \tag{236}
$$

The second order assumption (88)–(89) generates the three determining equations

$$
\begin{aligned}
& E_2 \equiv -2aSC - 6W_t + 6CW_x = 0, \\
& E_3 \equiv aS_t + 2aSC_x - 6C_x W_x = 0, && (237) \\
& E_4 \equiv \mathrm{pHS}(W) - a(S^2C + S_{xt}/2 + SC_{xx}) - 3SW_t + 3(SC + C_{xx})W_x = 0.
\end{aligned}
$$

In the generic case $C_x \neq 0$, their general solution is unknown, in particular we have not succeeded to perform the elimination of (S, C) to find the constraint(s) satisfied by W. It is easy to eliminate W but this gives rise to two equations for (S, C)

$$6W_x = 2aS + a\frac{S_t}{C_x},\ 6W_t = a\frac{CS_t}{C_x}, \tag{238}$$

$$M_{23} \equiv \left(\frac{CS_t}{C_x}\right)_x - \left(2S + \frac{S_t}{C_x}\right)_t = 0, \tag{239}$$

$$M_4 \equiv 1 - 6a^{-1}F(t) - C_x^{-1}(4CSS_t + CS_{xxt} + 2C_{xx}S_t) \\ - C_x^{-2}(2CS_t^2 + C^2 S_t S_x - 2CC_{xx}S_{xt}) - 2C_x^{-3}CC_{xx}^2 S_t = 0, \tag{240}$$

and their possible functional dependence is unsettled. Anyhow, the field W cannot be a second solution of (235) [38].

The third order assumption (117)–(118), with the link (146) and the truncated expansion

$$w = W + a\partial_x \operatorname{Log} \tau, \tag{241}$$

generates seven determining equations (98). They are easily solved [38] and their unique solution defines the Lax pair (268)–(269), with W a second solution of (236).

The Tzitzéica equation

The equation is defined by (168), in the case $(a_1 = 0, a_2 \neq 0)$. It posseses two families, the first one defined by (169), the second one by

$$e^{-u} \sim \sqrt{(1/a_2)\varphi_x\varphi_t}(\varphi - \varphi_0)^{-1}, \text{ indices } (-1, 2). \tag{242}$$

These two families are not opposite, but the second family is irrelevant because the Tzitzéica equation has a one-to-one correspondence [23] with a one-family equation, namely the potential form (236) of the Hirota-Satsuma PDE in the particular case $F(t) = 0$. This correspondence is obtained by the elimination of a_2 in equation (168)

$$\left(F(t) = 0, e^u = \frac{2}{a\alpha} w_t\right) \Longrightarrow \left(e^{-2u}(e^{2u}\mathrm{Tzi}(u))_x = \left(\frac{\mathrm{pHS}(w)}{w_t}\right)_t\right). \tag{243}$$

The irrelevance of the second family is confirmed by the negative result of Weiss [126, 6] obtained when performing a truncation on e^{-u}.

All the truncations will accordingly take the same form (170) as for the Liouville equation, which implies that τ is an object invariant under the permutation (131). Depending on the Lax pair assumption, the link between τ and ψ will be either the identity (case of a scalar ψ invariant under the permutation (131)) or not (if the scalar ψ is not invariant, e.g. because the Lax pair itself is not invariant), as detailed below.

Let us first assume a second order Lax pair. To the author's knowledge, one cannot define a scalar ψ, linked to such a Lax pair, which, like τ, would be invariant under (131). This is probably the reason why the assumption

$\tau = \psi$ with ψ solution of the noninvariant Lax pair (88)–(89) generates so intricate determining equations that their general solution has not yet been obtained [40]; these equations are however consistent in the sense that one easily finds the particular exact solution

$$\alpha e^u = 2c\wp(x - ct - x_1, g_2, A + \frac{a_2\alpha^2}{8c^3}) - 2c\wp(x + ct - x_2, g_2, A - \frac{a_2\alpha^2}{8c^3}), \quad (244)$$

depending on five arbitrary constants (x_1, x_2, c, g_2, A) and representing the superposition of two traveling waves of opposite velocities.

From this second-order WTC truncation, and with appropriate assumptions, one can also find a particular solution which represents a binary Darboux transformation [116].

Let us now turn to the third order assumption. One can postulate either a Lax pair invariant under (131), such as the matrix pair (132)–(133), or a noninvariant Lax pair such as the scalar pair (117)–(118). In the first case, one must assume the identity link $\tau = \psi$, while in the second case the assumed link must be noninvariant. Both assumptions lead to a success [9]. Let us detail here the invariant assumption, i.e. *a priori* the simpler one.

The truncation is defined by (170), the link (146), and the matrix Lax pair (132)–(133)

$$E(u) \equiv \sum_{k=0}^{3}\sum_{l=0}^{3-k} E_{kl}(f_j, g_j, h_j, U) Y_1^k Y_2^l, \ \forall k, l \ : \ E_{kl} = 0, \quad (245)$$

in which (Y_1, Y_2) are the two components of the projective Riccati pseudopotential (135)–(138) equivalent to the Lax pair. To these ten determining equations in U and the nine unknown coefficients, one must add the six cross-derivative conditions $X_j = 0$ (139)–(140).

During their resolution, one first proves that the product $f_2 h_1$ cannot vanish (otherwise a_2 would be zero). This makes the sixteen equations algebraically independent and equivalent to the fifteen differential relations

$$f_{j,t}, g_{j,x}, g_{j,t}, h_{j,x}, g_{j,xt} = P(\{f_k, g_k, h_k\}, k = 1,2,3), \ j = 1,2,3, \quad (246)$$

with P polynomials whose coefficients depend on U, U_x, U_t, U_{xt}, plus the single algebraic relation

$$E_{00} \equiv a_2 - \frac{4}{\alpha^2}\left(g_3 + g_1 g_2 + (\alpha/2)e^U\right)^2 = 0. \quad (247)$$

They are solved successively as [equations are referenced as in (246)–(247)]

$$
\begin{array}{ll}
g_{3,xt} - (g_{3,x})_t & : E(U) = 0, \\
g_{1,x} - g_{2,t} & : \exists\ g_0(x,t) : g_1 = g_{0,t},\ g_2 = g_{0,x}, \\
g_{2,t} & : g_3 = -\alpha e^U - g_{0,x} g_{0,t} - g_{0,xt}, \\
E_{00} & : \exists\ f_0(x,t) \ne 0 : f_2 = \sqrt{a_2} W^{-1} f_0,\ h_1 = \sqrt{a_2} W^{-1} f_0^{-1}, \\
 & \text{notation } W = e^U + (2/\alpha) g_{0,xt}, \\
g_{2,x} & : f_3 = -\sqrt{a_2} W^{-1} f_0 g_{0,t} - f_1 g_{0,x} - g_{0,x}^2 + g_{0,xx}, \\
g_{3,x} & : f_1 = W_x/W + 2 g_{0,x}, \\
f_{2,t} & : h_2 = W_t/W + 2 g_{0,t} - f_{0,t}/f_0, \\
h_{1,x} & : f_{0,x} = 0, \\
g_{1,t} & : h_3 = g_{0,t}(f_{0,t}/f_0 - W_t/W - g_{0,t}) + g_{0,tt} - \sqrt{a_2} W^{-1} g_{0,x}/f_0, \\
g_{3,t} & : f_{0,t} = 0, \\
h_{2,x} & : g_{0,xt} = 0.
\end{array}
\tag{248}
$$

The irrelevant arbitrary function g_0 reflects the freedom in the definition (170) of τ and can be absorbed by redefining τ as τe^{-g_0}. Thus the solution is unique: the field U must satisfy the Tzitzéica PDE, and f_0 is an arbitrary nonzero complex constant λ. Accordingly, one has obtained a Lax pair and a singular part transformation. The equivalent projective Riccati representation of the matrix Lax pair is

$$Y_{1,x} = -Y_1^2 + U_x Y_1 + \sqrt{a_2}\lambda e^{-U} Y_2, \tag{249}$$
$$Y_{2,x} = -Y_1 Y_2 - \alpha e^U, \tag{250}$$
$$Y_{1,t} = -Y_1 Y_2 - \alpha e^U, \tag{251}$$
$$Y_{2,t} = -Y_2^2 + U_t Y_2 + \sqrt{a_2}\lambda^{-1} e^{-U} Y_1, \tag{252}$$

with cross-derivative conditions proportional to the Tzitzéica equation

$$(Y_{1,x})_t - (Y_{1,t})_x = Y_1 E(U),\ (Y_{2,x})_t - (Y_{2,t})_x = Y_2 E(U). \tag{253}$$

This Lax pair is the rewriting in matrix form of the scalar triplet given by Tzitzéica [57]

$$-\tau_{xx} + U_x \tau_x + \sqrt{a_2}\lambda e^{-U} \tau_t = 0, \tag{254}$$
$$-\tau_{tt} + U_t \tau_t + \sqrt{a_2}\lambda^{-1} e^{-U} \tau_x = 0, \tag{255}$$
$$-\tau_{xt} - \alpha e^U \tau = 0. \tag{256}$$

The Lax pair admits by construction the involution [58, 55]

$$(\tau, e^U, \lambda) \to \left(\frac{1}{\tau}, -e^U - \frac{2}{\alpha}\frac{\tau_x \tau_t}{\tau^2}, -\lambda\right), \tag{257}$$

equivalent to

$$(\tau, e^U, \lambda) \to (1/\tau, e^U + \mathcal{D} \operatorname{Log} \tau, -\lambda), \tag{258}$$

which defines another, equivalent, writing of the singular part transformation

$$e^u = -e^U - \frac{2}{\alpha}\frac{\tau_x \tau_t}{\tau^2}. \tag{259}$$

Remark. Knowing these results, one can also write this singular part transformation [130, 3] as a difference of the two fields $u - U$ in terms of the two components of a projective Riccati pseudopotential

$$u = U + \mathrm{Log}(-2\lambda^2 y_1 y_2 - 1),\ y_j = \alpha^{-1/2}\lambda^{-1}e^{-U/2}Y_j, \tag{260}$$

in a quite similar manner to the singular part transformation of Liouville and sine-Gordon (393). However, the field u is multivalued.

In order to find the BT, one must now eliminate one of the two equivalent projective components, and this defines two possible, different, eliminations.

In the first elimination, one takes Y_2 from (249) and substitutes it into the three remaining equations, which results in

$$Y_2 = (Y_{1,x} + Y_1^2 - U_x Y_1)e^U/(\sqrt{a_2}\lambda), \tag{261}$$
$$\mathrm{ODE} \equiv Y_{1,xx} + 3Y_1 Y_{1,x} + Y_1^3 - e^{-U}(e^U)_{xx}Y_1 + \alpha\sqrt{a_2}\lambda = 0, \tag{262}$$
$$\mathrm{PDE} \equiv Y_{1,t} + e^U\left((Y_1 Y_{1,x} + Y_1^3) - Y_1^2 U_x\right)/(\sqrt{a_2}\lambda) + \alpha e^U = 0, \tag{263}$$
$$(252) \equiv -Y_1 E(U) - \frac{e^U Y_1}{\sqrt{a_2}\lambda}\mathrm{ODE} + (2Y_1 - U_x + \partial_x)\mathrm{PDE} = 0, \tag{264}$$
$$[\mathrm{ODE},\mathrm{PDE}] = (Y_{1,xx})_t - (Y_{1,t})_{xx} = Y_1(e^{2U}E(U))_x. \tag{265}$$

Only two of them are functionally independent, as shown by relation (264), but the commutator (265) of equations (262)–(263) shows that this elimination fails to generate the auto-BT of Tzitzéica equation.

However, it does provide another result, which we now derive. The ODE (262) belongs to the classification of Gambier – this is the number 5, see Sect. 5.6 –, it is linearizable by the transformation $Y_1 = \partial_x \mathrm{Log}\,\psi$ into a third-order linear ODE, with the relation $\tau = \psi$ between the two functions. This transformation also linearizes the PDE (263), and the resulting linear system

$$\tau_{xxx} - (U_{xx} + U_x^2)\tau_x + \sqrt{a_2}\alpha\lambda\tau = 0, \tag{266}$$
$$-\sqrt{a_2}\lambda\tau_t + e^U\tau_{xx} - U_x e^U\tau_x = 0, \tag{267}$$

which cannot be a scalar Lax pair of the Tzitzéica equation, is, in fact, the scalar Lax pair of the Hirota-Satsuma equation (235), see Sect. 7.2,

$$\tau_{xxx} - (6/a)w_x\tau_x + \Lambda\tau = 0, \tag{268}$$
$$\Lambda\tau_t - (2/a)w_t\tau_{xx} + (2/a)w_{xt}\tau_x = 0, \tag{269}$$

under the change of variables (243).

In the second elimination, one takes Y_1 from (250) and substitutes it into the three remaining equations

$$\begin{aligned} Y_1 &= -(Y_{2,x} + \alpha e^U)/Y_2, && (270)\\ \text{ODE} &\equiv Y_2 Y_{2,xx} - 2Y_{2,x}^2 - (U_x Y_2 + 3\alpha e^U) Y_{2,x} \\ &\quad + \sqrt{a_2}\lambda e^{-U} Y_2^3 - \alpha^2 e^{2U} = 0, && (271)\\ \text{PDE} &\equiv Y_2 Y_{2,t} + Y_2^3 - U_t Y_2^2 + \sqrt{a_2}\lambda^{-1}(\alpha + e^{-U} Y_{2,x}) = 0, && (272)\\ (251) &\equiv E(U) + (\partial_x - \alpha e^U Y_2^{-1})\text{PDE} \\ &\quad - \sqrt{a_2}\lambda^{-1} e^{-U} Y_2^{-2}\text{ODE} = 0, && (273)\\ [\text{ODE},\text{PDE}] &= (Y_{2,xx})_t - (Y_{2,t})_{xx} \\ &= (3\alpha e^U + U_x Y_2 + 3Y_{2,x} - Y_2\partial_x) E(U). && (274) \end{aligned}$$

Only two of them are functionally independent, as shown by the relation (273), and the vanishing of the commutator (274) of equations (271)–(272) is equivalent to the vanishing of the Tzitzéica equation for U. This elimination therefore generates the auto-BT of Tzitzéica equation, by the substitution

$$Y_2 = (\alpha/2) \int \left(e^u - e^U\right) \mathrm{d}x \qquad (275)$$

into (271)–(272).

The ODE part (271) of the BT is equivalently written as [97]

$$\frac{w_{xx}}{w_x} - \frac{W_{xx}}{W_x} - 2\frac{w_x + W_x}{w - W} + \alpha\sqrt{a_2}\lambda\frac{(w-W)^2}{2w_x W_x} = 0, \qquad (276)$$

with the notation $Y_2 = (\alpha/2)(w - W), e^u = w_x, e^U = W_x$.

The nonlinear ODE (271) again belongs to the equivalence class of the fifth Gambier equation (G5), see Sect. 5.6, and its linearization

$$Y_2^{-1} = -\alpha^{-1} e^{-U} \partial_x \operatorname{Log}(e^U \psi) \qquad (277)$$

transforms the two equations (271)–(272) into the third-order scalar Lax pair of the Gel'fand and Dikii type (i.e. $f = 0$ in (117)–(118))

$$\begin{aligned} \mathcal{L}\psi &\equiv \psi_{xxx} + (2U_{xx} - U_x^2)\psi_x + ((2U_{xx} - U_x^2)_x/2 + \sqrt{a_2}\alpha\lambda)\psi = 0, && (278)\\ \mathcal{M}\psi &\equiv \psi_t + \sqrt{a_2}(\alpha\lambda)^{-1} e^{-2U}(\psi_{xx} + U_x\psi_x + U_{xx}\psi) \\ &\quad + (U_t + \int \left(\alpha e^U + a_2 e^{-2U}\right) \mathrm{d}x)\psi = 0, && (279)\\ [\mathcal{L},\mathcal{M}] &= 3E\partial_x^2 + (2(e^U)_x E + E_x)\partial_x \\ &\quad + (e^U)_x E_x + (3U_{xx} - U_x^2)E,\ E = E(U). && (280) \end{aligned}$$

Thus, the noninvariant (under (131)) link between τ and ψ that one would have had to postulate if one had chosen the scalar Lax pair (117)–(118) is *a posteriori* provided by the linearizing formula (277) and the Riccati equation (251), this is the invertible transformation

$$e^U \tau = (e^U \psi)_x,\ e^U \psi = -\alpha^{-1}\tau_t, \qquad (281)$$

and it clearly breaks the invariance under (131).

The Sawada-Kotera and Kaup-Kupershmidt equations

Because of their duality [27, 64], it is convenient to introduce simultaneously the Sawada-Kotera equation (SK) and the Kaup-Kupershmidt equation (KK). These are defined as

$$\mathrm{SK}(u) \equiv \beta u_t + \left(u_{xxxx} + \frac{30}{\alpha} u u_{xx} + \frac{60}{\alpha^2} u^3\right)_x = 0, \tag{282}$$

$$\mathrm{pSK}(v) \equiv \beta v_t + v_{xxxxx} + \frac{30}{\alpha} v_x v_{xxx} + \frac{60}{\alpha^2} v_x^3 = 0, \tag{283}$$

$$\mathrm{KK}(u) \equiv \beta u_t + \left(u_{xxxx} + \frac{30}{\alpha} u u_{xx} + \frac{45}{2\alpha} u_x^2 + \frac{60}{\alpha^2} u^3\right)_x = 0, \tag{284}$$

$$\mathrm{pKK}(v) \equiv \beta v_t + v_{xxxxx} + \frac{30}{\alpha} v_x v_{xxx} + \frac{45}{2\alpha} v_{xx}^2 + \frac{60}{\alpha^2} v_x^3 = 0, \tag{285}$$

in which u denotes the conservative field and v the potential one, with $u = v_x$.

Both equations have the Painlevé property [64]. Each of them has two families [64]

$$\mathrm{pSK, F1}: p = -1,\ v_0 = \alpha,\ \text{indices } -1, 1, 2, 3, 10, \tag{286}$$

$$\mathrm{pSK, F2}: p = -1,\ v_0 = 2\alpha,\ \text{indices } -2, -1, 1, 5, 12, \tag{287}$$

$$\mathrm{pKK, F1}: p = -1,\ v_0 = \alpha/2,\ \text{indices } -1, 1, 3, 5, 7, \tag{288}$$

$$\mathrm{pKK, F2}: p = -1,\ v_0 = 4\alpha,\ \text{indices } -7, -1, 1, 10, 12. \tag{289}$$

The singular part operator $\mathcal{D}$ attached to a given family is $\mathcal{D} = v_0 \partial_x$. The two families have residues which are not opposite, but fortunately each potential equation possesses in its hierarchy a "minus-one" equation [66]

$$\mathrm{pSK}_{-1}: v_{xxt} + \frac{6}{\alpha} v_x v_t = 0, \tag{290}$$

$$\mathrm{pKK}_{-1}: v_t v_{xxt} - \frac{3}{4} v_{xt}^2 + \frac{6}{\alpha} v_x v_t^2 = 0, \tag{291}$$

which has only one family (the first one is nothing else than the Hirota-Satsuma PDE, already processed in Sect. 7.2). The equations SK and KK are therefore to be considered as possessing the single family F1, Eqs. (147) and (149).

Let us assume the one-family singular part transformation (200) and, succesively, the second-order scalar Lax pair (88)–(89), then the third-order scalar one (117)–(118) with the Gel'fand-Dikii simplification $f = 0$. As to the link between τ and ψ, at second order this is the identity, while at third order it can be, as outlined in Sect. 5.6 and detailed in Musette lecture [91], either the linearizing transformation of the fifth Gambier equation or that of the twenty-fifth Gambier equation.

Therefore, at the fourth step of the singular manifold method, for each PDE, one has only three possibilities to examine: order two and Riccati, order three and (G5), order three and (G25). This is done in the next two sections.

The Sawada-Kotera equation

First truncation (order two and Riccati). The one-family truncation (95) with $\tau = \psi$ generates the three equations [20]

$$E_4 \equiv \beta C - 4S^2 + 9S_{xx} + 60SV_x/\alpha - 180(V_x/\alpha)^2 - 30V_{xxx}/\alpha = 0, \quad (292)$$
$$E_5 \equiv -\beta C_x - 2SS_x + S_{xxx} + 30S_xV_x/\alpha = 0, \quad (293)$$
$$E_6 \equiv \mathrm{pSK}(V) + (SE_4 - E_{5,x})/2 + 5S_x(3V_{xx}/\alpha - S_x/2) = 0. \quad (294)$$

These equations possess two solutions [20], a nongeneric one $S_x = 0$

$$\begin{aligned} &S = -k^2/2,\ C = c + c_0,\ c = k^4/\beta + 2c_0/3, \\ &V/\alpha = \zeta(x - (c - c_0)t, k^4/12 + \beta c_0/9, g_3) - k^2x/12 \\ &\qquad + ((5(k^4 + \beta c_0)k^2/36 - 12g_3)t)/\beta, \end{aligned} \quad (295)$$

in which ζ is the Weierstrass function and (k, c_0, g_3) are arbitrary constants, and a generic one $S_x \neq 0$ defined by the four equations

$$V_x = \alpha(\beta C_x + 2SS_x - S_{xxx})/(30S_x), \quad (296)$$
$$V_t = \dots \quad (297)$$
$$M_1 \equiv (G_x/S_x)_{xx} - G - S_x^2G_x^2/5 + 2SG_x/S_x = 0, \quad (298)$$
$$M_2 \equiv 29 \text{ terms} = 0, \text{ also vanishing if } G = 0, \quad (299)$$

in which G is defined by

$$G \equiv S_{xx} + 4S^2 - \beta C. \quad (300)$$

The general solution (S, C) of the system $M_1 = 0, M_2 = 0$ has not yet been obtained, for the elimination of S or C is difficult. This difficulty reflects the fact that second order is not the correct order. Nevertheless, these complicated equations admit the very simple *particular* solution [64] (it would be interesting to prove that this is the *general* solution),

$$G = 0,\ V_x = \alpha S/3,\ \mathrm{pKK}(V) = 0, \quad (301)$$

so the field V in the singular part transformation assumption (200) satisfies a different PDE, namely the potential KK equation. This defines a hetero-BT between the conservative forms of SK and KK [50, 23, 22]

$$\begin{aligned} &\alpha(v + V/2)_x + (v - V)^2 = 0,\ \beta(v + V/2)_t + \dots = 0, \\ &\mathrm{pSK}(v) = 0,\ \mathrm{pKK}(V) = 0 \end{aligned} \quad (302)$$

(see Ref. [22] for the exact expression of the t-part).

With $G = 0$, the linear system (88)–(89) is a degenerate Lax pair for KK, since it lacks a spectral parameter.

Still when (301) holds, the field χ^{-1} satisfies a fifth order PDE, the Fordy-Gibbons equation, and the explicit writing of its hetero-BT with the SK equation is left to the reader.

Second truncation (order three and (G5)). This assumption creates no *a priori* constraint on the coefficients (a, b) of the spectral problem (117), and the linearizing transformation of (G5) is just the identity $\tau = \psi$. This generates six determining equations (98). The process is successful [64, 126, 7] and V is found to be a second solution of pSK (notation $U = V_x$ as usual)

$$b = \lambda, \tag{303}$$

$$a = -6U/\alpha, \tag{304}$$

$$L_1 = \partial_x^3 + 6\frac{U}{\alpha}\partial_x - \lambda, \tag{305}$$

$$L_2 = \beta\partial_t + \left(18\frac{U_x}{\alpha} - 9\lambda\right)\partial_x^2 + \left(36\frac{U^2}{\alpha^2} - 6\frac{U_{xx}}{\alpha}\right)\partial_x - 36\lambda\frac{U}{\alpha}, \tag{306}$$

$$[L_1, L_2] = 6\beta^{-1}\alpha^{-1}\mathrm{SK}(U). \tag{307}$$

This is the Lax pair given by Satsuma and Kaup [53].

The BT results from the elimination of Y_2, which provides Eqs. (152)–(153) for $Y_1 = Y$,

$$Y_{xx} + 3YY_x + Y^3 + 6(U/\alpha)Y - \lambda = 0, \tag{308}$$

$$\begin{aligned}\beta Y_t - 9[(\lambda - 2U_x/\alpha)(Y_x + Y^2)\\ + 4(\lambda U/\alpha - (U/\alpha)^2 Y) + (2/3)(U_{xx}/\alpha)Y]_x = 0,\end{aligned} \tag{309}$$

$$\beta((Y_{xx})_t - (Y_t)_{xx})/Y = -(6/\alpha)\mathrm{SK}(U), \tag{310}$$

followed by the substitution $Y = (v - V)/\alpha$,

$$(v - V)_{xx}/\alpha + 3(v - V)(v + V)_x/\alpha^2 + (v - V)^3/\alpha^3 - \lambda = 0, \tag{311}$$

$$\begin{aligned}&\beta(v - V)_t/\alpha - (3/2)[(v - V)_{xxxx}/\alpha\\ &+(5(v - V)(v + V)_{xxx} + 15(v + V)_x(v - V)_{xx})/\alpha^2\\ &+(15(v - V)^2(v - V)_{xx} + 30(v - V)(v + V)_x^2)/\alpha^3\\ &+30(v - V)^3(v + V)_x/\alpha^4 + 6(v + V)^5/\alpha^5]_x = 0,\end{aligned} \tag{312}$$

a result due to Satsuma and Kaup [53].

The Kaup-Kupershmidt equation

First truncation (order two and Riccati). The one-family truncation (95) with $\tau = \psi$ generates the three equations [41]

$$E_2 \equiv 15(S/4 - 3V_x/\alpha) = 0, \tag{313}$$

$$E_4 \equiv \beta C/2 + 7S^2/4 + 3S_{xx}/4 - 15(SV_x + V_{xxx})/\alpha - 90(V_x/\alpha)^2 = 0, \tag{314}$$

$$\begin{aligned}E_6 \equiv (S/2)E_4/\alpha + (4\beta(CS + C_{xx}) - S^3 + (21/2)S_x^2 + 14SS_{xx} - 4S_{xxxx})/16\\ +(15/4)(3S^2 - 2S_{xx})V_x/\alpha - 45S(V_x/\alpha)^2 + \mathrm{pKK}(V) = 0.\end{aligned} \tag{315}$$

As opposed to the (difficult) SK case, these equations are easy to solve and possess the unique solution [64]

$$V_x = \alpha S/12,\ \mathrm{pSK}(V) = 0,\ S_{xx} + S^2/4 - \beta C = 0. \tag{316}$$

This is a strong indication that the particular solution (301) of (296)–(299) should be the general one. One again recovers, by a nice duality, the hetero-BT between KK and SK.

Second truncation (order three and (G5)). This generates thirteen determining equations (98). This truncation fails and provides no solution at all (one determining equation is $E_{2,2} \equiv 45/8 = 0$), not even the one-soliton solution. Indeed, the one-soliton solution of Kaup corresponds to constant coefficients for the scalar Lax pair (117)–(118) with $f = 0$, and, with the above procedure, the only way to obtain it [32, 49] is to enforce the two first integrals K_1 and K_2 which result from the zero value of b,

$$K_1 = \psi_{xx} - a\psi, K_2 = \psi_x^2 - a\psi^2 - 2(\psi_{xx} - a\psi)\psi. \tag{317}$$

Third truncation (order three and (G25)). This assumption implies, noindent among the coefficients (a, b) of the spectral problem (117), the *a priori* constraint [91]

$$b - a_x/2 = \lambda(t), \tag{318}$$

and the linearizing transformation of (G25) defines the link between τ and ψ

$$\frac{\tau_x}{\tau} = \frac{\lambda(t)}{Y_{1,x} + (1/2)Y_1^2 - a/2},\ Y_1 = \frac{\psi_x}{\psi}. \tag{319}$$

This generates fourteen determining equations (98) (i.e. the same order of magnitude as for the (G5) assumption) in the basis $(\psi_x/\psi, \psi_{xx}/\psi - ((\psi_x/\psi)^2 + b_x\psi_x/(b\psi) - a)/2)$; they are solved as follows (g_k denotes an arbitrary integration function, λ an arbitrary integration constant) [41]

$$\begin{aligned}
&E_{4,4} : c = 9\lambda(t)/\beta,\\
&E_{3,5} : a = -6V_x/\alpha,\\
&E_{1,5} : d = (3V_{xxx}/\alpha - (6V_x/\alpha)^2)/\beta,\\
&E_{0,5} : \lambda(t) = \lambda \text{ independent of } t,\\
&E_{0,6} : \mathrm{pKK}(V) = 0,\\
&E_{1,4} : e = g_2(t) + (36\lambda V_x/\alpha + 72V_xV_{xx}/\alpha^2 + 3V_{xxxx}/\alpha)/\beta,\\
&X_1\ \ : b = g_4(t) - 3V_{xx}/\alpha + V_{xxx}/(3\alpha)),\\
&X_0\ \ : g_4 = \lambda,
\end{aligned} \tag{320}$$

and the result is, with $U = V_x$,

$$b = \lambda - 3U_x/\alpha, \quad (321)$$

$$a = -6U/\alpha, \quad (322)$$

$$\partial_x \operatorname{Log} \tau = \frac{\lambda}{\psi_{xx}/\psi - (1/2)(\psi_x/\psi)^2 + 3(U/\alpha)}. \quad (323)$$

$$L_1 = \partial_x^3 + 6\frac{U}{\alpha}\partial_x + 3\frac{U_x}{\alpha} - \lambda, \quad (324)$$

$$L_2 = \beta\partial_t - 9\lambda\partial_x^2 + \left(3\frac{U_{xx}}{\alpha} + 36\frac{U^2}{\alpha^2}\right)\partial_x - 3\frac{U_{xxx}}{\alpha} -72\frac{UU_x}{\alpha^2} - 36\lambda\frac{U}{\alpha}, \quad (325)$$

$$\beta[L_1, L_2] = (6/\alpha)\mathrm{KK}(U)\partial_x + (3/\alpha)\mathrm{KK}(U)_x. \quad (326)$$

This is the Lax pair given by Kaup [27]. The integration of the first-order ODE (159) *modulo* the Lax pair yields the singular part transformation given by Levi and Ragnisco [32]:

$$\tau = \psi\psi_{xx} - (1/2)\psi_x^2 + 3(U/\alpha)\psi^2, \ \tau_x = \lambda\psi^2. \quad (327)$$

Although the relation $\tau_x/\psi^2 =$ constant is the same as in the case of KdV (see Eq. (4.14) in Ref. [65]), it cannot be taken as an *a priori* assumption, it is the result of the method.

Starting from the *vacuum* solution $U = 0$, the general solution ψ of $L_1\psi = 0, L_2\psi = 0$,

$$\psi = c_1 e^{Kx+9K^5t/\beta} + c_2 e^{jKx+9j^2K^5t/\beta} + c_3 e^{j^2Kx+9jK^5t/\beta},$$
$$j^2 + j + 1 = 0, \ K^3 = \lambda, \quad (328)$$

in which c_1, c_2, c_3, K are arbitrary complex constants, leads by (3) to the one-soliton solution of Kaup [27]

$$u = (\alpha/2)\partial_x^2 \operatorname{Log}(2 + \cosh(k/2)(x - (k/2)^4 t/\beta)), \ k \in \mathcal{R} \quad (329)$$

for the choice $(c_1, c_2, c_3) = (0, j^2, -j), K^2 = -k^2/12$, which corresponds to the entire function

$$\tau = -(k^2/12)(2 + \cosh(k/2)(x - (k/2)^4 t/\beta))e^{(k/2)(x+(k/2)^4 t/\beta)}. \quad (330)$$

Let us now obtain the auto-BT of KK, by an elimination. In order to perform this elimination easily, it is convenient to choose one of the two components of the pseudopotential $\mathbf{Y}$ so as to characterize the singular part transformation,

$$\mathrm{KK}: \frac{2(v - V)}{\alpha} = \frac{\tau_x}{\tau} = Z. \quad (331)$$

The chosen equivalent system is the system satisfied by (Y_1, Z)

$$
\begin{aligned}
Y_1 &= \frac{\psi_x}{\psi},\ Z = \frac{\tau_x}{\tau}, & (332)\\
Y_{1,x} &= -Y_1^2/2 + \lambda Z^{-1} - 3U/\alpha, & (333)\\
Z_x &= 2Y_1 Z - Z^2, & (334)\\
\beta Y_{1,t} &= [9\lambda Y_1^2/2 - (3U_{xx}/\alpha + 36(U/\alpha)^2)Y_1 + 9\lambda^2 Z^{-1} & \\
&\quad +3U_{xxx}/\alpha + 72UU_x/\alpha^2 + 9\lambda U/\alpha]_x, & (335)\\
\beta Z_t &= \big[18\lambda U/\alpha + 9\lambda^2 Z^{-1} + 9\lambda Y_1^2 & \\
&\quad +(45(U/\alpha)^2 + 6(U_{xx}/\alpha) - 18(U_x/\alpha)Y_1 & \\
&\quad +27(U/\alpha)Y_1^2 + (9/4)Y_1^4)Z\big]_x. & (336)
\end{aligned}
$$

The BT then arises from the elimination of Y_1 between (166), (167) and (336) (Eq. (335) must be discarded), which results in the two equations for $Z = Y$,

$$
\begin{aligned}
& Y_{xx} - (3/4)Y_x^2/Y + 3YY_x/2 + Y^3/4 + 6(U/\alpha)Y - 2\lambda = 0, & (337)\\
& \beta Y_t - (3/16)[3Y^5 + 15YY_x^2 + 30Y^2 Y_{xx} + 8Y_{xxxx} & \\
& +30(Y^3 + 2Y_{xx})(Y_x + 4V_x/\alpha) + 60Y(Y_x + 4V_x/\alpha)^2 & \\
& +30Y_x(Y_{xx} + 4V_{xx}/\alpha) + 20Y(Y_{xxx} + 4V_{xxx}/\alpha)]_x = 0, & (338)\\
& \beta((Y_{xx})_t - (Y_t)_{xx})/Y = -(6/\alpha)\mathrm{KK}(U), & (339)
\end{aligned}
$$

followed by the substitution $Y = 2(v - V)/\alpha$, [41]

$$
\begin{aligned}
& (v-V)_{xx}/\alpha - (3/4)(v-V)_x^2/(\alpha(v-V)) & \\
& \quad + 3(v-V)(v+V)_x/\alpha^2 + (v-V)^3/\alpha^3 - \lambda = 0, & (340)\\
& \beta(v-V)_t/\alpha - (3/2)[2(v-V)_{xxxx}/\alpha + 60(v-V)^3(v+V)_x/\alpha^4 & \\
& \quad + 12(v-V)^5/\alpha^5 + (10(v-V)(v+V)_{xxx} + 30(v+V)_x(v-V)_{xx} & \\
& \quad + 15(v-V)_x(v+V)_{xx})/\alpha^2 + (30(v-V)^2(v-V)_{xx} & \\
& \quad + 60(v-V)(v+V)_x^2 + 15(v-V)(v-V)_x^2)/\alpha^3]_x = 0. & (341)
\end{aligned}
$$

The simple form of the conservative equations (312) and (341) results from the addition of suitable differential consequences of (169) and (171). The x-part of the BT has already be given by Rogers and Carillo [50] for $\lambda = 0$.

If we write (166)–(167) in the variables $(Y_1, Z_2 = Z^{-1})$,

$$
\begin{aligned}
Y_{1,x} &= -Y_1^2/2 + \lambda Z_2 - 3U/\alpha, & (342)\\
Z_{2,x} &= -2Y_1 Z_2 + 1, & (343)
\end{aligned}
$$

both systems (124)–(125) and (342)–(343) are coupled Riccati systems, with the difference that the transformation from Y_1 to ψ_x/ψ is a point transformation while the one from Z_2 to ψ_x/ψ is a contact one. Thus, the Riccati system (124)–(125) is in the classification of linearizable coupled Riccati systems given by Lie (this is the projective one), while the Riccati system (342)–(343) is outside it.

A minor open problem is to find a bilinear BT for SK equation (the one given in Ref. [41] contains a nonbilinear term).

7.3 Nonintegrable equations, second scattering order

Strictly speaking, nonintegrable equations have no associated scattering order. What is meant in the title of this section is that one assumes a given scattering order to process some nonintegrable PDEs.

For algebraic PDEs in two variables, particular solutions in which (S, C) are constant are quite easy to find. They correspond to solutions u polynomial in $\tanh \frac{k}{2}(x - ct - x_0)$. The privilege of tanh is to be the general solution of the *unique* first order first degree nonlinear ODE with the PP, namely the Riccati equation $\tanh' + \tanh^2 + S/2$, in the particular case $S =$ constant.

A characteristic feature of nonintegrable equations is the absence of a BT. Therefore, the iteration of Sect. 2 can only generate a finite number of new solutions [30, 9], this will be seen on examples.

The Kuramoto-Sivashinsky equation

It is worth to handle this example in detail because it exhibits all the features of what should be done and more importantly of what should *not* be done when solving truncation equations.

The equation of Kuramoto and Sivashinsky (3) (notation $\mu = 19\mu'$) possesses a single family [100, 53]

$$p = -3,\ q = -7,\ u_0 = 120\nu,\ \text{indices}\ -1, 6, \frac{13 \pm i\sqrt{71}}{2}, \mathcal{D} = 60\nu\partial_x^3 + 60\mu'\partial_x. \tag{344}$$

and the orthogonality condition at index 6 is satisfied. Since equation (3) is a conservation law, we therefore study it on its potential form

$$E \equiv v_t + \frac{v_x^2}{2} + \mu v_{xx} + \nu v_{xxxx} + G(t) = 0,\ u = v_x, \tag{345}$$

which has the unique family

$$p = -2,\ q = -6,\ v_0 = -60\nu,\ \ \text{indices}\ -1, 2, \frac{13 \pm i\sqrt{71}}{2},\ \mathcal{D} = 60\nu\partial_x^2 + 60\mu'. \tag{346}$$

Although the no-log condition at index $i = 2$ is not satisfied, the ψ-series

$$v = -60\nu\chi^{-2} + 60\mu' \operatorname{Log} \psi + v_2 + 0(\chi),\ v_2 \text{ arbitrary function}, \tag{347}$$

in which the gradients of ψ and χ are given by (68)–(69) and (64)–(65), contains one logarithm only, which cancels by derivation.

The one-family truncation assumption is

$$v_T = v_0 \chi^{-2} + v_1 \chi^{-1} + v_{02} \operatorname{Log} \psi + v_2, \; v_{02} \text{ constant} \tag{348}$$

equivalent to a truncated series $(-3 : 0)$ for u. Substituting (348) in (345) and eliminating any derivative of χ and ψ, one obtains

$$E = \sum_{j=0}^{6} E_j \chi^{j-6}. \tag{349}$$

Together with the identity (78), this defines a system of eight equations in the six unknowns $(v_0, v_1, v_{02}, v_2, S, C)$.

Equations $j = 0, 1, 2$ are solved for v_0, v_1, v_{02}, v_2 exactly as in the Painlevé test and yield the values in (347). The next five equations ($j = 3, 4, 5, 6$ and (78)) now read [30]

$$E_3 \equiv 120\nu \left(-C + 15\nu S_x + v_{2,x}\right) = 0, \tag{350}$$

$$E_4 \equiv 60 \left(-6\nu^2 S_{xx} - 4\nu^2 S^2 - 20\mu'\nu S + 2\nu C_x + 11\mu'^2\right) = 0, \tag{351}$$

$$E_5 \equiv \frac{S}{2} E_3 + 60 \left(-\mu' C + 20\mu'\nu S_x - 2\nu^2 S S_x + \nu^2 S_{xxx} \right. \\ \left. -\nu C_{xx} + \mu' v_{2,x}\right) = 0, \tag{352}$$

$$E_6 \equiv E(v_2) + 30 \left(\mu' C_x - 19\mu'^2 S - \mu'\nu(20 S^2 + S_{xx}) \right. \\ \left. -\nu^2 (4S^3 + 3S_x^2 + 4 S S_{xx})\right) = 0,$$

$$X \equiv S_t + C_{xxx} + 2C_x S + C S_x = 0. \tag{353}$$

The principles to be obeyed during the resolution are the following.

1. Never increase the differential order of a given variable. On the contrary, solve for the higher derivatives in terms of the lower ones, and substitute the result, as well as its differential consequences, in the remaining equations.
2. Never integrate a differential equation, unless it is just a total derivative. On the contrary, perform an *algebraic* resolution.
3. Never solve for a function of, say, one variable as an expression in several variables.
4. Close the solution by exhausting all Schwarz cross-derivative conditions.

This computation is systematic, and its algorithmic version is known as the construction of a differential Groebner basis [85, 6].

The full system is split into (E_3, E_4, E_5), independent of ∂_t, and (E_6, X), explicitly depending on ∂_t. The subsystem (E_3, E_4) is first solved, according to rule 1, as a Cramer system for $(v_{2,x}, S_{xx})$. After substitution of $(v_{2,x}, S_{xx})$ and their derivatives in all the other equations, equation E_5 is solved, according to rule 1, for C_{xx} and the result is recognized as being an x−derivative. This allows us to solve for C_x after the introduction of an arbitrary integration function λ of t only. As to (E_6, X), they are solved, according to rule 1, as a Cramer system in variables involving only t-derivatives, namely $(v_{2,t}, S_t)$, for expressions independent of ∂_t. To summarize this first stage, the original system is now equivalent to

$$S_{xx} = -\frac{3}{2}S^2 - \frac{5\mu'}{2\nu}S + \frac{\mu'^2}{8\nu^2}(\lambda + 22), \tag{354}$$

$$C_x = -\frac{5\nu}{2}S^2 + \frac{5\mu'}{2}S + \frac{\mu'^2}{8\nu}(3\lambda + 22), \tag{355}$$

$$v_{2,x} = C - 15\nu S_x, \tag{356}$$

$$\frac{v_{2,t}}{\nu} = -\frac{1}{\nu}G(t) + \frac{1243\mu'^3}{2\nu^2} + 16\frac{\mu'^3}{\nu^2}\lambda + 10\frac{\mu'^2}{\nu}(\tag{357}$$

$lambda - 2)S$

$$+110\mu' S^2 - \frac{1}{2\nu}C^2 + 15CS_x - \frac{125\nu}{2}S_x^2, \tag{358}$$

$$S_t = -\frac{5\mu'^3}{16\nu^2}(\lambda + 22) - \frac{\mu'^2}{8\nu}(\lambda - 116)S - \frac{55\mu'}{4}S^2 - \frac{5\nu}{2}S^3$$
$$-CS_x + 5\nu S_x^2. \tag{359}$$

One equation, and only one, namely (354), is an ODE. Integrating it as an elliptic ODE for S [30] would create useless subcases and complications and should, according to rule 2, *not* be done. This ODE should also *not* be replaced by its first integral, because the integrating factor S_x could be, and will indeed be, zero. According to rule 3, it is also forbidden to eliminate $\lambda(t)$ by solving e.g. (355) for it. The only thing to do is (rule 4) to close this solution by cross-differentiation. There are two such conditions:

$$(S_{xx})_t - (S_t)_{xx} \equiv -\frac{\mu'^4}{64\nu^3}(3\lambda^2 + 308\lambda + 5324) + \frac{5\mu'^3}{8\nu^2}(15\lambda + 374)S$$
$$+\frac{25\mu'^2}{8\nu}(\lambda - 16)S^2 - \frac{165\mu'}{2}S^3 - \frac{75\nu}{4}S^4$$
$$+\frac{\mu'^2}{8\nu^2}\lambda' + \frac{85\mu'}{2}S_x^2 + 25\nu S S_x^2 = 0, \tag{360}$$

$$(v_{2,t})_x - (v_{2,x})_t \equiv -C_t - \frac{\mu'^2}{8\nu}(3\lambda + 22)C - \frac{5\mu'}{2}SC + \frac{5\nu}{2}S^2C$$
$$+\frac{15\mu'^2}{4}(3\lambda + 71)S_x - 255\mu'\nu S S_x$$
$$-150\nu^2 S^2 S_x = 0. \tag{361}$$

The latter is solved for C_t and provides a third cross-derivative condition

$$
\begin{aligned}
\frac{(C_x)_t - (C_t)_x}{\nu} \equiv & -\frac{\mu'^4}{64\nu^3}(81\lambda^2 + 4028\lambda + 47476) \\
& + \frac{5\mu'^3}{8\nu^2}(101\lambda + 2322)S + \frac{5\mu'^2}{8\nu}(55\lambda + 96)S^2 \\
& - \frac{1415\mu'}{2} S^3 - \frac{825\nu}{4} S^4 \\
& + \frac{3\mu'^2}{8\nu^2}\lambda' + \frac{535\mu'}{2} S_x^2 + 275\nu S S_x^2 = 0.
\end{aligned} \tag{362}
$$

This ends the linear part of the resolution, and now comes the nonlinear part (algebraic Groebner). The two remaining equations (360) and (362), considered as nonlinear in the two unknowns (S_x, S), imply without computation $S_x = 0$, which then allows one to solve (354) for the monomial S^2 as a polynomial in S of a smaller degree. Equations (360) and (362) thus become linear in S and λ', and their resultant in λ' factorizes as a product of *linear* factors

$$
(\lambda + 33)(\lambda - 11 + \frac{40\nu}{\mu'} S) = 0. \tag{363}
$$

The first factor yields no solution. The second one provides the unique solution

$$
\left(S - \frac{\mu}{38\nu}\right)\left(S + \frac{11\mu}{38\nu}\right) = 0,\ C = \text{ arbitrary constant } c \tag{364}
$$

and it leads to the two-parameter (c, x_0) solution (4). The two equations (364) represent the SME.

If one performs the iteration of Sect. 2, starting from $u = c$, one generates the solitary wave (4) and no more [30].

The reduction $u(x,t) = c + U(\xi), \xi = x - ct$ of the PDE (3) yields the ODE

$$
\nu U''' + \mu U' + U^2/2 + K = 0,\ K \text{ arbitrary}, \tag{365}
$$

with the indices $-1, (13 \pm i\sqrt{71})/2$. Due to the two irrational indices, the *general analytic solution* (see definition in Sect. 9.3) can only depend on one arbitrary constant. This one-parameter solution, whose local expansion contains no logarithm, is known globally only for $K = -450\nu k^2/(19^2\mu)$, this is (4), but its closed form expression for any K is still an open problem.

Being autonomous, the ODE (365) is equivalent to the nonautonomous second order ODE for $V(U)$

$$
V = \frac{\mathrm{d}U}{\mathrm{d}\xi}\ :\ \nu \frac{\mathrm{d}^2(V^2)}{\mathrm{d}U^2} + 2\mu + \frac{U^2 + 2K}{V} = 0, \tag{366}
$$

an equation which has been studied from the Hamiltonian point of view [7].

7.4 Nonintegrable equations, third scattering order

An example is given in the Sect. 9.3.

8 Two common errors in the one-family truncation

Two errors are frequently made in the method of Sect. 7.

8.1 The constant level term does not define a BT

Consider the one-family truncation as done by WTC (the subscript T means "truncated")

$$u_T^{\mathrm{WTC}} = \sum_{j=0}^{-p} u_j^{\mathrm{WTC}} \varphi^{j+p} \tag{367}$$

in which φ is the function defining the singularity manifold.

In the WTC truncation, one considers three solutions of the PDE

1. the lhs u_T^{WTC} of the truncation (367),
2. the "constant level" coefficient u_{-p}^{WTC},
3. the field U which appears in the Lax pair after the successful completion of the method.

The frequently encountered argument "The constant level coefficient u_{-p}^{WTC} also satisfies the PDE, therefore one has obtained a BT" is wrong. This is obvious, since nonintegrable PDEs, which have no BT, nevertheless have this property. One can check it by taking the explicit example of a nonintegrable PDE [30].

A hint that the above argument might be wrong is the fact, observed on all successful truncations, that the U in the Lax pair is *never* u_{-p}^{WTC}. Let us prove this fact, with the homographically invariant analysis [6]. The truncation of the same variable in the invariant formalism is

$$u_T = \sum_{j=0}^{-p} u_j \chi^{j+p}, \tag{368}$$

in which χ is given by (58). This u_T depends on the movable constant φ_0 and one has

$$\begin{cases} u_T^{\mathrm{WTC}} = u_T(\varphi_0 = 0) \\ u_{-p}^{\mathrm{WTC}} = u_T(\varphi_0 = \infty). \end{cases} \tag{369}$$

Since the results of the truncation do not depend on the movable constant φ_0, this proves that the lhs u_T^{WTC} of the truncation and the constant level coefficient u_{-p}^{WTC} are not considered as distinct by the singular manifold method. Since the U in the Lax pair cannot be the truncated u (otherwise one would not have a singular part transformation), this ends the proof.

8.2 The WTC truncation is suitable iff the Lax order is two

We mean the truncation as originally introduced, not its updated version of Sect. 7.

When the Lax pair has second order, everything is consistent. When the Lax pair has a higher order, e.g. three, the original method, as well as its original invariant version [38], presents the following inconsistency. In a first stage, it generates the $-q+p$ equations $E_j(S,C,U) = 0$ of formula (95), which intrinsically correspond to a *second*-order scattering problem (and this is precisely the inconsistency), and in a second stage it injects in each of these $-q+p$ equations a link between (S,C) and the scalar field ψ of the Lax pair of *higher* order, thus generating determining equations which are hybrid between the second order and the higher one. The first nearly correct treatment has been made in Ref. [39].

For the same reason, in order to obtain the Lax pair when its order is higher than two, it is also inconsistent to consider the so-called *singular manifold equation* (SME) [65, 6, 49], defined in Sect. 7.1. When the Lax order is three, the correct extension of this SME notion would be the set of three relations on (a,b,c,d) resulting from the elimination of U between the four coefficients of the Lax pair (e is derivable from (122) so we discard it), but this seems of little interest.

Although these inconsistencies may still provide the full result for some "robust" equations (Boussinesq [125], Sawada-Kotera [64], Hirota-Satsuma [38]), there do exist equations for which it leads to a failure, and the Kaup-Kupershmidt equation [41] is one of them.

9 The singular manifold method applied to two-family PDEs

By two-family, we mean two opposite families. This includes also the one-family truncation as a particular case.

When the base member of the hierarchy of integrable equations has more than a single family, these families usually come by pairs of opposite singular part operators, just like P2–P6. Examples are enumerated at the end of Sect. 5.2. Then the sum of the two opposite singular parts

$$\mathcal{D} \operatorname{Log} \tau_1 - \mathcal{D} \operatorname{Log} \tau_2 \tag{370}$$

only depends on the variable

$$Y = \frac{\tau_1}{\tau_2}. \tag{371}$$

The current status of the method [40, 49], which used to be called the two-singular manifold method [40], is as follows. Most of the method for one-family equations still applies, with the difference that it is much more convenient to represent the Lax pairs in a Riccati form than in a scalar linear form. Let us restrict here to second-order scattering problems (for the third order case, see Sect. 9.2) and to identity links (146) between the two τ and the two ψ functions. Then Y satisfies a Riccati system and, as explained in Sect. 6.1, its most general expression is given by (66).

In the first step, τ is simply replaced by Y in the assumption (3) for a singular part transformation.

In the second step, the scattering problem is represented by the Riccati system satisfied by Y, whose coefficients depend on (S, C, A, B).

The fourth step contains the main difference. Rather than truncating u at the level $j = -p$, one truncates it at the level $j = -2p$ [40, 49], in order to implement the two movable singularities $\tau_1 = 0$ and $\tau_2 = 0$. So the truncation is [49] (for second order Lax pairs only)

$$u = \mathcal{D} \operatorname{Log} Y + U, \tag{372}$$

$$Y^{-1} = B(\chi^{-1} + A), \tag{373}$$

$$E(u) = \sum_{j=0}^{-2q} E_j(S, C, A, B, U) Y^{j+q}, \tag{374}$$

$$\forall j \; E_j(S, C, A, B, U) = 0, \tag{375}$$

in which nothing is imposed on U.

Let us remark that the relation $A \neq 0$ does not characterize two-family PDEs, see the Liouville case in Sect. 9.1.

9.1 Integrable equations with a second order Lax pair

The sine-Gordon equation

The sine-Gordon equation is defined for convenience as the case $a_1 \neq 0, a_2 = 0$ of the equation (168). Although not algebraic in u, it becomes algebraic in e^u and it possesses two opposite families (opposite in the field u), both with $p = -2, q = -2$

$$e^u \sim -(2/\alpha)\varphi_x\varphi_t(\varphi - \varphi_0)^{-2}, \text{ indices } (-1, 2), \; \mathcal{D} = (2/\alpha)\partial_x\partial_t. \tag{376}$$

$$e^{-u} \sim (2/a_1)\varphi_x\varphi_t(\varphi - \varphi_0)^{-2}, \text{ indices } (-1, 2), \; \mathcal{D} = -(2/a_1)\partial_x\partial_t. \tag{377}$$

The resulting singular part transformation assumption

$$e^u + (a_1/\alpha)e^{-u} = (2/\alpha)\partial_x\partial_t \operatorname{Log} Y + \tilde{W}, \; E(u) = 0 \tag{378}$$

with Y defined by (371), can be integrated twice due to the special form of the PDE, resulting in

$$u = -2 \operatorname{Log} Y + W, \; E(u) = 0, \tag{379}$$

in which nothing is imposed on W (we use W to reserve the symbol U for future use). For $a_1 = 0$, this truncation is what was called in Sect. 7.1 the second truncation of Liouville equation.

The five determining equations in the unknowns (S, C, A, B, W) are [49, 9]

$$E_0 \equiv \alpha B^2 e^W - 2C = 0, \tag{380}$$
$$E_1 \equiv 2(C_x + 2AC) = 0, \tag{381}$$
$$E_2 \equiv 0, \text{ (Fuchs index)} \tag{382}$$
$$E_3 \equiv -\sigma_t - \sigma(C_x + 2AC) = 0, \tag{383}$$
$$E_4 \equiv \sigma\left(C\sigma + (C_x + 2AC)_x\right)/2 + a_1 B^{-2} e^{-W} = 0, \tag{384}$$

with the abbreviation

$$\sigma = S + 2A^2 - 2A_x, \tag{385}$$

and, together with the cross-derivative condition (78), they are solved as usual by ascending values of j

$$E_0 : \ B^2 e^W = \frac{2}{\alpha} C, \tag{386}$$
$$E_1 : \ A = -\frac{1}{2}(\operatorname{Log} C)_x, \tag{387}$$
$$E_3 : \ S = -F(x) + \frac{C_x^2}{2C^2} - \frac{C_{xx}}{C}, \tag{388}$$
$$E_4 : \ CC_{xt} - C_x C_t + F(x)C^3 + a_1 \alpha F(x)^{-1} C = 0, \tag{389}$$
$$X : \ a_1 F'(x) = 0. \tag{390}$$

in which F is a function of integration. For sine-Gordon, $F(x)$ must be a constant

$$F(x) = 2\lambda^2. \tag{391}$$

In the Liouville case, for which the truncation imposes no restriction on $F(x)$, let us also require that $F(x)$ be a constant. Then, for both equations, $\operatorname{Log} C$ is proportional to a second solution U of the PDE

$$C = \frac{\alpha}{2} \lambda^{-2} e^U, \ E(U) = 0, \tag{392}$$

and one has obtained the singular part transformation

$$u = -2 \operatorname{Log} y + U, \ y = \lambda B Y, \tag{393}$$

in which y satisfies the Riccati system

$$y_x = \lambda + U_x y - \lambda y^2, \tag{394}$$
$$y_t = -\frac{\alpha}{2} \lambda^{-1} (e^U + (a_1/\alpha) e^{-U} y^2), \tag{395}$$
$$(\operatorname{Log} y)_{xt} - (\operatorname{Log} y)_{tx} = E(U). \tag{396}$$

The linearization

$$y = \psi_1/\psi_2 \tag{397}$$

yields the second-order matrix Lax pair

$$(\partial_x - L)\begin{pmatrix}\psi_1\\ \psi_2\end{pmatrix} = 0,\ L = \begin{pmatrix}U_x/2 & \lambda\\ \lambda & -U_x/2\end{pmatrix}, \tag{398}$$

$$(\partial_t - M)\begin{pmatrix}\psi_1\\ \psi_2\end{pmatrix} = 0,\ M = -(\alpha/2)\lambda^{-1}\begin{pmatrix}0 & e^U\\ -(a_1/\alpha)e^{-U} & 0\end{pmatrix}. \tag{399}$$

The auto-BT (classical for sine-Gordon, Ref. [87] for Liouville) results from the substitution $y = e^{-(u-U)/2}$ into (136)–(137)

$$(u+\tilde{U})_x = -4\lambda \sinh\frac{u-\tilde{U}}{2}, \tag{400}$$

$$(u-\tilde{U})_t = \lambda^{-1}\left(\alpha e^{(u+\tilde{U})/2} + a_1 e^{-(u+\tilde{U})/2}\right). \tag{401}$$

It coincides in the sine-Gordon case with the one given earlier, equations (7)–(8). The ODE part (136) of the BT is a Riccati equation.

The SME is [49]

$$S + C^{-1}C_{xx} - \frac{1}{2}C^{-2}C_x^2 + 2\lambda^2 = 0, \tag{402}$$

and it coincides, but this is not generic, with the one [124, 6] obtained from the (incorrect) truncation in χ.

Remarks.

1. The reason for the presence of the apparently useless parameter B in the definition (66) is to allow the precise correspondence (146)

$$\tau_1 = \psi_1,\ \tau_2 = \psi_2 \tag{403}$$

for some choice of B, namely

$$B = \lambda^{-1},\ y = Y,\ W = U. \tag{404}$$

2. In the Liouville case $a_1 = 0$, this is an example of a PDE with only one family and a nonzero value of A.

The modified Korteweg-de Vries equation

This PDE has the same scattering problem as sine-Gordon, so the computation should be, and indeed is, quite similar to that for sine-Gordon.

Since this PDE has the conservative form

$$\text{mKdV}(w) \equiv bw_t + \left(w_{xx} - 2(w-\beta)^3/\alpha^2 + 6\nu w\right)_x = 0,\ w = r_x, \quad (405)$$

it is technically cheaper to process its potential form

$$\text{p-mKdV}(r) \equiv br_t + r_{xxx} - 2(r_x - \beta)^3/\alpha^2 + 6\nu(r_x - \beta) + F(t) = 0. \quad (406)$$

Its invariance under the involution $w - \beta \mapsto -(w - \beta)$ provides an elegant way [14] to derive the BT of the KdV equation and its hierarchy. Although the constants β and ν could be set to zero by a transformation on (r, x, t) preserving the PP, it is convenient to keep them nonzero, for reasons explained at the end of this Section. This last PDE admits the two opposite families (α is any square root of α^2)

$$p = 0^-,\ q = -3,\ r \sim \alpha \operatorname{Log} \psi,\ \text{indices } (-1, 0, 4),\ \mathcal{D} = \alpha. \quad (407)$$

The truncation is defined by

$$r = \alpha \operatorname{Log} Y + R, \quad (408)$$

with (372)–(375), and this generates five equations $E_j = 0$ [49], with the notation (385) for σ

$$\begin{aligned}
E_1 &\equiv 6\alpha A - 6((R - \alpha \operatorname{Log} B)_x - \beta) = 0, & (409)\\
E_2 &\equiv \alpha(2A_x + 4A^2 - bC - 2\sigma + 6\nu) - \alpha^{-1}(E_1 + 6\alpha A)^2/6 = 0, & (410)\\
E_3 &\equiv \text{p-mKdV}(R - \alpha \operatorname{Log} B) - (3/2)\alpha^{-1}\sigma_x & \\
&\quad +(\sigma - 4A^2 - (1/3)\alpha^{-1}E_{1,x} - 2A_x)E_1 - 2AE_{1,x} - 2AE_2 - E_{2,x} & (411)\\
E_4 &\equiv \text{expression vanishing with } E_1, E_2, E_3, E_5, & (412)\\
E_5 &\equiv (3/4)\alpha\sigma\sigma_x + (1/4)\sigma^2 E_1 = 0, & (413)\\
X &\equiv S_t + C_{xxx} + 2C_x S + CS_x = 0. & (414)
\end{aligned}$$

They depend on (R, B) only through the combination $R - \alpha \operatorname{Log} B$. Equation $j = 4$ is a differential consequence of equations $j = 1, 2, 3, 5$, because 4 is a Fuchs index, and the other equations have been written so as to display how they are solved:

$$\begin{aligned}
E_1 &: \ A = \alpha^{-1}((R - \alpha \operatorname{Log} B)_x - \beta), & (415)\\
E_5 &: \ \sigma = -2(\lambda(t)^2 - \nu),\ \lambda \text{ arbitrary function}, & (416)\\
E_2 &: \ bC = 2A_x - 2A^2 + 4\lambda(t)^2 + 2\nu, & (417)\\
E_3 &: \ \text{p-mKdV}(R - \alpha \operatorname{Log} B) = 0, & (418)\\
X &: \ \lambda'(t) = 0. & (419)
\end{aligned}$$

Thus, their general solution can be expressed in terms of a second solution W of the mKdV equation (405) and an arbitrary complex constant λ [49]

$$\begin{aligned}
W &= (R - \alpha \operatorname{Log} B)_x,\ A = (W - \beta)/\alpha,\\
bC &= 2W_x/\alpha - 2(W - \beta)^2/\alpha^2 + 2\nu + 4\lambda^2,\\
S &= 2W_x/\alpha - 2(W - \beta)^2/\alpha^2 + 2\nu - 2\lambda^2, \qquad (420)
\end{aligned}$$

and the cross-derivative condition $X_1 = 0$ (Eq. (106)), equivalent to the mKdV equation (405) for W, proves that one has obtained a singular part transformation and a Lax pair.

The SME, obtained by the elimination of W between S and C,

$$bC - S - 6\lambda^2 = 0, \tag{421}$$

is identical to that of the KdV equation (216).

The auto-BT of mKdV is obtained by the substitution

$$\mathrm{Log}(BY) = \alpha^{-1} \int (w - W)\mathrm{d}x \tag{422}$$

in the two equations for the gradient of $y = BY$

$$\frac{y_x}{y} = \lambda \left(\frac{1}{y} - y\right) - 2\frac{W - \beta}{\alpha}, \tag{423}$$

$$b\frac{y_t}{y} = \frac{1}{y}\left(-4\lambda\frac{W-\beta}{\alpha} + (2\frac{(W-\beta)^2}{\alpha^2} + 2\frac{W_x}{\alpha} - 4\lambda^2)y\right)_x. \tag{424}$$

In the same manner as in the KdV truncation, these two Riccati equations can also be interpreted as the hetero-BT between the mKdV equation and the PDE satisfied by the pseudopotential y, called the Chen-Calogero-Degasperis-Fokas PDE.

The nonlinear Schrödinger equation

For the AKNS system of two second order equations in (u, v) (whose reduction $\bar{u} = v$ is NLS, see (39)), no two-family truncation has yet been defined which strictly follows the method and provides the desired result. It should be noted that the fourth order equation for u resulting from the elimination of v, known as the Broer-Kaup equation or classical Boussinesq system, admits a two-family truncation without any problem [42].

The full result (singular part transformation, BT) can be found [35] for the AKNS system by performing the one-family truncation [125] and then applying four involutions to the result of Weiss.

A second open problem for this PDE is that its bilinear BT is not yet known.

9.2 Integrable equations with a third order Lax pair

In principle, there is no additional difficulty to extend the method to a scattering order higher than two. A good equation to process would be the modified Boussinesq equation

$$E \equiv \begin{cases} -u_t + (v - (3/2)a^2u^2)_x = 0, \\ -v_t - 3a^2(u_{xx} - uv + a^2u^3)_x = 0, \end{cases} \tag{425}$$

which has two opposite families

$$u \sim (2/a)\chi^{-1}, \ v \sim 6\chi^{-2} \tag{426}$$

and a third order Lax pair like the Boussinesq equation.

The one-family assumption [51]

$$u = U + (2/a)\partial_x \operatorname{Log} \tau, \ v = V - 6\partial_x^2 \operatorname{Log} \tau, \ E(u,v) = E(U,V) = 0, \tag{427}$$

with the identity link $\tau = \psi$ and the choice of the scalar Lax pair (117)–(118), already leads to the solution

$$\begin{aligned} &f = -(3/2)aU, \ a = (V - 3a^2U^2 - 3aU_x)/4, \ b = \lambda, \\ &c = -3a, \ d = -3a^2U, \ e = 0. \end{aligned} \tag{428}$$

Despite this success, it would be more consistent with the two-family structure to process this PDE with a two-family assumption, removing in passing the restriction $E(U,V) = 0$ in (427). This could make the coefficients (f, a, b) linear in (U, V), which is not the case in (428).

Table 1 summarizes, for a sample of PDEs, the currently best method to obtain its Lax pair, singular part transformation and Bäcklund transformation from a truncation.

9.3 Nonintegrable equations, second and third scattering order

A nonintegrable equation has no determined scattering order, so this section cannot be split according to the scattering order.

The KPP equation

The KPP equation (77) possesses the two opposite families (78)–(80) and it fails the test at index 4, so there can only exist particular solutions. Let us first review all the known solutions to this equation.

In addition to the notation (81), it is convenient to introduce the symmetric constant

$$a_1 = (2e_1 - e_2 - e_3)(2e_2 - e_3 - e_1)(2e_3 - e_1 - e_1)/(3d)^3 \tag{429}$$

and the entire function

$$\Psi_3 = \sum_{n=1}^{3} C_n e^{k_n (x + (3/b) k_n t)}, \ k_n = \frac{3e_n - s_1}{3d}, \ C_n \text{ arbitrary}, \tag{430}$$

i.e. the general solution of the third order linear system (117)–(118) with constant coefficients [32]

Table 1. The relevant truncation for some 1+1-dimensional PDEs. The successive columns are: the usual name of the PDE (a p means the potential equation), its number of families (a * indicates that only one family is relevant, see details in Ref), the order of its Lax pair, the truncation variable(s), the link between τ and ψ, the singularity orders of u and $E(u)$, the Fuchs indices (without the ever present -1), the number of determining equations, the reference to the place where the right method was first applied (earlier references may be found in it). The "?" in the AKNS system entry (the one whose NLS is a reduction) means that the method has not yet been applied to it, see text.

Name	f	Lax	Trunc. var.	τ	$-p:-q$	indices	nb.det.ẽq.	Ref
Liouville	1		τ		$0:2$	2	3	[9]
KdV	1	2	χ	ψ	$2:5$	$4,6$	2	[65]
AKNS eq.	1	2	χ	ψ	$1:5$	$4,6$	3	[89]
p-mKdV	2	2	Y	ψ	$0:3$	$0,4$	4	[49]
sine-Gordon	2	2	Y	ψ	$0:2$	2	4	[49]
Broer-Kaup	2	2	Y	ψ	$0:4$	$0,3,4$	4	[49]
pp-Boussinesq	1	3	$(\psi_x/\psi,\psi_{xx}/\psi)$	ψ	$0:4$	$0,1,6$	6	[39]
p-SK	1*	3	$(\psi_x/\psi,\psi_{xx}/\psi)$	ψ	$1:6$	$1,2,3,10$	6	[41]
p-KK	1*	3	$(\psi_x/\psi,\psi_{xx}/\psi)$	G25(ψ)	$1:6$	$1,3,5,7$	14	[91]
Tzitzéica	1*	3	$(\psi_x/\psi,\psi_t/\psi)$	ψ	$2:6$	2	10	[9]
AKNS system	4	2	?		$1:3,1:3$	$0,3,4$		[35]

$$(S) \equiv \begin{cases} \psi_{xxx} - 3a_2\psi_x - a_1\psi = 0, \\ b\psi_t - 3\psi_{xx} = 0. \end{cases} \tag{431}$$

Let us also denote (j,l,m) any permutation of $(1,2,3)$. Three distinct solutions are presently known.

The first solution is trigonometric, this is a *collision of two fronts* [72]

$$u = \frac{s_1}{3} + d\partial_x \operatorname{Log} \Psi_3,\ C_1C_2C_3 \neq 0 \tag{432}$$

which depends on two arbitrary constants $C_1/C_3, C_2/C_3$. For $C_j = 0, C_lC_m \neq 0$, it degenerates into three heteroclinic (i.e. with different limits at both infinities) *propagating fronts* which depend on one arbitrary constant x_0

$$u = \frac{e_l + e_m}{2} + d\frac{k}{2}\tanh\frac{k}{2}(x - ct - x_0), \tag{433}$$
$$k^2 = (k_l - k_m)^2,\ c = -3(k_l + k_m)/b.$$

The second solution is elliptic [10],

$$u = s_1/3 + d\psi_x\sqrt{\wp(\psi)},\ \psi = \Psi_3,\ g_3 = 0,\ g_2 \text{ arbitrary},\ a_1 = 0, \tag{434}$$

it only exists under the constraint (codimension is one) that one root e_j be at the middle of the two others and it depends on the four arbitrary constants

C_1, C_2, C_3, g_2. Its degeneracy $g_2 = 0$ (i.e. $\wp(\psi) = \psi^{-2}$) is the degeneracy $a_1 = 0$ of the collision of two fronts solution (432).

The third and last solution is the stationary elliptic solution $u(x)$

$$u(x): \ -u'' + 2d^{-2}(u-e_1)(u-e_2)(u-e_3) = 0. \tag{435}$$

A trigonometric degeneracy bounded at infinity is made of the three homoclinic *stationary pulses*

$$u = e_j + \frac{e_l - e_m}{\sqrt{2}} \operatorname{sech} i \frac{e_l - e_m}{d\sqrt{2}}(x - x_0),\ a_1 = 0,\ 2e_j - e_l - e_m = 0, \tag{436}$$

it has codimension one and it depends on the arbitrary constant x_0.

Let us now apply the various methods we have seen, in order to retrieve these solutions, namely

1. enforcement of one of the two no-log conditions (82),
2. enforcement of the two no-log conditions (82),
3. one-family truncation with a second order assumption,
4. one-family truncation with a third order assumption,
5. two-family truncation with a second order assumption,
6. two-family truncation with a third order assumption.

The single no-log condition (82) has two solutions. The first one $C = 0$ implies $u_t = 0$ and thus defines the reduction $u(x)$, i.e. the elliptic equation (435). The second one is a first order nonlinear PDE for $C(x,t)$, integrated by the method of characteristics as [9]

$$\begin{aligned} &F(I_1, I_2) = 0,\ c_n = 3k_n/b = (3e_n - s_1)/(bd),\\ &I_1 = e^x (C - c_1)^{p_1} (C - c_2)^{p_2} (C - c_3)^{p_3},\\ &I_2 = e^t (C - c_1)^{q_1} (C - c_2)^{q_2} (C - c_3)^{q_3}, \end{aligned} \tag{437}$$

in which p_n, q_n depend on e_j. Unless some specific choice of the arbitrary function F is made, or (x,t) are no more taken as the independent variables, one cannot integrate further the system (64)–(65) for χ and S.

The two no-log conditions (82) together, apart the already encountered solution $C = 0$, provide the two relations

$$\begin{aligned} &a_1 = 0,\ e_1 = (e_2 + e_3)/2,\\ &b^2 d^2 C^3 - (9/4)(e_2 - e_3)^2 C - 3bd^2 (C_t + C C_x) = 0, \end{aligned} \tag{438}$$

whose solution is similarly

$$\begin{aligned} &a_1 = 0,\ F(I_1, I_2) = 0,\\ &I_1 = \frac{bC - (3(e_2 - e_3)/(2d))}{bC + (3(e_2 - e_3)/(2d))} e^{(e_2 - e_3)x/d},\\ &I_2 = \frac{C^2}{(bC)^2 - (3(e_2 - e_3)/(2d))^2} e^{(e_2 - e_3)^2 t/(bd^2)}. \end{aligned} \tag{439}$$

The one-family truncation (142) with $\tau = \psi$ and the second order assumption (88)–(89) generates three determining equations $E_j(S, C, U) = 0, j = 1, 2, 3$. After solving the first one for U

$$U = s_1/3 - bdC/6, \tag{440}$$

the two remaining equations, with the ever present condition (78), are [21]

$$E_2 \equiv -6a_2 - S + (b^2/6)C^2 - bC_x = 0 \tag{441}$$

$$E_3 \equiv a_2 bC - 2a_1 + bSC/2 - b^3C^3/108$$
$$+S_x/2 - b^2C_t/6 + 2bC_{xx}/3 = 0. \tag{442}$$

The elimination of S and C_t yields a factorized equation

$$-b^2C_t + bC_{xx} - 2b^2CC_x + (4/9)(bC)^3 - 12a_2bC - 12a_1 = 0, \tag{443}$$

$$[bC_{xx} - b^2CC_x + b^3C^3/9 - 3a_2bC - 3a_1]C = 0. \tag{444}$$

The subcase $C = 0$, hence $S = -6a_2, a_1 = 0$, yields the degeneracy $a_1 = 0$ of the three fronts (433). In the other subcase, the system for C is linearizable into the third order system (S) (431) in which both ∂_x and ∂_t change sign, the generic solution (S, C) of (E_2, E_3) is therefore

$$bC = -3\partial_x \operatorname{Log} \Psi_3(-x, -t), \; S = -6a_2 + (bC)^2/6 - bC_x, \tag{445}$$

and there only remains to integrate (64)–(65) for χ or (60)–(61) for φ. Since the one-form $dx - Cdt$ possesses an integrating factor [9], the PDE (61) for φ can be integrated by the method of characteristics,

$$\varphi = \Phi(F), \; F = \frac{\Psi_x + k_2\Psi}{\Psi_x + k_3\Psi} \frac{e^{-k_2x - k_2^2t/b}}{e^{-k_3x - k_3^2t/b}}. \tag{446}$$

Note that the cyclic permutation of the roots e_j is broken when going from (S, C) to φ. With the classical identity on Schwarzians

$$\{\varphi; x\} \equiv \{\Phi; F\}F_x^2 + \{F; x\}, \tag{447}$$

the third order ODE (60) for φ becomes

$$\{\Phi; F\} = 0, \tag{448}$$

which integrates as

$$\varphi = \Phi(F) = \frac{A_1F + A_2}{A_3F + A_4}, \; A_j \text{ arbitrary constants, } A_1A_4 - A_2A_3 \neq 0. \tag{449}$$

The value of χ^{-1}

$$\chi^{-1} = \frac{F_x}{F - F_0} - \frac{F_{xx}}{2F}, \; F_0 \text{ arbitrary constant,} \tag{450}$$

is again invariant under a cyclic permutation of the roots e_j, and the solution u finally obtained is (432).

The one-family truncation (142) with $\tau = \psi$ and the third order assumption (117)–(118) generates five determining equations (98), their straightforward resolution yields

$$u = d\partial_x \operatorname{Log} \psi + s_1/3 + dU, \tag{451}$$
$$\psi_{xxx} + 3U\psi_{xx} - 3(a_2 - U^2 - U_x)\psi_x$$
$$- (a_1 + 3a_2 U - U^3 - bU_t/2 - 6UU_x + U_{xx}/2)\psi = 0, \tag{452}$$
$$b\psi_t - 3\psi_{xx} - 3U\psi_x - 6U^2\psi = 0, \tag{453}$$

in which U is constrained by two relations. But, since the coefficient f can be set to zero without loss of generality, the choice $U = 0$ represents the general solution (just like in [32] where constant values were assumed *ab initio* for the coefficients in (117)–(118)), and it represents again the collision of two fronts (432).

The contrast of difficulty between the second order assumption (laborious) and the third order assumption (immediate) is the signature that the good scattering order of KPP is three, despite the irrelevance of such a notion for nonintegrable equations.

The two-family truncation with a second order assumption (372)–(375) [31, 32, 108] generates five determining equations. Despite the factorized form of E_5, we have not yet found their general solution. The three particular solutions for which (S, C, A) are constant provide immediately the three pulses (436), for they belong to the class of polynomials in tanh and sech, generated by negative and positive powers of χ according to the elementary identities (157).

The elliptic solution (434) can also be written

$$u = s_1/3 + \partial_x \operatorname{Log}(\operatorname{ns}(\psi) - \operatorname{cs}(\psi)),\ \psi = \Psi_3,\ g_3 = 0,\ g_2 \text{ arb.},\ a_1 = 0, \tag{454}$$

a relation in which the argument of the logarithm is the ratio of two entire functions. Therefore it could be possible to find it by a suitable extension of a two-family truncation with a third order assumption.

This solution was first found by the following two-step procedure [10], which, unfortunately for this nice method, only works for a restricted class of PDEs (those with $p = -1, u_0 = c_0, u_1 = c_1 C + c_2, c_j =$ constant, see (80)). The first step is to define the truncation

$$u = d\partial_x \operatorname{Log}(\varphi - \varphi_0) + U,\ U = \text{constant},$$
$$E(u) = \sum_{j=0}^{3} E_j(\varphi_{xx}/\varphi_x, \varphi_t/\varphi_x) \left(\frac{\varphi - \varphi_0}{\varphi_x}\right)^{j-3},\ \forall j\ :\ E_j = 0, \tag{455}$$

whose general solution is $U = s_1/3, \varphi - \varphi_0 = \Psi_3$ (indeed, comparing with (80), this assumption *a priori* implies $b\varphi_t - 3\varphi_{xx} = 0$). The second step is not a truncation, but the change of function $u \mapsto f$

$$u = s_1/3 + (d\partial_x \operatorname{Log} \Psi_3)\, f(\Psi_3), \tag{456}$$

which transforms (77) into

$$U'' - 2U^3 + 2a_1 \Psi_{3,x}^{-3} = 0,\ U(\psi) = f(\psi)/\psi. \tag{457}$$

This is an ODE iff $a_1 = 0$, in which case its solution is (434). Therefore, the assumption (456) has defined a reduction of the PDE to an ODE. This subject will be further examined in Sect. 10.

The cubic complex Ginzburg-Landau equation

The cubic complex Ginzburg-Landau equation (CGL3)

$$E(u) \equiv iu_t + pu_{xx} + q|u|^2 u - i\gamma u = 0,\ pq \neq 0,\ (u,p,q) \in \mathcal{C},\ \gamma \in \mathcal{R}, \tag{458}$$

with p, q, γ constant, is a generic PDE describing the propagation of the signal in an optical fiber as well as superfluidity, spatiotemporal intermittency, pattern formation, etc.

One easily checks that $|u|$ generically behaves like a simple pole. The dominant behaviour

$$u \sim a_0 \chi^{-1+i\alpha},\ \overline{u} \sim \overline{a}_0 \chi^{-1-i\alpha}, \tag{459}$$

in which a_0 is a complex constant, α a real constant, is solution of the nonlinear algebraic system

$$p(-1+i\alpha)(-2+i\alpha) + qa_2 = 0, \tag{460}$$
$$\overline{p}(-1-i\alpha)(-2-i\alpha) + \overline{q}a_2 = 0, \tag{461}$$

with $a_2 = |a_0|^2$. This defines two families for $|u|^2$ (four for $|u|$) [9]

$$a_2 = \frac{9|p|^2}{2|q|^2 d_i^2}[d_r + \Delta],\ \alpha = \frac{3}{2d_i}(d_r + \Delta), \tag{462}$$
$$\frac{p}{q} = d_r - id_i,\ \Delta^2 = d_r^2 + (8/9)d_i^2. \tag{463}$$

To prevent these irrational expressions to mess up all subsequent computations (Fuchs indices, no-log conditions, truncations), the system (460)–(461) can equivalently be solved as a *linear* system on $\mathcal{C}$ [32, 36]

$$a_2 = -\frac{p}{q}(1-i\alpha)(2-i\alpha), \tag{464}$$
$$\overline{p} = Kp(1-i\alpha)(2-i\alpha),\ \overline{q} = Kq(1+i\alpha)(2+i\alpha), \tag{465}$$

in which K is an irrelevant arbitrary nonzero complex constant.

The indicial equation is the determinant [26] of the second order matrix

$$\mathbf{P}(j) = \begin{pmatrix} (2a_0\overline{a_0})q & a_0^2 q \\ \overline{q}\overline{a_0}^2 & (2a_0\overline{a_0})\overline{q} \end{pmatrix}$$
$$+ \mathrm{diag}(p(j-1+i\alpha)(j-2+i\alpha), \overline{p}(j-1-i\alpha)(j-2-i\alpha)), \quad (466)$$

and with the resolution (464)–(465) it evaluates to

$$\det \mathbf{P}(j) = (j+1)j(j^2 - 7j + 6\alpha^2 + 12) = 0. \quad (467)$$

For generic values of (p,q), two of the four indices are irrational.

Let us consider, for simplification, the solitary wave reduction

$$u(x,t) = U(\xi)e^{i(\omega t + \varphi(\xi))}, \ \xi = x - ct, \quad (468)$$

in which (U,φ) are functions of the reduced independent variable ξ, and let us restrict to the pure CGL3 case $\mathrm{Im}(p/q) \neq 0$. The general solution of the fourth order system of ODEs for (U,φ) *a priori* depends on six arbitrary constants, the four constants of integration plus the two reduction parameters (c,ω). From these six constants, one must subtract

1. the irrelevant origin ξ_0 of ξ (Fuchs index -1), which represents the invariance under a space translation,
2. the irrelevant origin φ_0 of the phase (Fuchs index 0), which represents the invariance under a phase shift,
3. and the number of irrational Fuchs indices, generically two. Indeed, these irrational indices represent the chaotic nature of CGL3 (see the expansion (48) in [32]) and they cannot contribute to any analytic solution.

Therefore only two relevant arbitrary constants are present in what can be called the *general analytic solution* of the reduction (468).

Presently, one only knows four particular solutions of the reduction $\xi = x - ct$ with a zero codimension (no constraint on (p,q,γ)). These are

1. a pulse or solitary wave [107]

$$u = -ia_0 k \,\mathrm{sech}\, kx e^{i[\alpha \,\mathrm{Log} \cosh kx + K_1 t]}, \ K_2 k^2 - \gamma = 0, \quad (469)$$

2. a front or shock [101]

$$u = a_0 \frac{k}{2}\left[\tanh \frac{k}{2}\xi \pm 1\right] e^{i[\alpha \,\mathrm{Log} \cosh \frac{k}{2}\xi + K_3 c\xi - K_4 c^2 t]}, \quad (470)$$

3. a source or propagating hole [5]

$$u = a_0 \left[\frac{k}{2} \tanh \frac{k}{2}\xi + (K_1 + iK_2)c\right]$$
$$\times e^{i[\alpha \,\mathrm{Log} \cosh \frac{k}{2}\xi + K_3 c\xi - (K_4 k^2 + K_5 c^2)t]}, \ K_6 k^2 + K_7 c^2 = \gamma, \quad (471)$$

4. an unbounded solution [32]

$$|u|^2 = a_2(\tan^2 \frac{k}{2}\xi + K^2),\ c = 0. \tag{472}$$

In the above expressions, all parameters (a_2, α, k, c, K_l) are real and only depend on (p, q, γ), except in (471) where the velocity c is arbitrary.

These four particular solutions are four different degeneracies [32] of the yet unknown general analytic solution.

In experiments or computer simulations, one has observed [5, 79, 12, 62] other regular patterns which should correspond to other degeneracies of the general analytic solution. One of them [62] is a homoclinic hole solution, complementing the heteroclinic hole (471). Another one, of the highest interest in fiber optics, is a propagating pulse, extrapolating (469) to $c \neq 0$ and reducing in the NLS limit (p, q real, $\gamma = 0$) to a "bright soliton" of arbitrary velocity.

Let us now address the question of retrieving these four solutions (and ideally of finding the unknown one ar at least other degeneracies) by some truncation. For a truncation to be successful, the truncated variables should be free of any multivaluedness in their dominant behaviour. This is not the case of the natural physical variables $(u, \overline{u})$ or $(\operatorname{Re} u, \operatorname{Im} u)$, which are *always* locally multivalued as seen from (459). A more detailed study [32] uncovers the best representation for this purpose, namely a *complex modulus* Z and a real argument Θ uniquely defined by

$$u = Ze^{i\Theta},\ \overline{u} = \overline{Z}e^{-i\Theta}, \tag{473}$$

and the above four exact solutions are written in this notation. For each family, if one excludes the contribution of the irrational Fuchs indices, the three fields $(Z, \overline{Z}, \operatorname{grad} \Theta)$ are locally singlevalued and they behave like simple poles. The physical variables $(|u|^2, \operatorname{grad} \arg u)$ also have this nice property of being locally singlevalued (they respectively behave like a double pole and a simple pole), but they are not as elementary as $(Z, \overline{Z}, \operatorname{grad} \Theta)$.

The one-family truncation of the third order ODE satisfied by $|u|^2$ (after elimination of φ), with a constant coefficient second order assumption, evidently captures all four solutions, since $|u|^2$ is a degree-two polynomial in $\tanh \kappa\xi$. Such a truncation generates cumbersome computations and provides no additional solution.

The one-family truncation of $(Z, \overline{Z}, \operatorname{grad} \Theta)$ with the same constant coefficient second order assumption is defined as [32, 34]

$$\begin{cases} Z = a_0(\chi^{-1} + X + iY), \\ \overline{Z} = \overline{a}_0(\chi^{-1} + X - iY), \\ \Theta = \omega t + \alpha \operatorname{Log} \psi + K\xi, \\ (\operatorname{Log} \psi)' = \chi^{-1},\ \chi' = 1 - (k^2/4)\chi^2, \\ Ee^{-i\Theta} = \sum_{j=0}^{3} E_j \chi^{j-3}, \end{cases} \tag{474}$$

in which χ and ψ are functions of $\xi = x-ct$, (ω, X, Y, K, k^2) are real constants. One has to solve the four complex (eight real) equations $E_j = 0$ in the eight real unknowns $(a_2, \alpha, \omega, X, Y, K, c, k^2)$, the two complex parameters (p, q), and the real parameter γ. If there exists a solution, the elementary building block functions evaluate to

$$\chi = \frac{k}{2} \tanh \frac{k\xi}{2}, \ \psi = \cosh \frac{k\xi}{2}. \tag{475}$$

The good procedure [32, 36] is again to select, among the eleven complex variables considered as equivalent, four variables which make the system a *linear* one of Cramer type. The system (E_0, E_1, E_2) is of Cramer type in (a_2, K, ω), and after its resolution the last equation E_3 is independent of (p, q, γ, c) and factorizes into a product of *linear* factors

$$E_3 \equiv [k^2 - 4(X + iY)^2](\alpha Y - 2X) = 0. \tag{476}$$

Finally, this one-family truncation recovers all four solutions except the pulse (469).

The two-family truncation of $(Z, \overline{Z}, \mathrm{grad}\, \Theta)$ with the same constant coefficient second order assumption retrieves the pulse solution (469), but finds nothing new.

Similar truncations for two coupled CGL3 equations can be found in [36, 37].

The nonintegrable Kundu-Eckhaus equation

The PDE for the complex field $U(x,t)$ [74, 8]

$$iU_t + \alpha U_{xx} + (\frac{\beta^2}{\alpha} |U|^4 + 2be^{i\gamma}(|U|^2)_x)U = 0, \ (\alpha, \beta, b, \gamma) \in \mathcal{R}, \tag{477}$$

with $\alpha\beta \cos\gamma \neq 0$, is linearizable when $b^2 = \beta^2$ into the Schrödinger equation

$$iV_t + \alpha V_{xx} = 0, \ U = \sqrt{\frac{\alpha}{2\beta \cos\gamma}} \frac{V}{\sqrt{\int |V|^2 \mathrm{d}x}}. \tag{478}$$

This suggests considering the PDE for $u = \int |U|^2 \mathrm{d}x$

$$\begin{aligned} &\frac{\alpha}{2}(u_{xxxx}u_x^2 + u_{xx}^3 - 2u_x u_{xx} u_{xxx}) + 2\frac{\beta^2 - (b\sin\gamma)^2}{\alpha} u_x^4 u_{xx} \\ &+2(b\cos\gamma)u_x^3 u_{xxx} + \frac{1}{2\alpha}(u_{tt}u_x^2 + u_{xx}u_t^2 - 2u_t u_x u_{xt}) = 0. \end{aligned} \tag{479}$$

When $b^2 \neq \beta^2$, this PDE fails the test [15] because, for each of the two families for u, one index is generically irrational. However, its one-family truncation with a second order assumption (i.e. the usual WTC truncation) is a very rich exercise [33] which yields quite unusual solutions, among them an elliptic one involving the ODE of class III of Chazy [13].

10 Singular manifold method *versus* reduction methods

In order to find exact solutions of PDEs, there exist two main classes of methods. The first class, which has been detailed in these lectures, is based on the structure of the movable singularities and it can be called, in short, the singular manifold method.

The second class, presented in another course of this school [129], basically relies on group theory and consists in finding the *reductions* to a PDE in a lesser number of independent variables, and at the end to an ODE. These reductions are obtained either by looking for the infinitesimal symmetries of the PDE (space translation, etc) and by integrating them, or by a direct search not involving any group theory. The main methods in this second class are known as (see references in [129]) the *classical method* (point symmetries), the *nonclassical method* (conditional point symmetries), the *direct method* (direct search).

The question of the comparison of these four methods by their results is an active research subject [48, 49, 57], and its current state is given in [17, 129].

Let us take as an example a second order nonintegrable PDE, this is enough to give an idea of the comparison. The KPP equation (77) has been studied in detail with the singular manifold method, Sect. 9.3. It has also been investigated with the three other methods, and the results are the following.

In the classical and nonclassical methods, let us denote

$$\tau(x,t,u)u_t + \xi(x,t,u)u_x - \eta(x,t,u) = 0 \tag{480}$$

the PDE for $u(x,t)$ which, after computation of the symmetries (τ,ξ,η), defines the constraint on u susceptible to yield a reduction if the constraint can be integrated.

Classical method. It yields only two reductions $u(x,t) \mapsto U(z)$ [9, 10], one noncharacteristic (i.e. conserving the differential order two)

$$z = x - ct,\ u = U,\ -U'' - bcU' + 2d^{-2}(U-e_1)(U-e_2)(U-e_3) = 0, \tag{481}$$

one characteristic (i.e. lowering the differential order two)

$$z = t,\ u = U,\ bU' + 2d^{-2}(U-e_1)(U-e_2)(U-e_3) = 0. \tag{482}$$

This is in fact a unique reduction $z = \lambda x + \mu t$, but the splitting according to the characteristic nature is relevant for the Painlevé property. None of these two ODEs has the Painlevé property, unless $c = (3e_j - s_1)/(bd)$ in (481).

Nonclassical method. It yields three sets of values for (τ,ξ,η) [102], two with $\tau \neq 0$ and one with $\tau = 0, \xi \neq 0$.

The first one [10] has codimension one

$$a_1 = 0,\ e_1 = (e_2+e_3)/2,\ bu_t - 3d^{-2}\left((\mathrm{Log}\,\psi)_x(u-e_1)\right)_x = 0,\ \psi = \Psi_3, \tag{483}$$

with Ψ_3 defined by (430), its integration defines the noncharacteristic reduction to an elliptic equation

$$a_1 = 0,\ z = \Psi_3,\ u = e_1 + dz_x U(z),\ U'' - 2U^3 = 0, \tag{484}$$

and one finds the solution (434).

The second one [102] has codimension zero

$$bu_t + d^{-1}(u - s_1/3)u_x + 3d^{-2}(u - e_1)(u - e_2)(u - e_3) = 0, \tag{485}$$

and, remarkably, this first order PDE, which fails the test, identifies to the no-log condition (82) with $3u - s_1 = bdC$. Its integration (437) cannot define a reduction unless some choice of the arbitrary function is made. Nevertheless, the common solution to the PDEs (77) and (485) is (432).

The third one [102] is

$$u_x - \eta = 0, \tag{486}$$

in which η satisfies the second order PDE

$$\begin{aligned} &\eta_{xx} + 2\eta\eta_{xu} + \eta^2\eta_{uu} + 2d^{-2}(u - e_1)(u - e_2)(u - e_3)\eta_u \\ &-2d^{-2}(3u^2 - 2s_1 u + s_1^2/3 - 3d^2 a_2)\eta + b\eta_t = 0. \end{aligned} \tag{487}$$

Integrating (487) is equivalent to integrating the original PDE (77), since the transformation (486) simply exchanges them, so one is stranded. The only way out is to put some additional constraints on η. The consistent way to do that ([17] page 634) is to eliminate u_x and its derivatives (in this case u_{xx} only) between (486) and (77), which results in a nonlinear first order ODE for the function $t \mapsto u(x,t)$ (i.e. with x as a parameter)

$$bu_t - (\eta_x + \eta\eta_u) + 2d^{-2}(u - e_1)(u - e_2)(u - e_3) = 0. \tag{488}$$

Requiring the invariance of this ODE under the infinitesimal transformation $(\tau = 0, \xi = 1, \eta)$ of the classical method creates constraints on η, an exercise which is left to the reader.

Direct method. The search for a reduction $u(x,t) \mapsto U(z)$ in the class

$$u(x,t) = \lambda(x,t)U(z(x,t)) + \mu(x,t), \tag{489}$$

apart from the characteristic reduction $z = t, u = U$, yields the noncharacteristic reduction [102]

$$U'' - 2U^3 + g_1(z)U' + g_2(z)U + g_3(z) = 0, \tag{490}$$

$$u = dz_x U + s_1/3,\ z_x \neq 0, \tag{491}$$

provided z, g_1, g_2, g_3 satisfy the system

$$\begin{cases} bz_{xt} - z_{xxx} - 6a_2 z_x + z_x^3 g_2 = 0, \\ bz_t - 3z_{xx} + z_x^2 g_1 = 0, \\ 2a_1 - z_x^3 g_3 = 0. \end{cases} \tag{492}$$

The constraint $z_x \neq 0$ splits the discussion into $a_1 = 0$ and $a_1 \neq 0$. The case a_1 arbitrary defines the reduction [102]

$$z = x - ct,\ u = dU + s_1/3,\ U'' + cU' - 2(U^3 - 3a_2 U - a_1) = 0, \tag{493}$$

identical to (481). In the case $a_1 = 0$, hence $g_3 = 0$, the system is solved for (z_{xxx}, z_t), and the condition $(z_{xxx})_t = (z_t)_{xxx}$ reads

$$\left(18\left(\frac{z_{xx}}{z_x^2}\right)^2 + 18\frac{z_{xx}}{z_x^2} + 2g_1'(z) - g_1(z)\partial_z + 3\partial_z^2\right) Q(z) = 0, \tag{494}$$

$$Q(z) = 9g_2 - 2g_1^2 - 3g_1'. \tag{495}$$

For $Q(z) \neq 0$, the condition integrates as

$$z = G(xf_1(t) + f_2(t)), \tag{496}$$

in which G is an arbitrary function, and (f_1, f_2) are further constrained by $f_1' = 0, f_2' = 0$. The result is $a_1 = 0, z = G(x)$, and the ODE (490) transforms to an elliptic equation under

$$U(z) = \frac{f(G^{-1}(z))}{G'(z)},\ f'' - 2f^3 + 6a_2 f = 0, \tag{497}$$

in which G^{-1} denotes the inverse function of G. This solution is not distinct from the stationary elliptic reduction (435).

For $Q(z) = 0$, if one defines the function G by

$$g_1(z) = -3(\mathrm{Log}(G'(z)))', \tag{498}$$

the system (492) is equivalent to

$$\begin{cases} g_2 = (2/9)g_1^2 + (1/3)g_1', \\ (\partial_x^3 - 3a_2\partial_x)G(z(x,t)) = 0, \\ (b\partial_t - 3\partial_x^2)G(z(x,t)) = 0, \\ f''(Z) - 2f(Z)^3 = 0,\ U(z) = G'(z)f(Z),\ Z = G(z), \end{cases} \tag{499}$$

and this proves that the particular solution $g_1 = g_2 = 0$ considered in [102] is the general solution, equivalent to the reduction $z = \Psi_3$ in (434).

11 Truncation of the unknown, not of the equation

When applied for instance to the second Painlevé equation P2

$$\text{P2} \quad E(u) \equiv u'' - 2u^3 - xu - \alpha = 0, \tag{500}$$

the one-family singular manifold method in the case of a second order scattering problem, i.e. the one originally performed by WTC, see Sects. 7.1 and 7.3, presents the following drawback. The P2 ODE has two families $u \sim \varepsilon\chi^{-1}, \varepsilon^2 = 1$. With the definition of grad χ

$$\chi' = 1 + (S/2)\chi^2, \tag{501}$$

the one-family truncated expansion for u is found to be [59]

$$u = \varepsilon\chi^{-1}, \ E(u) = \varepsilon(S - x)\chi^{-1} + (-\alpha - \varepsilon S'/2)\chi^0, \tag{502}$$

$$E_2 \equiv \varepsilon(S - x) = 0, \tag{503}$$

$$E_3 \equiv -\alpha - \varepsilon S'/2 = 0, \tag{504}$$

and its general solution is

$$S = x, \ 2\alpha + \varepsilon = 0, \ u = \varepsilon(\operatorname{Log}\operatorname{Ai}(x))', \ \operatorname{Ai}'' + (x/2)\operatorname{Ai} = 0. \tag{505}$$

One therefore finds only a one-parameter particular solution in terms of the Airy function, at the price of one constraint on the parameter α. This is unsatisfactory because the method fails to find the highest information on (P2) (highest in the context of these lectures), namely its *Schlesinger transformation.* Such a transformation is by definition a *birational transformation* between two different copies of P2, denoted $u(x, \alpha)$ and $U(X, A)$, and it reads [81]

$$x = X, \ u + U = \frac{-2A - 1}{2(U' + U^2) + X} = \frac{2\alpha - 1}{2(u' - u^2) - x}, \ \alpha = A + 1. \tag{506}$$

A method to remedy this drawback is the following [16] [1] We rephrase it in the homographically invariant formalism, which simplifies the exposition. Firstly, rather than splitting $E(u)$, defined in (502), into one equation per power of χ, one retains the single information $E(u) = 0$, and one eliminates u and χ between the three equations (501) and (502) to obtain the second order ODE for $S(x)$

$$2(S - x)S'' - S'^2 + 2S' + 2S^3 - 4xS^2 + 2x^2 S + 4\alpha(\alpha + \varepsilon) = 0. \tag{507}$$

This ODE for $S(x)$, which is birationally equivalent to P2 under the transformation

[1] After these lecture notes were first written, a more direct method has been proposed [39], it is briefly described in Sect. 12.

$$u = \varepsilon\chi^{-1},\ S = -2(\chi^{-1})' - 2\chi^{-2},\ \chi^{-1} = \frac{S' + 2\varepsilon\alpha}{2(S - x)}, \tag{508}$$

bears the number 34 in the classification of Painlevé and Gambier [20].

Secondly, despite the fact that one already knows the general solution $S(x)$ in terms of the P2 function $u(x, \alpha)$, one takes advantage of the two-family structure of P2 (the sign ε is ± 1) to perform an involution by representing $S(x)$ with another P2 function $U(X, A)$ as

$$S = -2V' - 2V^2,\ U = \varepsilon_2 V,\ \varepsilon_2^2 = 1,\ U'' - 2U^3 - XU - A = 0,\ X = x. \tag{509}$$

The elimination of S between (508) and (509) provides a relation between $(\varepsilon\alpha, \varepsilon_2 A)$ only

$$(A + \varepsilon_2/2)^2 = (\alpha + \varepsilon/2)^2. \tag{510}$$

The solution $A = -\varepsilon_2(\varepsilon\alpha + 1)$ is the Schlesinger transformation.

An equivalent presentation can be found in Ref. [61]. In the latter, one first computes the two coefficients u_0, u_1 of the Laurent expansion

$$u = u_0\chi^{-1} + u_1, \tag{511}$$

then the Schlesinger transformation is readily obtained by (more precisely, the computation of [61] reduces to) the elimination of the three variables u, Z, U'' between the four equations (u_0, u_1, u, U, Z, s are functions of $X = x$)

$$u = u_0 Z^{-1} + U, \tag{512}$$

$$Z' = 1 + 2\frac{U - u_1}{u_0} Z + \frac{s}{2} Z^2, \tag{513}$$

$$\mathrm{Pn}(u, x, \alpha, \beta, \gamma, \delta) = 0, \tag{514}$$

$$\mathrm{Pn}(U, X, A, B, \Gamma, \Delta) = 0. \tag{515}$$

Equation (512) is an assumption for a singular part transformation, and (513) defines a Riccati equation for the expansion variable Z which depends on a free function s. The elimination is differential for u and Z, algebraic for U'', and it results in

$$F(U', U; s, s', s'', \alpha, \beta, \gamma, \delta, A, B, C, D) = 0. \tag{516}$$

The algebraic independence of (U', U), consequence of the irreducibility of Pn, requires the identical vanishing of F as a polynomial of the two variables (U', U), and this provides two solutions: the identity $(u = U, Z^{-1} = 0)$ and, at least for P2 and P4, the Schlesinger transformation. The result for P2 is

$$\mathrm{P2}\ :\ \varepsilon Z^{-1} = u - U = \frac{\varepsilon(A - \alpha)}{2U' + \varepsilon(2U^2 + x)},\ \alpha + A + \varepsilon = 0,\ s = 0, \tag{517}$$

and the inverse transformation

$$\text{P2} \ : \ u - U = \frac{\varepsilon(A-\alpha)}{2u' + \varepsilon(2u^2 + x)} \tag{518}$$

follows from the elimination of U' between (517) and

$$(U-u)' + \varepsilon(U^2 - u^2) = 0, \tag{519}$$

itself obtained by the elimination of Z between (512) and (513). This Schlesinger transformation is identical, thanks to the parity invariance of P2, to (506).

The result for P4 is

$$\text{P4} \ u'' - u'^2/(2u) - (3/2)u^3 - 4xu^2 - 2x^2u + 2\alpha u - \beta/u = 0, \tag{520}$$

$$\begin{aligned} \varepsilon Z^{-1} = u - U &= \frac{4\varepsilon(\alpha - A)U}{3U' + \varepsilon(3U^2 + 6xU - 2A - 4\alpha) + 6}, \\ &= \frac{4\varepsilon(\alpha - A)u}{3u' + \varepsilon(3u^2 + 6xu - 2\alpha - 4A) + 6}, \end{aligned} \tag{521}$$

$$(U-u)' + \varepsilon(U^2 - u^2 + 2x(U-u) + 2(\alpha - A)/3) = 0, \tag{522}$$

$$9\beta + 2(\alpha + 2A - 3\varepsilon)^2 = 0, \ 9B + 2(A + 2\alpha - 3\varepsilon)^2 = 0, \tag{523}$$

$$s = 4(A - \alpha)/3. \tag{524}$$

12 Birational transformations of the Painlevé equations

The difficulty for the truncation (512)–(515) to handle P6 has been explained in [40], and the following general method has been given [39] to overcome it.

Consider an Nth order, first degree ODE with the Painlevé property. This is necessarily [104, pp. 396–409] a Riccati equation for $U^{(N-1)}$, with coefficients depending on x and the lower derivatives of U, e.g. in the case of P6,

$$U'' = A_2(U,x)U'^2 + A_1(U,x)U' + A_0(U,x), \tag{525}$$

$$\begin{aligned} A_2 &= \frac{1}{2}\left[\frac{1}{U} + \frac{1}{U-1} + \frac{1}{U-x}\right], \ A_1 = -\left[\frac{1}{x} + \frac{1}{x-1} + \frac{1}{U-x}\right], \\ A_0 &= \frac{U(U-1)(U-x)}{x^2(x-1)^2}\left[A + B\frac{x}{U^2} + \Gamma\frac{x-1}{(U-1)^2} + \Delta\frac{x(x-1)}{(U-x)^2}\right]. \end{aligned} \tag{526}$$

Then, assume its Lax pair to have second order, which is indeed the case for P6 [54]. This implies that the pseudopotential of the singular manifold method has only one component [38], which can be chosen so as to satisfy some Riccati ODE,

$$Z' = 1 + z_1 Z + z_2 Z^2, \ z_2 \neq 0, \tag{527}$$

in which (Z, z_1, z_2) are rational functions of $(U^{(N-1)}, ..., U, x)$ to be determined.

Since the group of invariance of a Riccati equation is the homographic group, the variables $U^{(N-1)}$ and Z, which both satisfy a Riccati ODE, are not independent, but they are linked by a homography, the three coefficients g_j of which are rational in $(U^{(N-2)}, ..., U, x)$, e.g. in the case $N = 2$ relevant for P6,

$$(U' + g_2)(Z^{-1} - g_1) - g_0 = 0, \ g_0 \neq 0. \tag{528}$$

This allows us to obtain the two coefficients z_j of the Riccati pseudopotential equation (527) as explicit expressions of $(g_j, \partial_U g_j, \partial_x g_j, A_2, A_1, A_0, U^{(N-1)})$. Indeed, eliminating U' between (525) and (528) defines a first order ODE for Z, whose identification with (527) *modulo* (528) provides the three relations

$$g_0 = g_2^2 A_2 - g_2 A_1 + A_0 + \partial_x g_2 - g_2 \partial_U g_2, \tag{529}$$
$$z_1 = A_1 - 2g_1 + \partial_U g_2 - \partial_x \operatorname{Log} g_0 + (2A_2 - \partial_U \operatorname{Log} g_0) U', \tag{530}$$
$$z_2 = -g_1 z_1 - g_1^2 - g_0 A_2 - \partial_x g_1 - (\partial_U g_1) U'. \tag{531}$$

Therefore, the natural unknowns in the present problem are the two coefficients g_1, g_2 of the homography, which are functions of the two variables (U, x), and not the two functions (z_1, z_2) of the three variables (U', U, x).

The truncation assumption is, for an ODE with movable simple poles such as P6,

$$u = u_0 Z^{-1} + U, \ u_0 \neq 0, \ x = X, \tag{532}$$

with u and U solutions of the same ODE but with different parameters, and it now defines a consistent truncation. This is achieved by the elimination of u, Z', U'', U' between (514), (515), (532), (527) and (528), followed by the elimination of (g_0, z_1, z_2) from (529)–(531) (q denotes the singularity order of Pn written as a differential polynomial in u, it is -6 for P6),

$$E(u) = \sum_{j=0}^{-q+2} E_j(U, x, u_0, g_1, g_2, \boldsymbol{\alpha}, \boldsymbol{A}) Z^{j+q-2} = 0, \tag{533}$$
$$\forall j \ : \ E_j(U, x, u_0, g_1, g_2, \boldsymbol{\alpha}, \boldsymbol{A}) = 0. \tag{534}$$

The nonlinear *determining equations* $E_j = 0$ are independent of U', and this is the main difference with previous work [61]. Another difference is the greater number ($-q+3$ instead of $-q+1$) of equations $E_j = 0$, which is due to the additional elimination of U' with (528). Their resolution for P6 [39] provides as the unique solution the unique first degree birational transformation of P6, first found by Okamoto [103], which can be written as [28]

$$\frac{N}{u-U} = \frac{x(x-1)U'}{U(U-1)(U-x)} + \frac{\Theta_0}{U} + \frac{\Theta_1}{U-1} + \frac{\Theta_x - 1}{U-x} \tag{535}$$

$$= \frac{x(x-1)u'}{u(u-1)(u-x)} + \frac{\theta_0}{u} + \frac{\theta_1}{u-1} + \frac{\theta_x - 1}{u-x}, \tag{536}$$

$$\forall j = \infty, 0, 1, x \; : \; (\theta_j^2 + \Theta_j^2 - (N/2)^2)^2 - (2\theta_j \Theta_j)^2 = 0, \tag{537}$$

$$N = \sum(\theta_k^2 - \Theta_k^2), \tag{538}$$

in which the two sets of monodromy exponents are only defined by their squares,

$$\theta_\infty^2 = 2\alpha,\ \theta_0^2 = -2\beta,\ \theta_1^2 = 2\gamma,\ \theta_x^2 = 1 - 2\delta, \tag{539}$$

$$\Theta_\infty^2 = 2A,\ \Theta_0^2 = -2B,\ \Theta_1^2 = 2\Gamma,\ \Theta_x^2 = 1 - 2\Delta. \tag{540}$$

The eight signs of the monodromy exponents are arbitrary and independent, and the equivalent affine representation of (537)–(538) is

$$\theta_j = \Theta_j - \frac{1}{2}\left(\sum \Theta_k\right) + \frac{1}{2},\ \Theta_j = \theta_j - \frac{1}{2}\left(\sum \theta_k\right) + \frac{1}{2}, \tag{541}$$

$$N = 1 - \sum \Theta_k = -1 + \sum s_k \theta_k = 2(\theta_j - \Theta_j),\ j = \infty, 0, 1, x, \tag{542}$$

in which $j, k = \infty, 0, 1, x$.

The well known confluence from P6 down to P2 then allows us to recover [38] all the first degree birational transformations of the five Painlevé equations (P1 admits no such transformation because it depends on no parameter), thus providing a unified picture of these transformations.

13 Conclusion, open problems

The singular manifold method, which is based on the singularity structure, is quite powerful to provide exact solutions or other analytic results. There still exist many challenging problems, in particular in nonlinear optics and spatiotemporal intermittency [5, 62], in which the equations, although nonintegrable, possess some regular "patterns" which could well be described by exact particular solutions. The difficulty to find them [32] comes from the good guess which must be made for the functions ψ, which do not necessarily satisfy a linear system any more. Methods from group theory usually provide complementary results, although they also fail in the two just quoted examples.

Acknowledgments

The author thanks the CIME Fundazione and l'Università di Palermo for invitation.

References

1. M. J. Ablowitz and P. A. Clarkson, *Solitons, nonlinear evolution equations and inverse scattering* (Cambridge University Press, Cambridge, 1991).
2. M. J. Ablowitz, D. J. Kaup, A. C. Newell and H. Segur, The inverse scattering transform-Fourier analysis for nonlinear problems, Stud. Appl. Math. **53** (1974) 249–315.
3. I. Ayari and A. M. Grundland, Conditional symmetries for kth order partial differential equations, 24 pages, preprint CRM–2488 (1997).
4. A. V. Bäcklund, Om ytor med konstant negativ krökning, Lunds Universitets Arsskrift Avd. 2 **19** (1883).
5. N. Bekki and K. Nozaki, Formations of spatial patterns and holes in the generalized Ginzburg-Landau equation, Phys. Lett. A **110** (1985) 133–135.
6. F. Boulier, *Étude et implantation de quelques algorithmes en algèbre différentielle* (Thèse, Université des sciences et technologies de Lille, 1994).
7. S. Bouquet, Hamiltonian structure and integrability of the stationary Kuramoto-Sivashinsky equation, J. Math. Phys. **36** (1995) 1242–1258.
8. F. Calogero and W. Eckhaus, Nonlinear evolution equations, rescalings, model PDEs and their integrability, I, Inverse problems **3** (1987) 229–262.
9. F. Cariello and M. Tabor, Painlevé expansions for nonintegrable evolution equations, Physica D **39** (1989) 77–94.
10. F. Cariello and M. Tabor, Similarity reductions from extended Painlevé expansions for nonintegrable evolution equations, Physica D **53** (1991) 59–70.
11. P. J. Caudrey, The inverse problem for the third order equation $u_{xxx}+q(x)u_x+r(x)u=-i\zeta^3 u$, Phys. Lett. A **79** (1980) 264–268.
12. H. Chaté, Spatiotemporal intermittency régimes of the one-dimensional complex Ginzburg-Landau equation, Nonlinearity **7** (1994) 185–204.
13. J. Chazy, Sur les équations différentielles du troisième ordre et d'ordre supérieur dont l'intégrale générale a ses points critiques fixes, Thèse, Paris (1910); Acta Math. **34** (1911) 317–385.
14. H. H. Chen, General derivation of Bäcklund transformations from inverse scattering problems, Phys. Rev. Lett. **33** (1974) 925–928.
15. P. A. Clarkson and C. M. Cosgrove, The Painlevé property and a generalised derivative nonlinear Schrödinger equation, J. Phys. A **20** (1987) 2003–2024.
16. P. A. Clarkson, N. Joshi, and A. Pickering, Bäcklund transformations for the second Painlevé hierarchy: a modified truncation approach, Inverse Problems **15** (1999) 175–187.
17. P. A. Clarkson and P. Winternitz, Symmetry reduction and exact solutions of nonlunear partial differential equations, *The Painlevé property, one century later*, 591–660, ed. R. Conte, CRM series in mathematical physics (Springer, New York, 1999).
18. R. Conte, Universal invariance properties of Painlevé analysis and Bäcklund transformation in nonlinear partial differential equations, Phys. Lett. A **134** (1988) 100–104.
19. R. Conte, Invariant Painlevé analysis of partial differential equations, Phys. Lett. A **140** (1989) 383–390.
20. R. Conte, Painlevé singular manifold equation and integrability, *Inverse methods in action*, pp. 497–504, ed. P. C. Sabatier, Springer-Verlag series "Inverse problems and theoretical imaging" (Springer-Verlag, Berlin, 1990).

21. R. Conte, Painlevé analysis and Bäcklund transformation of the nonintegrable KPP equation, *Nonlinear evolution equations: integrability and spectral methods*, 187–192, eds. A. Degasperis, A. P. Fordy, and M. Lakshmanan (Manchester University Press, Manchester, 1990).
22. R. Conte, Towards the equivalence between integrability and Painlevé test for partial differential equations, *Nonlinear and chaotic phenomena in plasmas, solids and fluids*, 94–101, eds. W. Rozmus and J. A. Tuszyński (World Scientific, Singapore, 1991).
23. R. Conte, Unification of PDE and ODE versions of Painlevé analysis into a single invariant version, *Painlevé transcendents, their asymptotics and physical applications*, 125–144, eds. D. Levi and P. Winternitz (Plenum, New York, 1992).
24. R. Conte, Exact solutions of nonlinear wave equations by singularity methods, Rendiconti del Circolo Matematico di Palermo, Suppl. **57** (1998) 165–175.
25. R. Conte (ed.), *The Painlevé property, one century later*, 810 pages, CRM series in mathematical physics (Springer, New York, 1999).
26. R. Conte, The Painlevé approach to nonlinear ordinary differential equations, *The Painlevé property, one century later*, 77–180, ed. R. Conte, CRM series in mathematical physics (Springer, New York, 1999). Solv-int/9710020.
27. R. Conte, Various truncations in Painlevé analysis of partial differential equations, 103–116, *Nonlinear dynamics: integrability and chaos*, eds. M. Daniel, K. M. Tamizhmani, and R. Sahadevan (Narosa publishing house, New Delhi, 2000). Solv-int/9812008.
28. R. Conte, Sur les transformations de Schlesinger de la sixième équation de Painlevé, C. R. Acad. Sc. Paris **332** (2001) 501–504. math.CA/0103165.
29. R. Conte, A. P. Fordy and A. Pickering, A perturbative Painlevé approach to nonlinear differential equations, Physica D **69** (1993) 33–58.
30. R. Conte and M. Musette, Painlevé analysis and Bäcklund transformation in the Kuramoto-Sivashinsky equation, J. Phys. A **22** (1989) 169–177.
31. R. Conte and M. Musette, Link between solitary waves and projective Riccati equations, J. Phys. A **25** (1992) 5609–5623.
32. R. Conte and M. Musette, Linearity inside nonlinearity: exact solutions to the complex Ginzburg-Landau equation, Physica D **69** (1993) 1–17.
33. R. Conte and M. Musette, Exact solutions to the partially integrable Eckhaus equation, *Nonlinear evolution equations and dynamical systems*, eds. L. Martina and A. K. Pogrebkov, Teoreticheskaya i Matematicheskaya Fizika **99** (1994) 226–233 [English: Theor. and Math. Phys. **99** (1994) 543–548].
34. R. Conte and M. Musette, Exact solutions to the complex Ginzburg-Landau equation of nonlinear optics, Pure Appl. Opt. **4** (1995) 315–320.
35. R. Conte and M. Musette, Beyond the two–singular manifold method, *Nonlinear physics: theory and experiment*, 67–74, eds. E. Alfinito, M. Boiti, L. Martina and F. Pempinelli (World Scientific, Singapore, 1996).
36. R. Conte and M. Musette, Analytic expressions of hydrothermal waves, Reports on mathematical physics **46** (2000) 77–88. nlin.SI/0009022
37. R. Conte and M. Musette, On the solitary wave of two coupled nonintegrable Ginzburg-Landau equations, *Nonlinear integrability and all that: twenty years after NEEDS '99*, 382–388, eds. M. Boiti, L. Martina, F. Pempinelli, B. Prinari, and G. Soliani (World Scientific, Singapore, 2000).

38. R. Conte and M. Musette, First degree birational transformations of the Painlevé equations and their contiguity relations, J. Phys. A **34** (2001) 10507–10522. nlin.SI/0110028.
39. R. Conte and M. Musette, New contiguity relation of the sixth Painlevé equation from a truncation, Physica D **161** (2002) 129–141. nlin.SI/0112007.
40. R. Conte and M. Musette, A truncation for obtaining all the first degree birational transformations of the Painlevé transcendents, J. Nonlinear Math. Phys. **9** Supp. 1 (2002) 14–28. nlin.SI/0110031. *Recent advances in integrable systems*, eds. Ma Wen-xiu, P. J. Olver, Yan Min, and Hu Ji-shan, First joint meeting of the American mathematical society and the Hong-Kong mathematical society, Hong-Kong, 13–16 December 2000.
41. R. Conte, M. Musette and A. M. Grundland, Bäcklund transformation of partial differential equations obtained from the Painlevé-Gambier classification, II. Tzitzéica equation, J. Math. Phys. **40** (1999) 2092–2106.
42. R. Conte, M. Musette and A. Pickering, The two–singular manifold method, II. Classical Boussinesq system, J. Phys. A **28** (1995) 179–185.
43. G. Contopoulos, B. Grammaticos and A. Ramani, Painlevé analysis for the mixmaster universe model, J. Phys. A **25** (1993) 5795–5799.
44. G. Darboux, Sur une proposition relative aux équations linéaires, C. R. Acad. Sc. Paris **94** (1882) 1456–1459.
45. G. Darboux, *Leçons sur la théorie générale des surfaces et les applications géométriques du calcul infinitésimal*, vol. III (Gauthier-Villars, Paris, 1894). Reprinted, *Théorie générale des surfaces* (Chelsea, New York, 1972). Reprinted (Gabay, Paris, 1993).
46. P. Di Francesco, 2-D quantum and topological gravities, matrix models, amd integrable differential systems, *The Painlevé property, one century later*, 229–285, ed. R. Conte, CRM series in mathematical physics (Springer, New York, 1999).
47. P. G. Estévez, P. R. Gordoa, L. Martínez Alonso, and E. Medina Reus, Modified singular manifold expansion: application to the Boussinesq and Mikhailov-Shabat systems, J. Phys. A **26** (1993) 1915–1925.
48. P. G. Estévez and P. R. Gordoa, Nonclassical symmetries and the singular manifold method: theory and six examples, Stud. Appl. Math. **95** (1995) 73–113.
49. P. G. Estévez and P. R. Gordoa, Nonclassical symmetries and the singular manifold method: a further two examples, J. Phys. A **31** (1998) 7511–7519.
50. A. P. Fordy and J. Gibbons, Some remarkable nonlinear transformations, Phys. Lett. A **75** (1980) 325–325.
51. A. P. Fordy and A. Pickering, *Symmetries and Integrability of Difference equations* London Mathematical Society LNS **255** (1999) 287–298.
52. A. R. Forsyth, *Theory of differential equations.* Part IV (vol. VI) Partial differential equations (Cambridge University Press, Cambridge, 1906). Reprinted (Dover, New York, 1959).
53. J.-D. Fournier, E. A. Spiegel and O. Thual, Meromorphic integrals of two non-integrable systems, *Nonlinear dynamics*, 366–?, eds. G. Servizi and G. Turchetti (World Scientific, Singapore, 1989).
54. R. Fuchs, Sur quelques équations différentielles linéaires du second ordre, C. R. Acad. Sc. Paris **141** (1905) 555–558.
55. B. Gaffet, A class of 1−d gas flows soluble by the inverse scattering transform, Physica D **26** (1987) 123–139.

56. B. Gambier, Sur les équations différentielles du second ordre et du premier degré dont l'intégrale générale est à points critiques fixes, Thèse, Paris (1909); Acta Math. **33** (1910) 1–55.
57. M.L. Gandarias and M.S. Bruzón, Nonclassical symmetries for a family of Cahn-Hilliard equations, Phys. Lett. A **263** (2000) 331–337.
58. T. I. Garagash, On a modification of the Painlevé test for the BLP equation, *Nonlinear evolution equations and dynamical systems*, 130–133, eds. V. G. Makhankov, I. V. Puzynin, and O. K. Pashaev (World Scientific, Singapore, 1993).
59. J. D. Gibbon, A. C. Newell, M. Tabor, and Zeng Y.-b, Lax pairs, Bäcklund transformations and special solutions for ordinary differential equations, Nonlinearity **1** (1988) 481–490.
60. J. D. Gibbon, P. Radmore, M. Tabor, and D. Wood, The Painlevé property and Hirota's method, Stud. Appl. Math. **72** (1985) 39–63.
61. P. Gordoa, N. Joshi, and A. Pickering, Mappings preserving locations of movable poles: a new extension of the truncation method to ordinary differential equations, Nonlinearity **12** (1999) 955–968.
62. M. van Hecke, Building blocks of spatiotemporal intermittency, Phys. Rev. Lett. **80** (1998) 1896–1899.
63. E. Hille, *Ordinary differential equations in the complex domain* (J. Wiley and sons, New York, 1976).
64. R. Hirota, Fundamental properties of the binary operators in soliton theory and their generalization, *Dynamical problems in soliton systems*, 42–49, ed. S. Takeno (Springer, Berlin, 1985)
65. R. Hirota and A. Ramani, The Miura transformations of Kaup's equation and of Mikhailov's equation, Phys. Lett. A **76** (1980) 95–96.
66. R. Hirota and J. Satsuma, N-soliton solutions of model equations for shallow water waves, J. Phys. Soc. Japan Letters **40** (1976) 611–612.
67. R. Hirota and J. Satsuma, Nonlinear evolution equations generated from the Bäcklund transformation for the Boussinesq equation, Prog. Theor. Phys. **57** (1977) 797–807.
68. M. Ito, An extension of nonlinear evolution equations of the K-dV (mK-dV) type to higher orders, J. Phys. Soc. Japan **49** (1980) 771–778.
69. M. Jimbo, M. D. Kruskal and T. Miwa, Painlevé test for the self-dual Yang-Mills equation, Phys. Lett. A **92** (1982) 59-60.
70. M. Jimbo and T. Miwa, Solitons and infinite dimensional Lie algebras, Publ. RIMS, Kyoto **19** (1983) 943–1001.
71. D. J. Kaup, On the inverse scattering problem for cubic eigenvalue problems of the class $\psi_{xxx} + 6Q\psi_x + 6R\psi = \lambda\psi$, Stud. Appl. Math. **62** (1980) 189–216.
72. T. Kawahara and M. Tanaka, Interactions of traveling fronts: an exact solution of a nonlinear diffusion equation, Phys. Lett. A **97** (1983) 311–314.
73. A. N. Kolmogorov, I. G. Petrovskii and N. S. Piskunov, The study of a diffusion equation, related to the increase of the quantity of matter, and its application to one biological problem, Bulletin de l'Université d'État de Moscou, série internationale, section A Math. Méc. **1** (1937) 1–26.
74. A. Kundu, Landau-Lifshitz and higher order nonlinear systems gauge generated from nonlinear Schrödinger type equations, J. Math. Phys. **25** (1984) 3433–3438.

75. Y. Kuramoto and T. Tsuzuki, Persistent propagation of concentration waves in dissipative media far from thermal equilibrium, Prog. Theor. Phys. **55** (1976) 356–369.
76. G. L. Lamb Jr, Propagation of ultrashort optical pulses, Phys. Lett. A **25** (1967) 181–182.
77. G. L. Lamb Jr, Bäcklund transformations for certain nonlinear evolution equations, J. Math. Phys. **15** (1974) 2157–2165.
78. P. D. Lax, Integrals of nonlinear equations of evolution and solitary waves, Comm. Pure Appl. Math. **21** (1968) 467–490.
79. J. Lega, B. Janiaud, S. Jucquois and V. Croquette, Localized phase jumps in wave trains Phys. Rev. A **45** (1992) 5596-5604.
80. D. Levi and O. Ragnisco, Non-isospectral deformations and Darboux transformations for the third-order spectral problem, Inverse Problems **4** (1988) 815–828.
81. N. A. Lukashevich, The second Painlevé equation, Differentsial'nye Uravneniya **7** (1971) 1124–1125 [English: Diff. equ. **7** (1971) 853–854].
82. F. Magri, G. Falqui, and M. Pedroni, The method of Poisson pairs in the theory of nonlinear PDEs, 55 pages, this volume. SISSA/135/1999/FM, nlin.SI/0002009.
83. V. G. Makhankov and O. K. Pashaev, Nonlinear Schrödinger equation with noncompact isogroup, Teoreticheskaya i Matematicheskaya Fizika **53** (1982) 55–67 [English: Theor. and Math. Phys. **53** (1982) 979–987].
84. P. Manneville, The Kuramoto-Sivashinsky equation: a progress report, *Propagation in systems far from equilibrium*, 265–280, eds. J. Weisfreid, H. R. Brand, P. Manneville, G. Albinet, and N. Boccara (Springer, Berlin, 1988).
85. E. Mansfield, *Differential Groebner bases* (Ph D Thesis, University of Sydney, 1991).
86. V. B. Matveev and M. A. Salle, *Darboux transformations and solitons* (Springer-Verlag, Berlin, 1991).
87. D. W. McLaughlin and A. C. Scott, A restricted Bäcklund transformation, J. Math. Phys. **14** (1973) 1817–1828.
88. H. C. Morris, Prolongation structures and a generalized inverse scattering problem, J. Math. Phys. **17** (1976) 1867–1869.
89. M. Musette, Insertion of the Darboux transformation in the invariant Painlevé analysis of nonlinear partial differential equations, *Painlevé transcendents, their asymptotics and physical applications*, 197–209, eds. D. Levi and P. Winternitz (Plenum Publishing Corp., New York, 1992).
90. M. Musette, Painlevé analysis for nonlinear partial differential equations, *The Painlevé property, one century later*, 517–572, ed. R. Conte, CRM series in mathematical physics (Springer, New York, 1999) Solv-int/9804003.
91. M. Musette, Chapter 1, this volume.
92. M. Musette and R. Conte, Algorithmic method for deriving Lax pairs from the invariant Painlevé analysis of nonlinear partial differential equations, J. Math. Phys. **32** (1991) 1450–1457.
93. M. Musette and R. Conte, Solitary waves and Lax pairs from polynomial expansions of nonlinear differential equations, *Nonlinear evolution equations and dynamical systems*, pp. 161–170, eds. M. Boiti, L. Martina and F. Pempinelli (World Scientific, Singapore, 1992).
94. M. Musette and R. Conte, The two–singular manifold method, I. Modified KdV and sine-Gordon equations, J. Phys. A **27** (1994) 3895–3913.

95. M. Musette and R. Conte, Non-Fuchsian extension to the Painlevé test, Phys. Lett. A **206** (1995) 340–346.
96. M. Musette and R. Conte, Bäcklund transformation of partial differential equations obtained from the Painlevé-Gambier classification, I. Kaup-Kupershmidt equation, J. Math. Phys. **39** (1998) 5617–5630.
97. M. Musette, R. Conte, and C. Verhoeven, Bäcklund transformation and nonlinear superposition formula of the Kaup-Kupershmidt and Tzitzéica equations, *Bäcklund and Darboux transformations: the geometry of soliton theory*, eds. A. Coley, D. Levi, C. Rogers, and P. Winternitz, CRM Proceedings and Lecture Notes **29** (2001) 345–362. American Mathematical Society, Providence, R.I. AARMS-CRM workshop (Halifax, 5–9 June 1999).
98. A. Newell, M. Tabor, and Zeng Y.-b., A unified approach to Painlevé expansions, Physica D **29** (1987) 1–68.
99. A. C. Newell and J. A. Whitehead, Finite bandwidth, finite amplitude convection, J. Fluid Mech. **38** (1969) 279–303.
100. K. Nozaki, Hirota's method and the singular manifold expansion, J. Phys. Soc. Japan **56** (1987) 3052–3054.
101. K. Nozaki and N. Bekki, Exact solutions of the generalized Ginzburg-Landau equation, J. Phys. Soc. Japan **53** (1984) 1581–1582.
102. M. C. Nucci and P. A. Clarkson, The nonclassical method is more general than the direct method for symmetry reductions. An example of the Fitzhugh-Nagumo equation Phys. Lett. A **164** (1992) 49–56.
103. K. Okamoto, Isomonodromic deformation and Painlevé equations, and the Garnier system, J. Fac. Sci. Univ. Tokyo, Sect. IA **33** (1986) 575–618.
104. P. Painlevé, *Leçons sur la théorie analytique des équations différentielles* (Leçons de Stockholm, 1895) (Hermann, Paris, 1897). Reprinted, *Œuvres de Paul Painlev'e*, vol. I (Éditions du CNRS, Paris, 1973).
105. P. Painlevé, Sur les équations différentielles du second ordre et d'ordre supérieur dont l'intégrale générale est uniforme, Acta Math. **25** (1902) 1–85.
106. P. Painlevé, Sur les équations différentielles du second ordre à points critiques fixes, C. R. Acad. Sc. Paris **143** (1906) 1111–1117.
107. N. R. Pereira and L. Stenflo, Nonlinear Schrödinger equation including growth and damping, Phys. Fluids **20** (1977) 1733–1743.
108. A. Pickering, A new truncation in Painlevé analysis, J. Phys. A **26** (1993) 4395–4405.
109. A. Pickering, The singular manifold method revisited, J. Math. Phys. **37** (1996) 1894–1927.
110. A. Ramani, B. Grammaticos, and T. Bountis, The Painlevé property and singularity analysis of integrable and nonintegrable systems, Physics Reports **180** (1989) 159–245.
111. C. Rogers and S. Carillo, On reciprocal properties of the Caudrey-Dodd-Gibbon and Kaup-Kupershmidt hierarchies, Physica Scripta **36** (1987) 865–869.
112. C. Rogers and W. F. Shadwick, *Bäcklund transformations and their applications* (Academic press, New York, 1982).
113. J. Satsuma, Exact solutions of Burgers' equation with reaction terms, *Topics in soliton theory and exact solvable nonlinear equations*, 255–262, eds. M. J. Ablowitz, B. Fuchssteiner, and M. D. Kruskal (World Scientific, Singapore, 1987).

114. J. Satsuma and D. J. Kaup, A Bäcklund transformation for a higher order Korteweg-de Vries equation, J. Phys. Soc. Japan **43** (1977) 692–697.
115. C. Scheen and J. Demaret, Analytic structure and chaos in Bianchi type-IX relastivistic dynamics, Phys. Lett. B **399** (1997) 207–214.
116. W. Schief, The Tzitzéica equation: a Bäcklund transformation interpreted as truncated Painlevé expansion, J. Phys. A **29** (1996) 5153–5155.
117. K. M. Tamizhmani and M. Lakshmanan, Linearization and Painlevé property of Liouville and Cheng equations, J. Math. Phys. **27** (1986) 2257–2258.
118. G. Tzitzéica, Sur une nouvelle classe de surfaces, Rendiconti del Circolo Matematico di Palermo **25** (1908) 180–187.
119. G. Tzitzéica, Sur une nouvelle classe de surfaces, C. R. Acad. Sc. Paris **150** (1910) 955–956.
120. H. D. Wahlquist and F. B. Estabrook, Prolongation structures of nonlinear evolution equations, J. Math. Phys. **16** (1975) 1–7.
121. J. Weiss, The Painlevé property for partial differential equations. II: Bäcklund transformation, Lax pairs, and the Schwarzian derivative, J. Math. Phys. **24** (1983) 1405–1413.
122. J. Weiss, Bäcklund transformation and linearizations of the Hénon-Heiles system, Phys. Lett. A **102** (1984) 329–331.
123. J. Weiss, On classes of integrable systems and the Painlevé property, J. Math. Phys. **25** (1984) 13–24.
124. J. Weiss, The sine–Gordon equations: Complete and partial integrability, J. Math. Phys. **25** (1984) 2226–2235.
125. J. Weiss, The Painlevé property and Bäcklund transformations for the sequence of Boussinesq equations, J. Math. Phys. **26** (1985) 258–269.
126. J. Weiss, Bäcklund transformation and the Painlevé property, J. Math. Phys. **27** (1986) 1293–1305.
127. J. Weiss, M. Tabor and G. Carnevale, The Painlevé property for partial differential equations, J. Math. Phys. **24** (1983) 522–526.
128. R. Willox, private communication (1995).
129. P. Winternitz, Chapter 5, this volume.
130. Yang Huan-xiong and Li You-quan, Prolongation approach to Bäcklund transformation of Zhiber-Mikhailov-Shabat equation, J. Math. Phys.**37** (1996) 3491–3497.
131. V. E. Zakharov, On stochastization of one-dimensional chains of nonlinear oscillators, Zh. Eksp. Teor. Fiz. **65**, 219–225 (1973) [English: Soviet Physics JETP **38**, 108–110 (1974)].
132. V. E. Zakharov and A. B. Shabat, A scheme for integrating the nonlinear equations of mathematical physics by the method of the inverse scattering problem. I, Funktsional'nyi Analiz i Ego Prilozheniya **8**, 43–53 (1974) [English: Funct. Anal. Appl. **8**, 226–235 (1974)].
133. A. V. Zhiber and A. B. Shabat, Klein-Gordon equations with a nontrivial group, Dokl. Akad. Nauk SSSR **247**, 1103–1107 (1979) [English: Sov. Phys. Dokl. **24**, 607–609 (1979)].

The method of Poisson pairs in the theory of nonlinear PDEs*

Franco Magri[1], Gregorio Falqui[2], and Marco Pedroni[3]

[1] Dipartimento di Matematica e Applicazioni, Università di Milano–Bicocca, Via degli Arcimboldi 8, I-20126 Milano, Italy; *magri@vmimat.mat.unimi.it*
[2] SISSA, Via Beirut 2/4, I-34014 Trieste, Italy; *falqui@sissa.it*
[3] Dipartimento di Matematica, Università di Genova, Via Dodecaneso 35, I-16146 Genova, Italy; *pedroni@dima.unige.it*

Summary. The aim of these lectures is to show that the methods of classical Hamiltonian mechanics can be profitably used to solve certain classes of nonlinear partial differential equations. The prototype of these equations is the well-known Korteweg–de Vries (KdV) equation.
In these lectures we touch the following subjects:

i) the birth and the role of the method of Poisson pairs inside the theory of the KdV equation;
ii) the theoretical basis of the method of Poisson pairs;
iii) the Gel'fand–Zakharevich theory of integrable systems on bi-Hamiltonian manifolds;
iv) the Hamiltonian interpretation of the Sato picture of the KdV flows and of its linearization on an infinite–dimensional Grassmannian manifold.
v) the reduction technique(s) and its use to construct classes of solutions;
vi) the role of the technique of separation of variables in the study of the reduced systems;
vii) some relations intertwining the method of Poisson pairs with the method of Lax pairs.

1 Introduction: The tensorial approach and the birth of the method of Poisson pairs

This lecture is an introduction to the Hamiltonian analysis of PDEs from an "experimental" point of view. This means that we are more concerned in unveiling the spirit of the method than in working out the theoretical details. Therefore the style of the exposition will be informal, and proofs will be mainly omitted. We shall follow, step by step, the birth and the evolution of the Hamiltonian analysis of the KdV equation

$$u_t = \frac{1}{4}(u_{xxx} - 6uu_x) \ , \tag{1}$$

* Work partially supported by the M.U.R.S.T. and the G.N.F.M. of the Italian C.N.R.

F. Magri, G. Falqui, and M. Pedroni, The method of Poisson pairs in the theory of nonlinear PDEs, Lect. Notes Phys. **632**, 85–136 (2003)
http://www.springerlink.com/

from its "infancy" to the final representation of the KdV flow as a linear flow on an infinite-dimensional Grassmannian due to Sato [27]. The route is long and demanding. Therefore the exposition is divided in two parts, to be carried out in this and in the fourth lecture (see Sect. 4). Here our primary aim is to show the birth of the method of Poisson pairs. It is reached by means of a suitable use of the well-known methods of tensor analysis. We proceed in three steps. First, by using the transformation laws of vector fields, we construct the Miura map and the so called *modified* KdV equation (mKdV). This result leads quite simply to the theory of (elementary) Darboux transformations and to the concept of Poisson pair. Indeed, a peculiarity of mKdV is to possess an elementary Hamiltonian structure. By means of the transformation law of Poisson bivectors, we are then able to transplant this structure to the KdV equation, unraveling its "bi-Hamiltonian structure". This structure can be used in turn to define the concept of *Lenard chain* and to plunge the KdV equation into the "KdV hierarchy". This step is rather important from the point of view of finding classes of solutions to the KdV equation. Indeed the hierarchy is a powerful instrument to construct finite-dimensional invariant submanifolds of the equation and, therefore, finite-dimensional reductions of the KdV equation. The study of this process of restriction and of its use to construct solutions will be one of the two leading themes of these lectures. It is intimately related to the theory of separation of variables dealt with in the last two lectures. The second theme is that of the linearization of the full KdV flow on the infinite-dimensional Sato Grassmannian. The starting point of this process is surprisingly simple, and once again based on a simple procedure of tensor calculus. By means of the transformation laws of one–forms, we pull back the KdV hierarchy from its phase space onto the phase space of the mKdV equation. In this way we obtain the "mKdV hierarchy". In the fourth lecture we shall show that this hierarchy can be written as a flow on an infinite–dimensional Grassmann manifold, and that this flow can be linearized by means of a (generalized) Darboux transformation.

1.1 The Miura map and the KdV equation

As an effective way of probing the properties of equation (1) we follow the tensorial approach. Accordingly, we regard equation (1) as the *definition* of a vector field

$$u_t = X(u, u_x, u_{xx}, u_{xxx}) \tag{2}$$

on a suitable function space, and we investigate how it transforms under a point transformation in this space. Since our "coordinate" u is a function and not simply a number, we are allowed to consider transformations of coordinates depending also on the derivatives of the new coordinate function of the type[4]

[4] For further details on these kind of transformations, see [8].

$$u = \Phi(h, h_x) \ . \tag{3}$$

We ask whether there exists a transformed vector field

$$h_t = Y(h, h_x, h_{xx}, h_{xxx}) \tag{4}$$

related to the KdV equation according to the transformation law for vector fields,

$$X(\Phi(h)) = \Phi'_h \cdot Y(h), \tag{5}$$

where Φ'_h is the (Fréchet) derivative of the operator Φ defining the transformation. This condition gives rise to a (generally speaking) over-determined system of partial differential equations on the unknown functions $\Phi(h, h_x)$ and $Y(h, h_x, h_{xx}, h_{xxx})$. In the specific example the over-determined system can be solved. Apart form the trivial solution $u = h_x$, we find the *Miura transformation* [23, 16],

$$u = h_x + h^2 - \lambda \ , \tag{6}$$

depending on an arbitrary parameter λ. The transformed equation is the *modified* KdV equation:

$$h_t = \frac{1}{4}(h_{xxx} - 6h^2 h_x + 6\lambda h_x) \ . \tag{7}$$

Exercise 1.1. Work out in detail the transformation law (5), checking that X, Φ, and Y, defined respectively by equations (1), (6) and (7) do satisfy equation (5). ◇

The above result is plenty of consequences. The first one is a simple method for constructing solutions of the KdV equation. It is called the method of (elementary) Darboux transformations [22]. It rests on the remark that the mKdV equation (7) admits the discrete symmetry

$$h \mapsto h' = -h \ . \tag{8}$$

Let us exploit this property to construct the well-known one–soliton solution of the KdV equation. We notice that the point $u = 0$ is a very simple (singular) invariant submanifold of the KdV equation. Its inverse image under the Miura transformation is the 1–dimensional submanifold S_1 formed by the solutions of the special Riccati equation

$$h_x + h^2 = \lambda \ . \tag{9}$$

This submanifold, in its turn, is invariant with respect to equation (7). A straightforward computation shows that, on this submanifold,

$$\frac{1}{4}(h_{xxx} - 6h^2 h_x + 6\lambda h_x) = \lambda h_x \ . \tag{10}$$

Therefore, on S_1 the mKdV equation takes the simple form $h_t = \lambda h_x$. Solving the first order system formed by this equation and the Riccati equation $h_x + h^2 = \lambda$, and setting $\lambda = z^2$, we find the general solution

$$h(x,t) = z\,\tanh(zx + z^3 t + c) \tag{11}$$

of the mKdV equation on the invariant submanifold S_1. At this point we use the symmetry property and the Miura map. By the symmetry property (8) the function $-h(x,t)$ is a new solution of the modified equation, and by the Miura map the function

$$u'(x,t) = -h_x + h^2 - z^2 = 2z^2 \mathrm{sech}^2(zx + z^3 t + c) \tag{12}$$

is a new solution of the KdV equation. It is called the *one soliton* solution[5]. It can also be interpreted in terms of invariant submanifolds. To this end, we have to notice that the Miura map (6) transforms the invariant submanifold S_1 of the modified equation into the submanifold formed by the solutions of the first order differential equation

$$\frac{1}{2}\left(-\frac{1}{2}u_x^2 + u^3\right) + \lambda u^2 = 0\ . \tag{13}$$

As one can easily check, this set is preserved by the KdV equation, and therefore it is an invariant one–dimensional submanifold of the KdV equation, built up from the singular manifold $u = 0$. On this submanifold, the KdV equation takes the simple form $u_t = \lambda u_x$, and the flow can be integrated to recover the solution (12).

This example clearly shows that the Darboux transformations are a mechanism to build invariant submanifolds of the KdV equation. Some of these submanifolds will be examined in great detail in the present lectures. The purpose is to show that the reduced equations on these submanifolds are classical Hamiltonian vector fields whose associated Hamilton–Jacobi equations can be solved by separation of variables. In this way, we hope, the interest of the Hamiltonian analysis of the KdV equations can better emerge.

1.2 Poisson pairs and the KdV hierarchy

We shall now examine a more deep and far reaching property of the Miura map. It is connected with the concept of Hamiltonian vector field. From Analytical Mechanics, we know that the Hamiltonian vector fields are the images of exact one–forms through a suitable linear map, associated with a so–called Poisson bivector. We shall formally define these notions in the next lecture. These definition can be easily extended to vector fields on infinite–dimensional manifolds. Let us give an example, by showing that the mKdV

[5] For a very nice account of the origin and of the properties of the KdV equation and of other soliton equations and their solutions, see, e.g., [24].

equation is a Hamiltonian vector field. This requires a series of three consecutive remarks. First we notice that equation (7) can be factorized as

$$h_t = \left[\frac{1}{2}\partial_x\right] \cdot \left[\frac{1}{2}h_{xx} - h^3 + 3\lambda h\right] . \tag{14}$$

Then, we notice that the linear operator in the first bracket, $\frac{1}{2}\partial_x$, is a constant skewsymmetric operator which we can recognize as a Poisson bivector. Finally, we notice that in the differential polynomial appearing in the second bracket in the right hand side of equation (14), we can easily recognize an exact one–form. Indeed,

$$\int \left(\frac{1}{2}h_{xx} - h^3 + 3\lambda h\right) \dot{h}\, dx = \frac{d}{dt}\int \left(-\frac{1}{4}h_x^2 - \frac{3}{4}h^4 + \frac{3}{2}\lambda h^2\right) dx \tag{15}$$

for any tangent vector $\dot{h}$. These statements are true under suitable boundary conditions, as explained in, e.g., [8]. Here and in the rest of these lectures we will tacitly use periodic boundary conditions.

The Hamiltonian character of the mKdV equation is obviously independent of the existence of the Miura map. However, this map finely combines this property from the point of view of tensor analysis. Let us recall that a Poisson bivector is a skewsymmetric linear map from the cotangent to the tangent spaces satisfying a suitable differential condition (see Lecture 2). It obeys the transformation law

$$Q_{\Phi(h)} = \Phi'_h P_h \Phi'^*_h \tag{16}$$

under a change of coordinates (or a map between two different manifolds). In this formula the point transformation is denoted (in operator form) by $u = \Phi(h)$. The operators Φ'_h and Φ'^*_h are the Fréchet derivative of Φ and its adjoint operator. The symbols P_h and Q_u denote the Poisson bivectors in the space of the functions h and u, respectively. Since the Miura map $u = h_x + h^2 - \lambda$ is not invertible, it is rather nontrivial that there exists a Poisson bivector Q_u, on the phase space of the KdV equation, which is Φ–related (in the sense of equation (16)) to the Poisson bivector $P_h = \frac{1}{2}\partial_x$ associated with the modified equation. Surprisingly, this is the case. One can check that the operator Q_u is defined by

$$Q_u = -\frac{1}{2}\partial_{xxx} + 2(u+\lambda)\partial_x + u_x . \tag{17}$$

Exercise 1.2. Verify the above claim by computing the product (in the appropriate order) of the operators $\Phi'_h = \partial_x + 2h$, $P_h = \frac{1}{2}\partial_x$, and $\Phi'^*_h = -\partial_x + 2h$, and by expressing the results in term of $u = h_x + h^2 - \lambda$. ◇

This exercise shows that the Miura map is a peculiar Poisson map. Since it depends on the parameter λ, the final result is that the phase space of the KdV equation is endowed with a one–parameter family of Poisson bivectors,

$$Q_\lambda = Q_1 - \lambda Q_0 \ , \tag{18}$$

which we call a Poisson pencil. The operators (Q_1, Q_0) defining the pencil are said to form a *Poisson pair*, a concept to be systematically investigated in the next lecture.

These operators enjoy a number of interesting properties, and define new geometrical structures associated with the equation. One of the simplest but far–reaching is the concept of *Lenard chain.* The idea is to use the pair of bivectors to define a recursion relation on one–forms:

$$Q_0\alpha_{j+1} = Q_1\alpha_j \ . \tag{19}$$

In the applications a certain care must be taken in dealing with this recursion relation, since it does not define uniquely the forms α_j (the operator Q_0 is seldom invertible). Furthermore, it is still less apparent that it can be solved in the class of *exact* one–forms. However, in the KdV case we *bonafide* proceed and we find

$$\begin{aligned}
\alpha_0 &= 1 \\
\alpha_1 &= -\frac{1}{2}u \\
\alpha_2 &= \frac{1}{8}(-u_{xx} + 3u^2) \\
\alpha_3 &= \frac{1}{32}(-10u^3 + 10\,uu_{xx} - u_{xxxx} + 5u_x^2)
\end{aligned} \tag{20}$$

as first terms of the recurrence. The next step is to consider the associated vector fields (the meaning of numbering them with odd integers will be explained in Lecture 4):

$$\begin{aligned}
\frac{\partial u}{\partial t_1} &= Q_1\alpha_0 = Q_0\alpha_1 = u_x \\
\frac{\partial u}{\partial t_3} &= Q_1\alpha_1 = Q_0\alpha_2 = \frac{1}{4}(u_{xxx} - 6uu_x) \\
\frac{\partial u}{\partial t_5} &= Q_1\alpha_2 = Q_0\alpha_3 = \frac{1}{16}(u_{xxxxx} - 10uu_{xxx} - 20u_xu_{xx} + 30u^2u_x) \ .
\end{aligned} \tag{21}$$

They are the first members of the KdV hierarchy. In the fourth lecture, we shall show that it is a special instance of a general concept, the Gel'fand–Zakharevich hierarchy associated with any Poisson pencil of a suitable class.

1.3 Invariant submanifolds and reduced equations

The introduction of the KdV hierarchy has important consequences on the problem of constructing solutions of the KdV equation. The hierarchy is indeed a basic supply of invariant submanifolds of the KdV equation. This is due to the property of the vector fields of the hierarchy to commute among themselves. From this property it follows that the set of singular point of any linear combination (with constant coefficients) of the vector fields of the hierarchy is a finite–dimensional invariant submanifold of the KdV flow. These submanifold can be usefully exploited to construct classes of solutions of this equation.

As a first example of this technique we consider the submanifold defined by the condition

$$\frac{\partial u}{\partial t_3} = \lambda \frac{\partial u}{\partial t_1} , \tag{22}$$

that is, the submanifold where the second vector field of the hierarchy is a constant multiple of the first one. It is formed by the solutions of the third order differential equation

$$\frac{1}{4}(u_{xxx} - 6uu_x) - \lambda u_x = 0 \, . \tag{23}$$

Therefore it is a three dimensional manifold, which we denote by M_3. We can use as coordinates on M_3 the *values* of the function u and its derivatives u_x and u_{xx} at *any* point x_0. To avoid cumbersome notations, we will continue to denote these three *numbers* with the same symbols, u, u_x, u_{xx}, but the reader should be aware of this subtlety. To perform the reduction of the first equation of the hierarchy (21) on M_3, we consider the first two differential consequences of the equation $\dfrac{\partial u}{\partial t_1} = u_x$ and we use the constraint (22) to eliminate the third order derivative. We obtain the system

$$\frac{\partial u}{\partial t_1} = u_x , \qquad \frac{\partial u_x}{\partial t_1} = u_{xx} , \qquad \frac{\partial u_{xx}}{\partial t_1} = 6uu_x + 4\lambda u_x \, . \tag{24}$$

We call X_1 the vector field defined by these equations on M_3. It shares many of the properties of the KdV equation. For instance, it is related to a Poisson pair. The simplest way to display this property is to remark that X_1 possesses two integrals of motion,

$$\begin{aligned} H_0 &= u_{xx} - 3u^2 - 4\lambda u \\ H_1 &= -\frac{1}{2}u_x^2 + u^3 + 2\lambda u^2 + uH_0 \, . \end{aligned} \tag{25}$$

Then we notice that on M_3 there exists a unique Poisson bracket $\{\cdot,\cdot\}_0$ with the following two properties:

i) The function H_0 is a Casimir function, that is, $\{F, H_0\}_0 = 0$ for every smooth function F on M_3.
ii) X_1 is the Hamiltonian vector field associated with the function H_1.

Such a Poisson bracket $\{\cdot,\cdot\}_0$ is defined by the relations

$$\{u, u_x\}_0 = -1, \quad \{u, u_{xx}\}_0 = 0, \quad \{u_x, u_{xx}\}_0 = 6u + 4\lambda \ . \tag{26}$$

Similarly, one can notice that on M_3 there exists a unique Poisson bracket $\{\cdot,\cdot\}_1$ with the following "dual" properties:

i') The function H_1 is a Casimir function, that is, $\{F, H_1\}_1 = 0$ for every smooth function F on M_3.
ii') X_1 is the Hamiltonian vector field associated with the function H_0.

This second Poisson bracket $\{\cdot,\cdot\}_1$ is defined by the relations

$$\begin{aligned} &\{u, u_x\}_1 = u, \quad \{u, u_{xx}\}_1 = u_x, \\ &\{u_x, u_{xx}\}_1 = u_{xx} - u(6u + 4\lambda) \ . \end{aligned} \tag{27}$$

Exercise 1.3. Verify the stated properties of the Poisson pair (26) and (27). ◇

We now exploit the previous remarks to understand the geometry of the flow associated with X_1. First we use the Hamiltonian representation

$$\frac{dF}{dt} = X_1(F) = \{F, H_1\}_0 \ . \tag{28}$$

It entails that the level surfaces of the Casimir function H_0 are two–dimensional (symplectic) manifolds to which X_1 is tangent. Let us pick up any of these symplectic leaves, for instance the one passing through the origin $u = 0, u_x = 0, u_{xx} = 0$. Let us call S_2 this leaf. The coordinates (u, u_x) are *canonical* coordinates on S_2. The level curves of the Hamiltonian H_1 define a Lagrangian foliation of S_2. Our problem is to find the flow of the vector field X_1 along these Lagrangian submanifold. We have already given the solution of this problem in the particular case of the Lagrangian submanifold passing through the origin. This submanifold is the one–dimensional invariant submanifold (13) previously discussed in connection with Darboux transformations. The relative flow is the one–soliton solution to KdV. To deal with the generic Lagrangian submanifolds on an equal footing, it is useful to change strategy and to use the Hamilton–Jacobi equation

$$H_1(u, \frac{dW}{du}) = E \ . \tag{29}$$

In this rather simple example, there is almost nothing to say about this equation (which is obviously solvable), and the second Poisson bracket (26) seems not to play any role in the theory.

This (wrong!) impression is promptly corrected by the study of a more elaborated example. Let us consider the five–dimensional submanifold M_5 of the singular points of the third vector field of the KdV hierarchy. It is defined by the equation

$$u_{xxxxx} - 10uu_{xxx} - 20u_xu_{xx} + 30u^2u_x = 0 \ . \tag{30}$$

On this submanifold we can consider the restrictions of the first *two* vector fields of the same hierarchy. To compute the reduced equation we proceed as before. We regard the Cauchy data

$$(u, u_x, u_{xx}, u_{xxx}, u_{xxxx})$$

as coordinates on M_5. Then we compute the time derivatives

$$\frac{\partial u}{\partial t_1}, \frac{\partial u_x}{\partial t_1}, \frac{\partial u_{xx}}{\partial t_1}, \frac{\partial u_{xxx}}{\partial t_1}, \frac{\partial u_{xxxx}}{\partial t_1}$$

by taking the differential consequences of $\frac{\partial u}{\partial t_1} = u_x$, and by using (30) and its differential consequences as a constraint to eliminate all the derivatives of degree higher than four. We obtain the equations

$$\begin{aligned}
\frac{\partial u}{\partial t_1} &= u_x \\
\frac{\partial u_x}{\partial t_1} &= u_{xx} \\
\frac{\partial u_{xx}}{\partial t_1} &= u_{xxx} \\
\frac{\partial u_{xxx}}{\partial t_1} &= u_{xxxx} \\
\frac{\partial u_{xxxx}}{\partial t_1} &= 10uu_{xxx} + 20u_x u_{xx} - 30u^2 u_x
\end{aligned} \tag{31}$$

In the same way, for the reduction of the KdV equation, we get

$$\begin{aligned}
\frac{\partial u}{\partial t_3} &= \frac{1}{4}(u_{xxx} - 6uu_x) \\
\frac{\partial u_x}{\partial t_3} &= \frac{1}{4}(u_{xxxx} - 6uu_{xx} - 6u_x^2) \\
\frac{\partial u_{xx}}{\partial t_3} &= \frac{1}{4}(4uu_{xxxx} + 2u_x u_{xx} - 30u^2 u_x) \\
\frac{\partial u_{xxx}}{\partial t_3} &= \frac{1}{4}(2u_{xx}^2 + 6u_x u_{xxx} + 4uu_{xxxx} - 60uu_x^2 - 30u^2 u_{xx}) \\
\frac{\partial u_{xxxx}}{\partial t_3} &= \frac{1}{4}(10u_{xx}u_{xxx} + 10u_x u_{xxxx} - 120u^3 u_x - 100uu_x u_{xx} \\
&\quad + 10u^2 u_{xxx} - 60u_x^3) \, .
\end{aligned} \tag{32}$$

Exercise 1.4. Verify the previous computations. ◇

To find the corresponding solutions of the KdV equation, regarded as a partial differential equation in x and t, we have to:

1. Construct a common solution to the ordinary differential equations (31) and (32);

2. Consider the first component $u(t_1, t_3)$ of such a solution;
3. Set $t_1 = x$ and $t_3 = t$.

The function $u(x,t)$ obtained in this way is the solution we were looking for. What makes this procedure worth of interest is that the ODEs (31)–(32) can be solved by means of a variety of methods. In particular, they can be solved by means of the method of separation of variables[6]. It can be shown that they are rather special equations: They are Hamiltonian with respect to a Poisson pair; this Poisson pair allows to foliate the manifold M_5 into four–dimensional symplectic leaves with special properties; each symplectic leaf S_4 carries a Lagrangian foliation to which the vector fields (31) and (32) are tangent; the Poisson pair defines a special set of coordinates on each S_4; in these coordinates the Hamilton–Jacobi equations associated with the Hamiltonian equations (31) and (32) can be simultaneously solved by additive separation of variables. Most of these properties will be proved in the next lecture.

1.4 The modified KdV hierarchy

We leave for the moment the theme of the reduction, and come back to the KdV hierarchy in its general form. We notice that the first equations (21) can also be written in the form

$$\begin{aligned}\frac{\partial u}{\partial t_1} &= (Q_1 - \lambda Q_0)\alpha_0 \\ \frac{\partial u}{\partial t_3} &= (Q_1 - \lambda Q_0)(\lambda\alpha_0 + \alpha_1) \\ \frac{\partial u}{\partial t_5} &= (Q_1 - \lambda Q_0)(\lambda^2\alpha_0 + \lambda\alpha_1 + \alpha_2)\end{aligned} \tag{33}$$

This representation shows that these equations are Hamiltonian with respect to the whole Poisson pencil. This elementary property can be exploited to simply construct the modified KdV hierarchy. Let us write in general

$$\frac{\partial u}{\partial t_{2j+1}} = (Q_1 - \lambda Q_0)\alpha^{(j)}(\lambda)\ , \tag{34}$$

where

$$\alpha^{(j)}(\lambda) = \lambda^j\alpha_0 + \lambda^{j-1}\alpha_1 + \cdots + \alpha_j\ . \tag{35}$$

By means of the Miura map (6) we pull–back the one–forms $\alpha^{(j)}$ to one–form $\beta^{(j)}$ defined on the phase space of the modified equation according to the transformation law of one-forms,

[6] The fact that the stationary reductions of KdV can be solved by separation of variables is well-known (see, e.g., [9]). This classical method has recently found a lot of interesting new applications, as shown in the survey [26].

$$\beta(h) = {\Phi'_h}^* \alpha(\Phi(h)) \ . \tag{36}$$

We then define the equations

$$\frac{\partial h}{\partial t_{2j+1}} = P_h \beta^{(j)}(\lambda) \ . \tag{37}$$

They are Φ–related to the corresponding equations of the KdV hierarchy, exactly as the mKdV *equation* is Φ–related to the KdV *equation*. Indeed,

$$\frac{\partial u}{\partial t_{2j+1}} = \Phi'_h \frac{\partial h}{\partial t_{2j+1}} = \Phi'_h P_h {\Phi'_h}^* \alpha^{(j)}(\Phi(h); \lambda) = Q_u \alpha^{(j)}(u; \lambda) \ . \tag{38}$$

It is therefore natural to call equations (37) the *modified KdV hierarchy*. By using the explicit form of the operators P_h and ${\Phi'_h}^*$, it is easy to check that the modified hierarchy is defined by the conservation laws

$$\frac{\partial h}{\partial t_{2j+1}} = \partial_x H^{(2j+1)}, \tag{39}$$

where

$$H^{(2j+1)} = -\frac{1}{2}\alpha^{(j)}_x + \alpha^{(j)} h \ . \tag{40}$$

Exercise 1.5. Write down explicitly the first three equations of the modified hierarchy. ◇

The above formulas are basic in the Sato approach. In the fourth lecture, after a more accurate analysis of the Hamiltonian structure of the KdV equations, we shall be led to consider the currents $H^{(2j+1)}$ as defining a point of an infinite–dimensional Grassmannian. This point evolves in time as the point u moves according to the KdV equation. We shall determine the equation of motion of the currents $H^{(2j+1)}$. They define a "bigger" hierarchy called the *Central System*. This system contains the KdV hierarchy as a particular reduction. It enjoys the property of being linearizable. In this way, by a continuous process of extension motivated by the Hamiltonian structure of the equations (from the single KdV equation to the KdV hierarchy and to the Central System), we arrive at the result that the KdV flow can be linearized. At this point the following picture of the possible strategies for solving the KdV equations emerges: either we pass to the Sato infinite–dimensional Grassmannian and we use a linearization technique, or we restrict the equation to a finite–dimensional invariant submanifold and we use a technique of separation of variables. The two strategies complement themselves rather well. Our attitude is to see the Grassmannian picture as a compact way of defining the equations, and the "reductionist" picture as an effective way for finding interesting classes of solutions.

The plan of the lectures

This is the web of ideas which we would like to make more precise in the following lectures. As cornerstone of our presentation we choose the concept of Poisson pairs. In the second lecture, we develop the theory of these pairs up to the point of giving a sound basis to the concept of Lenard chain. In the third lecture we exhibit a first class of examples, and we explain a reduction technique allowing to construct the Poisson pairs of the reduced flows. In the fourth lecture we give a second look at the KdV theory, and we explain the reasons which, according to the Hamiltonian standpoint, suggest to pass on the infinite–dimensional Sato Grassmannian. In the fifth lecture we better explore the relation between the two strategies, and we touch the concept of Lax representation. Finally, the last lecture is devoted to the method of separation of variables. The purpose is to show how the geometry of the reduced Poisson pairs can be used to define the separation coordinates.

2 The method of Poisson pairs

In the previous lecture we have outlined the birth of the method of Poisson pairs and its main purpose: To define integrable hierarchies of vector fields. In this lecture we dwell on the theoretical basis of this construction presenting the concept of Gel'fand–Zakharevich system.

The starting point is the notion of *Poisson manifold.* A manifold is said to be a Poisson manifold[7] if a composition law on scalar functions has been defined obeying the usual properties of a Poisson bracket: bilinearity, skewsymmetry, Jacobi identity and Leibnitz rule. The last condition means that the Poisson bracket is a derivation in each entry:

$$\{fg,h\} = \{f,h\}g + f\{g,h\} \ . \tag{41}$$

Therefore, by fixing the argument of one of the two entries and keeping free the remaining one, we obtain a vector field,

$$X_h = \{\cdot, h\} \ . \tag{42}$$

It is called the *Hamiltonian vector field* associated with the function h with respect to the given Poisson bracket. Due to the remaining conditions on the Poisson bracket, these vector fields are closed with respect to the commutator. They form a Lie algebra homeomorphic to the algebra of functions defined by the Poisson bracket:

$$[X_f, X_g] = X_{\{f,g\}} \ . \tag{43}$$

Therefore a Poisson bracket on a manifold has a twofold role: it defines a Lie algebra structure on the ring of C^∞–functions, and provides a representation of this algebra on the manifold by means of the Hamiltonian vector fields.

[7] The books [17] and [31] are very good references for this topic.

Instead of working with the Poisson bracket, it is often suitable to work (especially in the infinite–dimensional case) with the associated Poisson tensor. It is the bivector field P on M defined by

$$\{f, g\} = \langle df,\, P\, dg\rangle \ . \tag{44}$$

In local coordinates, its components $P^{jk}(x^1, \ldots, x^n)$ are the Poisson brackets of the coordinate functions,

$$P^{jk}(x^1, \ldots, x^n) = \{x^j, x^k\} \ . \tag{45}$$

By looking at this bivector field as a linear skewsymmetric map $P : T^*M \to TM$, we can define the Hamiltonian vector fields X_f as the images through P of the exact one-forms,

$$X_f = P\, df \ . \tag{46}$$

In local coordinates this means

$$X_f^j = P^{jk} \frac{\partial f}{\partial x^k} \ . \tag{47}$$

Exercise 2.1. Show that the components of the Poisson tensor satisfy the cyclic condition

$$\sum_l \left(P^{jl} \frac{\partial P^{km}}{\partial x^l} + P^{kl} \frac{\partial P^{mj}}{\partial x^l} + P^{ml} \frac{\partial P^{jk}}{\partial x^l} \right) = 0 \ . \tag{48}$$

◇

Exercise 2.2. Suppose that M is an affine space A. Call V the vector space associated with A. Define a bivector field on A as a mapping $P : A \times V^* \to V$ which satisfies the skewsymmetry condition

$$\langle \alpha,\, P_u \beta\rangle = -\langle \beta,\, P_u \alpha\rangle \tag{49}$$

for every pair of covector (α, β) in V^* and at each point $u \in A$. Denote the directional derivative at the point u of the mapping $u \mapsto P_u \alpha$ along the vector v by

$$P'_u(\alpha; v) = \frac{d}{dt} P_{u+tv} \alpha|_{t=0} \ . \tag{50}$$

Show that the bivector P is a Poisson bivector if and only if it satisfies the cyclic condition

$$\langle \alpha,\, P'_u(\beta; P_u\gamma)\rangle + \langle \beta,\, P'_u(\gamma; P_u\alpha)\rangle + \langle \gamma,\, P'_u(\alpha; P_u\beta)\rangle = 0 \ . \tag{51}$$

◇

Exercise 2.3. Check that the bivector Q_λ of equation (17), associated with the KdV equation, fulfills the conditions (49) and (51). ◇

No condition is usually imposed on the rank of the Poisson bracket, that is, on the dimension of the vector space spanned by the Hamiltonian vector fields at each point of the manifold. If these vector fields span the whole tangent space the bracket is said to be regular, and the manifold M turns out to be a symplectic manifold. Indeed there exists, in this case, a unique symplectic 2-form ω such that

$$\{f,g\} = \omega(X_f, X_g) \ . \tag{52}$$

More interesting is the case where the bracket is singular. In this case, the Hamiltonian vector fields span a proper distribution D on M. It is involutive but, generically, not of constant rank. Nonetheless, this distribution is *completely integrable*: at each point there exists an integral submanifold of maximal dimension which is tangent to the distribution. These submanifolds are symplectic manifolds, and are called the symplectic leaves of the Poisson structure. The symplectic form is still defined by equation (52). Indeed, even if there is not a 1–1 correspondence between (differentials of) functions and Hamiltonian vector fields, this formula keeps its meaning, since the value of the Poisson bracket does not depend on the particular choice of the Hamiltonian function associated with a given Hamiltonian vector field. We arrive thus at the following conclusion: a Poisson manifold is either a symplectic manifold or a *stratification* of symplectic manifolds possibly of different dimensions. It can be proven that, in a sufficiently small open set where the rank of the Poisson tensor is constant, these symplectic manifolds are the level sets of some smooth functions $F_1, \ldots, F_k$, whose differentials span the kernel of the Poisson tensor. They are called *Casimir functions* of P (see below).

Exercise 2.4. Let $\{x_1, x_2, x_3\}$ be Cartesian coordinates in $M = \mathbb{R}^3$. Prove that the assignment

$$\{x_1, x_2\} = x_3, \quad \{x_1, x_3\} = -x_2, \quad \{x_2, x_3\} = x_1 \tag{53}$$

defines a Poisson tensor on M. Find its Casimir function, and describe the symplectic foliation associated with it. ◇

After these brief preliminaries on Poisson manifolds as natural settings for the theory of Hamiltonian vector fields, we pass to the theory of *bi-Hamiltonian* manifolds. Our purpose is to provide evidence that they are a suitable setting for the theory of *integrable* Hamiltonian vector fields. The simplest connection between the theory of integrable Hamiltonian vector fields and the theory of bi-Hamiltonian manifold is given by the Gel'fand–Zakharevich (GZ) theorem [13, 14] we shall discuss in this lecture.

A bi-Hamiltonian manifold is a Poisson manifold endowed with a *pair* of *compatible* Poisson brackets. We shall denote these brackets with $\{f,g\}_0$ and $\{f,g\}_1$. They are compatible if the Poisson pencil

$$\{f,g\}_\lambda := \{f,g\}_1 - \lambda\{f,g\}_0 \tag{54}$$

verifies the Jacobi identity for any value of the continuous (say, real) parameter λ. By means of this concept we catch the main features of the situation first met in the KdV example of Lecture 1. The new feature deserving attention is the dependence of the Poisson bracket (54) on the parameter λ. It influences all the objects so far introduced on a Poisson manifold: Hamiltonian fields and symplectic foliation. In particular, this foliation changes with λ. The useful idea is to extract from this moving foliation the invariant part. It is defined as the intersection of the symplectic leaves of the pencil when λ ranges over $\mathbb{R} \cup \{\infty\}$. The GZ theorem describes the structure of these intersections in particular cases.

Let us suppose that the dimension of M is odd, $\dim M = 2n+1$, and that the rank of the Poisson pencil is maximal. This means that the dimension of the characteristic distribution spanned by the Hamiltonian vector field is $2n$ for almost all the values of the parameter λ, and almost everywhere on the manifold M. In this situation the generic symplectic leaf of the pencil has accordingly dimension $2n$ and the intersection of all the symplectic leaves are submanifolds of dimension n. For brevity, we shall call this intersection the *support* of the pencil. The GZ theorem displays an important property of the leaves of the support of the pencil.

Theorem 2.1. *On a $(2n+1)$–dimensional bi-Hamiltonian manifold, whose Poisson pencil has maximal rank, the leaves of the support are generically Lagrangian submanifolds of dimension n contained on each symplectic leaf of dimension $2n$.*

This theorem contains two different statements. First of all it states that the dimension of the support is exactly half of the dimension of the generic symplectic leaf. It is the "hard" part of the theorem. Then it claims that the leaves of the support are Lagrangian submanifolds. Contrary to the appearances, this is the easiest part of the theorem, as we shall see. To better understand the content of the GZ theorem, we deem suitable to look at it from a different and, so to say, more constructive, point of view. It requires the use of the concept of *Casimir function*, defined as a function which commutes with all the other functions with respect to the Poisson bracket. Equivalently, it can be defined as a function whose differential belongs to the kernel of the Poisson tensor, i.e., a function generating the null vector field. In the case of a Poisson pencil, the Casimir functions depend on the parameter λ. If the rank of the Poisson pencil is maximal, the Casimir function is essentially unique (two Casimir functions are functionally dependent). The main content of the GZ theorem is that there exists a Casimir function which is a power series in the parameter λ. We suppose that it is a polynomial whose degree is exactly n if $\dim M = 2n + 1$. Thus we can write the Casimir function in the form

$$C(\lambda) = C_0\lambda^n + C_1\lambda^{n-1} + \cdots + C_n \,. \tag{55}$$

This result means that the Poisson pencil selects $n+1$ distinguished functions $(C_0, C_1, \dots C_n)$. Generically these functions are independent. Their common level surfaces are the leaves of the support of the pencil. Indeed, on the support the function $C(\lambda)$ must be constant independently of the particular value of λ. Thus all the coefficients $(C_0, C_1, \dots C_n)$ must be separately constant. Furthermore, as a consequence of the fact that $C(\lambda)$ is a Casimir function, it is easily seen that the coefficients C_k verify the Lenard recursion relations,

$$\{\cdot, C_k\}_1 = \{\cdot, C_{k+1}\}_0 \ , \tag{56}$$

together with the vanishing conditions

$$\{\cdot, C_0\}_0 = \{\cdot, C_n\}_1 = 0 \ . \tag{57}$$

In the language of the previous lecture, the functions

$$(C_0, C_1, \dots, C_n)$$

form a *Lenard chain.* A typical property of these functions is to be mutually in involution:

$$\{C_j, C_k\}_0 = \{C_j, C_k\}_1 = 0 \ . \tag{58}$$

This is proved by repeatedly using the recursion relation (56) to go back and forth along the chain. It follows that the leaves of the support are isotropic submanifolds, but since they are of maximal dimension n they are actually Lagrangian submanifolds. These short remarks should give a sufficiently detailed idea of the meaning of the GZ theorem.

Exercise 2.5. Check that that the integrals of motion H_0 and H_1 of the reduced flow X_1 on the invariant submanifold M_3 considered in Sect. 1.3 are the coefficients of the Casimir function $C(\lambda) = \lambda H_0 + H_1$ of the Poisson pencil defined on M_3. ◇

Exercise 2.6. Prove the claim (58) about the involutivity of the coefficients of a Casimir polynomial. ◇

From our standpoint, the above results are worthwhile of interest for two different reasons: First of all they show how the Lenard recursion relations, characteristic of the theory of "soliton equations", arise in a theoretically sound way in the framework of bi-Hamiltonian manifold. Secondly, they highlight the existence of a distinguished set of Hamiltonian $(C_0, C_1, \dots C_n)$ on the manifold M. Let us now choose one of the brackets of the pencil, say the bracket $\{\cdot,\cdot\}_0$. The function C_0 is a Casimir function for this bracket, and therefore its level surfaces are the symplectic leaves of the bracket $\{\cdot,\cdot\}_0$. Let us call ω_0 the symplectic 2–form defined on these submanifolds. As a consequence of the involution relation (58), the restrictions of the n functions $(C_1, \dots C_n)$ to the symplectic leaf are in involution with respect to ω_0. According to the Arnol'd–Liouville theorem, they define a family (or "hierarchy") of n completely integrable Hamiltonian vector fields on the symplectic leaf.

Definition 2.1. *The family of completely integrable Hamiltonian systems defined by the functions* $(C_1, \dots C_n)$ *on each symplectic leaf of the Poisson bracket* $\{\cdot,\cdot\}_0$ *will be called the GZ hierarchy associated with the Poisson pencil* $\{\cdot,\cdot\}_\lambda$ *defined on the bi-Hamiltonian manifold* M.

We shall be particularly interested in the study of this hierarchy for two reasons. First we want to show that the previous simple concepts allow to reconstruct a great deal of the KdV hierarchy, up to the linearization process on the infinite–dimensional Sato Grassmannian. In other words, we want to show that the theory of Poisson pairs is a natural gate to the theory of infinite-dimensional Hamiltonian systems described by partial differential equations of evolutionary type. Secondly, in a finite-dimensional setting, we want to show that the GZ vector fields are often more than integrable in the Liouville sense. Indeed, under some mild additional assumptions on the Poisson pencil, they are separable, and the separation coordinates are dictated by the geometry of the bi-Hamiltonian manifold. This result strenghtens the connection between Poisson pairs and integrability.

3 A first class of examples and the reduction technique

The aim of this lecture is to present a first class of nontrivial examples of GZ hierarchies. The examples are constructed to reproduce the reduced KdV flows discussed in the first lecture. The relation, however, will not be immediately manifest, and the reader has to wait until the fifth lecture for a full understanding of the motivations for some particular choice herewith made.

This lecture is split into three parts. In the first one we introduce a simple class of bi-Hamiltonian manifolds called Lie–Poisson manifolds. They are duals of Lie algebras endowed with a special Poisson pencil of Lie-theoretical origin. The Hamiltonian vector fields defined on these manifolds admit a *Lax representation* with a Lax matrix depending linearly on the parameter λ. In the second part we show how to combine several copies of these manifolds, in such a way to obtain Hamiltonian vector fields admitting a Lax representation depending polynomially on the parameter λ. Finally, in the third part, we introduce the geometrical technique of reduction of Marsden and Ratiu. It will allow us to specialize the form of the Lax matrix. The contact with the KdV theory, to be done in the fifth lecture, will then consist in showing that the reduced KdV flows admit exactly the Lax representation of the Hamiltonian vector fields considered in this lecture. This will ascertain the bi-Hamiltonian character of the reduced KdV flows. The lecture ends with an example worked out in detail.

3.1 Lie–Poisson manifolds

In this section $M = \mathfrak{g}^*$ is the dual of a Lie algebra $\mathfrak{g}$. We denote by S a point in M, and by $\dfrac{\partial F}{\partial S}$ the differential of a function $F : M \to \mathbb{R}$. This differential is the unique element of the algebra $\mathfrak{g}$ such that

$$\frac{dF}{dt} = \left\langle \frac{\partial F}{\partial S}, \dot{S} \right\rangle \tag{59}$$

along any curve passing through the point S. The Poisson pencil on M is defined by

$$\{F, G\}_\lambda = \left\langle S + \lambda A, \left[\frac{\partial F}{\partial S}, \frac{\partial G}{\partial S} \right] \right\rangle , \tag{60}$$

where A is any fixed element in $\mathfrak{g}^*$. In all the examples related to the KdV theory, $\mathfrak{g} = \mathfrak{sl}(2)$, S and $\dfrac{\partial F}{\partial S}$ are traceless 2×2 matrices, and

$$A = \begin{pmatrix} 0 & 0 \\ 1 & 0 \end{pmatrix} . \tag{61}$$

The Hamiltonian vector field X_F has the form

$$\dot{S} = \left[S + \lambda A, \frac{\partial F}{\partial S} \right] . \tag{62}$$

It is already in Lax form, with a Lax matrix given by

$$L(\lambda) = \lambda A + S . \tag{63}$$

Exercise 3.1. Compute the Poisson tensor and the Hamiltonian vector fields associated with the pencil (60). ◇

3.2 Polynomial extensions

We consider two copies of the algebra $\mathfrak{g}$. Accordingly, we denote by (S_0, S_1) a point in M and by $\left(\dfrac{\partial F}{\partial S_0}, \dfrac{\partial F}{\partial S_1} \right)$ the differential of the function $F : M \to \mathbb{R}$. By definition, along any curve $t \mapsto (S_0(t), S_1(t))$ we have

$$\frac{d}{dt} F = \left\langle \frac{\partial F}{\partial S_0}, \dot{S}_0 \right\rangle + \left\langle \frac{\partial F}{\partial S_1}, \dot{S}_1 \right\rangle . \tag{64}$$

The two copies of the algebra are intertwined by the Poisson brackets. As a Poisson pair on M we choose the following brackets

$$\begin{aligned} \{F, G\}_0 &= \left\langle A, \left[\frac{\partial F}{\partial S_0}, \frac{\partial G}{\partial S_1} \right] + \left[\frac{\partial F}{\partial S_1}, \frac{\partial G}{\partial S_0} \right] \right\rangle + \left\langle S_1, \left[\frac{\partial F}{\partial S_0}, \frac{\partial G}{\partial S_0} \right] \right\rangle \\ \{F, G\}_1 &= \left\langle A, \left[\frac{\partial F}{\partial S_1}, \frac{\partial G}{\partial S_1} \right] \right\rangle - \left\langle S_0, \left[\frac{\partial F}{\partial S_0}, \frac{\partial G}{\partial S_0} \right] \right\rangle \end{aligned} \tag{65}$$

The motivations can be found for instance in [20] (see also [25]). Later on we shall see how to extend this definition to the case of an arbitrary number of copies.

Exercise 3.2. Check that equations (65) indeed define a Poisson pair. ◇

Let us now study the Hamiltonian vector fields. Those defined by the brackets $\{\cdot,\cdot\}_0$ have the form

$$\begin{aligned} \dot{S}_0 &= \left[A, \frac{\partial F}{\partial S_1}\right] + \left[S_1, \frac{\partial F}{\partial S_0}\right] \\ \dot{S}_1 &= \left[A, \frac{\partial F}{\partial S_0}\right] . \end{aligned} \tag{66}$$

Those defined by the second bracket $\{\cdot,\cdot\}_1$ are

$$\begin{aligned} \dot{S}_0 &= -\left[S_0, \frac{\partial F}{\partial S_0}\right] \\ \dot{S}_1 &= \left[A, \frac{\partial F}{\partial S_1}\right] . \end{aligned} \tag{67}$$

It turns out that the Hamiltonian vector fields associated with the Poisson pencil are given by

$$\begin{aligned} \dot{S}_0 &= -\left[S_0 + \lambda S_1, \frac{\partial F}{\partial S_0}\right] - \left[\lambda A, \frac{\partial F}{\partial S_1}\right] \\ \dot{S}_1 &= -\left[\lambda A, \frac{\partial F}{\partial S_0}\right] + \left[A, \frac{\partial F}{\partial S_1}\right] . \end{aligned} \tag{68}$$

This computation allows to display an interesting property of these vector fields. If we multiply the second equation by λ and add the result to the first equation we find

$$(\lambda^2 A + \lambda S_1 + S_0)^\bullet = \left[\frac{\partial F}{\partial S_0}, \lambda^2 A + \lambda S_1 + S_0\right] . \tag{69}$$

This is a Lax representation with Lax matrix $L(\lambda) = \lambda^2 A + \lambda S_1 + S_0$. It depends polynomially on the parameter of the pencil. We have thus ascertained that all the Hamiltonian vector fields relative to the Poisson pencil (68) admit a Lax representation. The converse, however, is not necessarily true. Indeed, it must be noticed that the single Lax equation (69) is not sufficient to completely reconstruct the Poisson pencil (68). Additional constraints on the matrix $L(\lambda)$ are required to make the problem well-posed. The kind of constraints to be set are suggested by the geometric theory of reduction which we shall now outline.

3.3 Geometric reduction

We herewith outline a specific variant [5] of the reduction technique of Marsden and Ratiu [21] for Poisson manifolds. This variant is particularly suitable for bi-Hamiltonian manifolds.

Among the geometric objects defined by a Poisson pair (P_0, P_1) on a manifold M we consider:

i) a symplectic leaf S of one of the two Poisson bivectors, say P_0.
ii) the annihilator $(TS)^0$ of the tangent bundle of S, spanned by the 1–forms vanishing on the tangent spaces to S.
iii) the image $D = P_1(TS)^0$ of this annihilator according to the second Poisson bivector P_1. It is spanned by the Hamiltonian vector fields associated with the Casimir functions of P_0 by P_1.
iv) the intersection $E = D \cap TS$ of the distribution D with the tangent bundle of the selected symplectic leaf S.

It can be shown that E is an integrable distribution as a consequence of the compatibility of the Poisson brackets. Therefore we can consider the space of leaves of this distribution, $N = S/E$. We assume N to be a smooth manifold. By the Marsden–Ratiu theorem, N is a reduced bi-Hamiltonian manifold.

The reduced brackets on N can be computed by using the process of "prolongation of functions" from N to M. Given any function $f : N \to \mathbb{R}$, we consider it as a function on S, invariant along the leaves of E. Then we choose any function $F : M \to \mathbb{R}$ which annihilates D and coincides with f on S. This function is said to be a prolongation of f. It is not unique, but this fact is not disturbing. It can be shown that, if F and G are prolongations of f and g, their bracket $\{F, G\}_\lambda$ is an invariant function along E. Therefore it defines a function on N which is by definition the reduced bracket $\{f, g\}_\lambda$. The final result, of course, is independent of the particular choices of the prolongations F and G.

3.4 An explicit example

According to the spirit of these lectures, rather than discussing the proof of the reduction theorem stated in Sect. 3.3, we prefer to illustrate it on a concrete example. Let us thus perform the reduction of the Poisson pencil defined on two copies of $\mathfrak{g} = \mathfrak{sl}(2)$. The matrices S_0 and S_1 are traceless matrices whose entries we denote as follows:

$$S_0 = \begin{pmatrix} p_0 & r_0 \\ q_0 & -p_0 \end{pmatrix}, \qquad S_1 = \begin{pmatrix} p_1 & r_1 \\ q_1 & -p_1 \end{pmatrix}. \tag{70}$$

The space M has dimension six, and the entries of S_0 and S_1 are global coordinates on it. In these coordinates the differential of a function $F : M \to \mathbb{R}$ is represented by the pair of matrices

$$\frac{\partial F}{\partial S_0} = \begin{pmatrix} \frac{1}{2}\frac{\partial F}{\partial p_0} & 2 \\ \frac{\partial F}{\partial r_0} & -\frac{1}{2}\frac{\partial F}{\partial p_0} \end{pmatrix}, \qquad \frac{\partial F}{\partial S_1} = \begin{pmatrix} \frac{1}{2}\frac{\partial F}{\partial p_1} & 2 \\ \frac{\partial F}{\partial r_1} & -\frac{1}{2}\frac{\partial F}{\partial p_1} \end{pmatrix}. \tag{71}$$

Exercise 3.3. Define the pairing $\left\langle \frac{\partial F}{\partial S}, \dot{S} \right\rangle$ on $\mathfrak{g}$ as the trace of the product of the matrices $\frac{\partial F}{\partial S}$ and $\dot{S}$. Show that the matrices (71) verify the defining equation (64). ◇

The Hamiltonian vector fields (66) and (67) are consequently given by

$$\begin{aligned}
\dot{p_0} &= r_1 \frac{\partial F}{\partial r_0} - q_1 \frac{\partial F}{\partial q_0} - \frac{\partial F}{\partial q_1} \\
\dot{q_0} &= q_1 \frac{\partial F}{\partial p_0} - 2p_1 \frac{\partial F}{\partial r_0} + \frac{\partial F}{\partial p_1} \\
\dot{r_0} &= 2p_1 \frac{\partial F}{\partial q_0} - r_1 \frac{\partial F}{\partial p_0} \\
\dot{p_1} &= -\frac{\partial F}{\partial q_0} \\
\dot{q_1} &= \frac{\partial F}{\partial p_0} \\
\dot{r_1} &= 0
\end{aligned} \tag{72}$$

and by

$$\begin{aligned}
\dot{p_0} &= -r_0 \frac{\partial F}{\partial r_0} + q_0 \frac{\partial F}{\partial q_0} \\
\dot{q_0} &= 2p_0 \frac{\partial F}{\partial r_0} - q_0 \frac{\partial F}{\partial p_0} \\
\dot{r_0} &= -2p_0 \frac{\partial F}{\partial q_0} + r_0 \frac{\partial F}{\partial p_0} \\
\dot{p_1} &= -\frac{\partial F}{\partial q_1} \\
\dot{q_1} &= \frac{\partial F}{\partial p_1} \\
\dot{r_1} &= 0
\end{aligned} \tag{73}$$

respectively.

Step 1: The reduced space N

First we notice that the Hamiltonian vector fields (72) verify the constraints

$$\dot{r_1} = 0 , \qquad (r_0 + p_1^2 + r_1 q_1)^{\bullet} = 0 . \tag{74}$$

It follows that the submanifold $S \subset M$ defined by the equations

$$r_1 = 1 , \qquad r_0 + p_1^2 + r_1 q_1 = 0 \tag{75}$$

is a symplectic leaf of the first Poisson bivector P_0. Furthermore, it follows that the annihilator $(TS)^0$ is spanned by the exact 1–forms dr_1 and $d(r_0 +$

$p_1^2 + r_1 q_1)$. By computing the images of these forms according to the second Poisson bivector (73), we find the distribution D. It is spanned by the single vector field

$$\begin{aligned} \dot{p_0} &= -r_0 \\ \dot{q_0} &= 2p_0 \\ \dot{r_0} &= 0 \\ \dot{p_1} &= -1 \\ \dot{q_1} &= 2p_1 \\ \dot{r_1} &= 0 \end{aligned} \tag{76}$$

which verifies the five constraints

$$\begin{aligned} &(p_0 - r_0 p_1)^\bullet = 0, \qquad (q_0 + 2p_0 p_1 - r_0 p_1^2)^\bullet = 0, \\ &(q_1 + p_1^2)^\bullet = 0, \quad \dot{r_0} = 0, \quad \dot{r_1} = 0. \end{aligned} \tag{77}$$

They show that $D \subset TS$, and therefore $E = D$. Moreover they yield that the leaves of E on S are the level curves of the functions

$$\begin{aligned} u_1 &= q_1 + p_1^2 \\ u_2 &= p_0 + p_1 q_1 + p_1^3 \\ u_3 &= q_0 + 2p_0 p_1 + q_1 p_1^2 + p_1^4 \end{aligned} \tag{78}$$

We conclude that:

- $N = \mathbb{R}^3$;
- (u_1, u_2, u_3) are global coordinates on N;
- the canonical projection $\pi : S \to S/E$ is defined by equations (78).

Step 2: The reduced brackets

Consider any function $f : N \to \mathbb{R}$. The function

$$F := f(q_1 + p_1^2, p_0 + p_1 q_1 + p_1^3, q_0 + 2p_0 p_1 + q_1 p_1^2 + p_1^4) \tag{79}$$

is clearly a prolongation of f to M, since it coincides with f on S, and is invariant along D. We can thus use F to compute the first component of the reduced Hamiltonian vector field on N according to the following algorithm:

$$\begin{aligned} \dot{u_1} &\overset{(78)}{=} \dot{q_1} + 2p_1 \dot{p_1} \overset{(72)}{=} \frac{\partial F}{\partial p_0} - 2p_1 \frac{\partial F}{\partial q_0} \\ &\overset{(79)}{=} \left(\frac{\partial f}{\partial u_2} + 2p_1 \frac{\partial f}{\partial u_3} \right) - 2p_1 \frac{\partial f}{\partial u_3} = \frac{\partial f}{\partial u_2} . \end{aligned} \tag{80}$$

The other components are evaluated in the same way. The final result is that the Hamiltonian vector fields associated with the reduced Poisson pencil on N are defined by

$$\dot{u}_1 = (u_1 + \lambda)\frac{\partial f}{\partial u_2} + 2u_2\frac{\partial f}{\partial u_3}$$
$$\dot{u}_2 = -(u_1 + \lambda)\frac{\partial f}{\partial u_1} + (u_3 - 2\lambda u_1)\frac{\partial f}{\partial u_3} \qquad (81)$$
$$\dot{u}_3 = -2u_2\frac{\partial f}{\partial u_1} + (2\lambda u_1 - u_3)\frac{\partial f}{\partial u_2}$$

With this reduction process we passed from a six–dimensional manifold M to a three dimensional manifold N. Later on, we shall see that this manifold coincides with the invariant submanifold M_3 of KdV, defined by the constraint

$$u_{xxx} - 6uu_x = 0\ . \qquad (82)$$

Step 3: the GZ hierarchy

To compute the Casimir function of the pencil (81) we notice that these vector fields obey the constraint

$$(2\lambda u_1 - u_3)\dot{u}_1 + 2u_2\dot{u}_2 - (u_1 + \lambda)\dot{u}_3 = 0\ . \qquad (83)$$

Therefore, integrating this equation, we obtain that

$$C(\lambda) = \lambda(u_1^2 - u_3) + (u_2^2 - u_1u_3) = \lambda C_0 + C_1 \qquad (84)$$

is the Casimir sought for. It fulfills the scheme of the GZ theorem, and it defines a "short" Lenard chain

$$P_0dC_0 = 0 \qquad P_1dC_0 = P_0dC_1 = X_1 \qquad P_1dC_1 = 0\ . \qquad (85)$$

Therefore the GZ "hierarchy" consists of the single vector field

$$X_1\ :\quad \begin{aligned} \dot{u}_1 &= 2u_2 \\ \dot{u}_2 &= u_3 + 2u_1^2 \\ \dot{u}_3 &= 4u_1u_2 \end{aligned} \qquad (86)$$

As a last remark, we notice that this vector field coincides with the restriction of the first equation $\frac{\partial u}{\partial t_1} = u_x$ of the KdV hierarchy on the invariant submanifold (82). Indeed, by the procedure explained in Sect. 1, the reduced equation written in the "Cauchy data coordinates" (u, u_x, u_{xx}) is given by

$$\frac{\partial u}{\partial t_1} = u_x$$
$$\frac{\partial u_x}{\partial t_1} = u_{xx} \qquad (87)$$
$$\frac{\partial u_{xx}}{\partial t_1} = 6uu_x$$

We can now pass from (87) to (86) by the change of variables

$$u_1 = \frac{1}{2}u, \quad u_2 = \frac{1}{4}u_x, \quad u_3 = \frac{1}{4}u_{xx} - \frac{1}{2}u^2 \,. \tag{88}$$

This remark shows that the simplest reduced KdV flow is bi-Hamiltonian. In the fifth lecture we shall see that this property is general, and we shall explain the origin of the seemingly "ad hoc" change of variables (88).

3.5 A more general example

To deal with higher–order reduced KdV flows, we have to extend the class of bi-Hamiltonian manifolds to be considered. We outline the case of three copies of the algebra $\mathfrak{g}$. The formulas are similar to the ones of equation (65), albeit a little more involved. The brackets $\{F,G\}_0$ and $\{F,G\}_1$ are now given by

$$\begin{aligned}\{F,G\}_0 = &\left\langle A, \left[\frac{\partial F}{\partial S_0}, \frac{\partial G}{\partial S_2}\right] + \left[\frac{\partial F}{\partial S_1}, \frac{\partial G}{\partial S_1}\right] + \left[\frac{\partial F}{\partial S_2}, \frac{\partial G}{\partial S_0}\right]\right\rangle \\ &+ \left\langle S_2, \left[\frac{\partial F}{\partial S_0}, \frac{\partial G}{\partial S_1}\right] + \left[\frac{\partial F}{\partial S_1}, \frac{\partial G}{\partial S_0}\right]\right\rangle \\ &+ \left\langle S_1, \left[\frac{\partial F}{\partial S_0}, \frac{\partial G}{\partial S_0}\right]\right\rangle\end{aligned} \tag{89}$$

and

$$\begin{aligned}\{F,G\}_1 = &\left\langle A, \left[\frac{\partial F}{\partial S_1}, \frac{\partial G}{\partial S_2}\right] + \left[\frac{\partial F}{\partial S_2}, \frac{\partial G}{\partial S_1}\right]\right\rangle \\ &+ \left\langle S_2, \left[\frac{\partial F}{\partial S_1}, \frac{\partial G}{\partial S_1}\right]\right\rangle \\ &- \left\langle S_0, \left[\frac{\partial F}{\partial S_0}, \frac{\partial G}{\partial S_0}\right]\right\rangle.\end{aligned} \tag{90}$$

The comparison of the two examples allows to infer by induction the general rule for the Poisson pair, holding in the case of an arbitrary (finite) number of copies of $\mathfrak{g}$. The pencil (89)–(90) can be reduced according to the procedure shown before. If $\mathfrak{g} = \mathfrak{sl}(2)$ and A is still given by (61), the final result of the process is the following: We start from a nine–dimensional manifold M and, after reduction, we arrive at a five–dimensional manifold N. It fulfills the assumption of the GZ theorem. The GZ hierarchy consists of two vector fields, which are the reduced KdV flows given by (31) and (32).

Exercise 3.4. Perform the reduction of the pencil (89)–(90)) for $\mathfrak{g} = \mathfrak{sl}(2)$. ◇

4 The KdV theory revisited

In this lecture we consider again the KdV theory, but from a new point of view. Our purpose is twofold. The first aim is to show that the KdV hierarchy is another example of GZ hierarchy. The second aim is to explain in which sense the KdV hierarchy can be linearized. The algebraic linearization procedure dealt with in this lecture was suggested for the first time by Sato [27] (see also the developments contained in [17, 28, 29]), who exploited the so–called Lax representation of the KdV hierarchy in the algebra of pseudo–differential operators. Here we shall give a different description, strictly related to the Hamiltonian representation of the KdV hierarchy as a kind of infinite-dimensional GZ hierarchy. However, the presentation does not go beyond the limits of a simple sketch of the theory. We refer to [10] for full details.

4.1 Poisson pairs on a loop algebra

In this section we consider the infinite-dimensional Lie algebra M of C^∞–maps from the circle S^1 into $\mathfrak{g} = \mathfrak{sl}(2)$. A generic point of this manifold is presently a 2×2 traceless matrix

$$S = \begin{pmatrix} p(x) & r(x) \\ q(x) & -p(x) \end{pmatrix}, \tag{91}$$

whose entries are periodic functions of the coordinate x running over the circle. The three functions (p, q, r) play the role of "coordinates" on our manifold. The scalar-valued functions $F : M \to \mathbb{R}$ to be considered are local functionals

$$F = \int_{S^1} f(p, q, r; p_x, q_x, r_x; \dots)\, dx\ . \tag{92}$$

As before, their differentials are given by the matrices

$$\frac{\delta F}{\delta S} = \begin{pmatrix} \frac{1}{2}\frac{\delta f}{\delta p} & 2 \\ \frac{\delta f}{\delta r} & -\frac{1}{2}\frac{\delta f}{\delta p} \end{pmatrix}, \tag{93}$$

whose entries are the variational derivatives of the Lagrangian density f with respect to the functions (p, q, r). The Poisson pencil is similar to the first one considered in the previous lecture (see equation (60)). It is defined by

$$\{F, G\}_\lambda = \left\langle S + \lambda A, \left[\frac{\delta F}{\delta S}, \frac{\delta G}{\delta S}\right] \right\rangle + \omega \left(\frac{\delta F}{\delta S}, \frac{\delta G}{\delta S}\right). \tag{94}$$

It differs from the previous example by the addition of the nontrivial cocycle

$$\omega(a,b) = \int_{S^1} \mathrm{Tr}\left(\frac{da}{dx} b\right) dx \ . \tag{95}$$

This term is essential to generate partial differential equations. It is responsible for the appearance of the partial derivative in the expansion of the Hamiltonian vector fields

$$\dot{S} = \left(\frac{\delta F}{\delta S}\right)_x + \left[S + \lambda A, \frac{\delta F}{\delta S}\right] \ . \tag{96}$$

Exercise 4.1. Recall that a two–cocycle on $\mathfrak{g}$ is a bilinear skewsymmetric map $\omega : \mathfrak{g} \times \mathfrak{g} \to \mathbb{R}$ which verifies the cyclic condition

$$\omega(a,[b,c]) + \omega(b,[c,a]) + \omega(c,[a,b]) = 0 \ .$$

Using this identity and the periodic boundary conditions check that equation (96) defines a Poisson bivector. ◇

4.2 Poisson reduction

We apply the same reduction technique used in the previous lecture, avoiding to give all the details of the computations. They can be either worked out by exercise or found in [5, 19]

The first Poisson bivector P_0 is defined by

$$\dot{S} = \left[A, \frac{\delta F}{\delta S}\right], \tag{97}$$

where A is still defined by equation (61). These Hamiltonian vector fields obey the only constraint $\dot{r} = 0$. Therefore the submanifold $\mathcal{S}$ formed by the matrices

$$S = \begin{pmatrix} p & 1 \\ q & -p \end{pmatrix} \tag{98}$$

is a symplectic leaf of P_0. The annihilator $(T\mathcal{S})^0$ is spanned by the differentials of the functionals $F : M \to \mathbb{R}$ depending only on the coordinate function r. Consequently, the distribution D is spanned by the vector fields

$$\begin{aligned} \dot{p} &= \frac{\delta f}{\delta r} \\ \dot{q} &= \left(\frac{\delta f}{\delta r}\right)_x - 2p\frac{\delta f}{\delta r} \\ \dot{r} &= 0 \end{aligned} \tag{99}$$

The distribution D is thus tangent to $\mathcal{S}$ and E coincides with D. The vector field (99) verifies the constraint

$$\dot{q} + 2p\dot{p} - \dot{p_x} = (q + p^2 - p_x)^\bullet = 0 \, . \tag{100}$$

It follows that the leaves of the distribution E are the level sets of the function

$$u = q + p^2 - p_x \, . \tag{101}$$

Therefore the quotient space N is the space of scalar functions $u : S^1 \to \mathbb{R}$, and (101) is the canonical projection $\pi : \mathcal{S} \to \mathcal{S}/E$. We see that the manifold N is (isomorphic to) the phase space of the KdV equation.

We use the projection (101) to compute the reduced Poisson bivectors. The scheme of the computation is always the same. First we prolong any functional $\mathcal{F} = \int_{S^1} f(u, u_x, \cdots)dx$ on N into the functional

$$F(p, q, r) = \int_{S^1} f(q + p^2 - p_x, q_x + 2pp_x - p_{xx}; \ldots)\, dx \tag{102}$$

on $\mathcal{S}$. Then we compute its differential at the points of $\mathcal{S}$,

$$\frac{\delta F}{\delta S} = \begin{pmatrix} \frac{1}{2}\left(\frac{\delta f}{\delta u}\right)_x + p\frac{\delta f}{\delta u} & 2 \\ 0 & -\frac{1}{2}\left(\frac{\delta f}{\delta u}\right)_x - p\frac{\delta f}{\delta u} \end{pmatrix} . \tag{103}$$

Finally, we evaluate the reduced Hamiltonian vector fields on N according to the usual scheme:

$$\begin{aligned}
\dot{u} &\overset{(101)}{=} \dot{q} - \dot{p_x} + 2p\dot{p} \\
&\overset{(96)}{=} \left[\left(\frac{\delta f}{\delta r}\right)_x + (q+\lambda)\frac{\delta f}{\delta p} - 2p\frac{\delta f}{\delta r}\right] \\
&\quad - \left[\frac{1}{2}\left(\frac{\delta f}{\delta p}\right)_x + \frac{\delta f}{\delta r} - (q+\lambda)\frac{\delta f}{\delta q}\right]_x \\
&\quad + 2p\left[\frac{1}{2}\left(\frac{\delta f}{\delta p}\right)_x + \frac{\delta f}{\delta r} - (q+\lambda)\frac{\delta f}{\delta q}\right] \\
&\overset{(103)}{=} (q+\lambda)\left[\left(\frac{\delta f}{\delta u}\right)_x + 2p\frac{\delta f}{\delta u}\right] \\
&\quad - \left[\frac{1}{2}\left(\frac{\delta f}{\delta u}\right)_{xx} + \left(p\frac{\delta f}{\delta u}\right)_x - (q+\lambda)\frac{\delta f}{\delta u}\right]_x \\
&\quad + 2p\left[\frac{1}{2}\left(\frac{\delta f}{\delta u}\right)_{xx} + \left(p\frac{\delta f}{\delta u}\right)_x - (q+\lambda)\frac{\delta f}{\delta u}\right] \\
&= -\frac{1}{2}\left(\frac{\delta f}{\delta u}\right)_{xxx} + 2(q - p_x + p^2 + \lambda)\left(\frac{\delta f}{\delta u}\right)_x \\
&\quad + (q_x - p_{xx} + 2pp_x)\frac{\delta f}{\delta u} \\
&\overset{(101)}{=} -\frac{1}{2}\left(\frac{\delta f}{\delta u}\right)_{xxx} + 2(u+\lambda)\left(\frac{\delta f}{\delta u}\right)_x + u_x\frac{\delta f}{\delta u}.
\end{aligned} \tag{104}$$

We obtain the Poisson pencil of the KdV equation. This pencil is therefore the reduction of the "canonical" pencil (94) over a loop algebra.

4.3 The GZ hierarchy

The simplest way for computing the Casimir function of the above pencil is to use the Miura map. Since this map relates the pencil to the simple bivector of the mKdV equation, it is sufficient to compute the Casimir of the latter bivector, and to transform it back to the phase space of the KdV equation.

We notice that the Casimir function of the mKdV hierarchy (39) is given by

$$\mathsf{H}(h) = 2\,z \int_{S^1} h\,dx \ , \tag{105}$$

where the constant z has been inserted for future convenience.

To obtain the Casimir function of the KdV equation, we must "invert" the Miura map by expressing h as a function of u. To do that we exploit the dependence of the Miura map on the parameter $\lambda = z^2$ of the pencil. We know that in the finite-dimensional case the Casimir function can be found as a polynomial in λ. In the infinite-dimensional case, we expect the Casimir function to be represented by a series. It is then natural to look at h in the form of a Laurent series in z,

$$h(z) = z + \sum_{l\geq 1} h_l z^{-l} \ , \tag{106}$$

whose coefficients h_l are scalar-valued periodic functions of x. In this way we change our point of view on the Miura map. Henceforth it must be looked at as a relation between a scalar function u and a Laurent series $h(z)$. This change of perspective deeply influences all the mKdV theory. It is a possible starting point for the Sato picture of the KdV theory, as we shall show later.

By inserting the expansion (106) into the Miura map $h_x + h^2 = u + z^2$ and equating the coefficients of different powers of z, we easily compute recursively the coefficients h_l as differential polynomial of the function u. The first ones are

$$\begin{aligned}
h_1 &= \tfrac{1}{2}u \\
h_2 &= -\tfrac{1}{4}u_x \\
h_3 &= \tfrac{1}{8}(u_{xx} - u^2) \\
h_4 &= -\tfrac{1}{16}(u_{xxx} - 4uu_x) \\
h_5 &= \tfrac{1}{32}(u_{xxxx} - 6uu_{xx} - 5u_x^2 + 2u^3) \ .
\end{aligned} \tag{107}$$

One can notice (see [1]) that all the even coefficients h_{2l} are total x–derivatives. This remark explains the "strange" enumeration with odd times used for the KdV hierarchy in the first lecture.

To compute concretely the GZ vector fields, besides the Casimir function

$$\mathsf{H}(u,z) = 2z \sum_{l \geq 1} \int_{S^1} h_l z^{-l} \, dx \ , \tag{108}$$

we need its differential. To simplify the notation we set

$$\alpha := \frac{\delta \mathsf{H}}{\delta u} = 1 + \sum_{l \geq 1} \alpha_l z^{-l} \ . \tag{109}$$

Once again, the simplest way for evaluating this series is to use the Miura map. We notice that $\beta = 2z$ is the differential of the Casimir of the mKdV equation. From the transformation law of 1-forms,

$$\Phi_h'^{\,*}(\alpha) = \beta \ , \tag{110}$$

we then conclude that that α solves the equation

$$-\alpha_x + 2\alpha h = 2z \ . \tag{111}$$

As before, the coefficients α_l can be computed recursively. One finds a Laurent series in $\lambda = z^2$,

$$\alpha = 1 - \frac{1}{2} u \lambda^{-1} + \frac{1}{8}(3u^2 - u_{xx})\lambda^{-2} + \cdots \ , \tag{112}$$

whose first coefficients have already appeared in (20). From α we can easily evaluate the Lenard partial sums $\alpha^{(j)} = \left(\lambda^j \alpha\right)_+$ and write the odd GZ equations in the form

$$\frac{\partial u}{\partial t_{2j+1}} = \Big(-\frac{1}{2}\partial_{xxx} + 2(u+\lambda)\partial_x + u_x \Big)\left(\alpha^{(j)}\right) \ . \tag{113}$$

The even ones are

$$\frac{\partial u}{\partial t_{2j}} = 0 \ . \tag{114}$$

The above equations completely and tersely define the KdV hierarchy from the standpoint of the method of Poisson pairs.

4.4 The central system

We shall now pursue a little further the far–reaching consequences of the change of point of view introduced in the previous subsection. According to this new point of view, the mKdV hierarchy is defined in the space $\mathcal{L}$ of the Laurent series in z truncated from above. This affects the whole picture.

Let us consider again the basic formulas of the mKdV theory. They are the Miura map,

$$h_x + h^2 = u + z^2 \,, \tag{115}$$

the formula for the currents (40),

$$H^{(2j+1)} = -\frac{1}{2}\alpha_x^{(j)} + \alpha^{(j)}h, \quad H^{(2j)} = 0 \,, \tag{116}$$

and the definition of the mKdV hierarchy

$$\frac{\partial h}{\partial t_j} = \partial_x H^{(j)} \ . \tag{117}$$

They were obtained in the first lecture. Presently they are complemented by the information that $h(z)$ is a Laurent series of the form (106). We shall now investigate the meaning of the above formulas in this new setting.

We start form the series $h(z)$, and we associate with it a new family of Laurent series $h^{(j)}(z)$ defined recursively by

$$h^{(j+1)} = h_x^{(j)} + hh^{(j)} \ , \tag{118}$$

starting from $h^{(0)} = 1$. They form a *moving frame* associated with the point h in the space of (truncated) Laurent series. The first three elements of this frame are explicitly given by

$$h^{(0)} = 1, \qquad h^{(1)} = h, \qquad h^{(2)} = h_x + h^2 \ . \tag{119}$$

We see the basic block $h_x + h^2$ of the Miura transformation appearing. We call $\mathcal{H}_+$ the linear span of the series $\{h^{(j)}\}_{j\geq 0}$. It is a linear subspace of $\mathcal{L}$, attached to the point h. We can now interpret the three basic formulas of the mKdV theory as properties of this linear space:

- The Miura map (115) tells us that the linear space $\mathcal{H}_+$ is invariant with respect to the multiplication by λ,

$$\lambda(\mathcal{H}_+) \subset \mathcal{H}_+ \ . \tag{120}$$

- The formula (116) for the currents then entails that the currents $H^{(j)}$, for $j \in \mathbb{N}$, belong to $\mathcal{H}_+$:

$$H^{(j)} \in \mathcal{H}_+ \ . \tag{121}$$

- Furthermore, in conjunction with equation (111), it entails that the asymptotic expansion of the currents $H^{(j)}$ has the form

$$H^{(j)} = z^j + \sum_{l\geq 1} H_l^i z^{-l} = z^j + O(z^{-1}) \ . \tag{122}$$

- Finally, the mKdV equations (117) can be seen as the commutativity conditions of the operators $(\partial_x + h)$ and $(\frac{\partial}{\partial t_j} + H^{(j)})$:

$$\left[\partial_x + h, \frac{\partial}{\partial t_j} + H^{(j)}\right] = 0 \ . \tag{123}$$

Used together, conditions (121) and (123) imply that the operators $(\frac{\partial}{\partial t_j} + H^{(j)})$ leave the linear space $\mathcal{H}_+$ invariant:

$$(\frac{\partial}{\partial t_j} + H^{(j)})(\mathcal{H}_+) \subset \mathcal{H}_+ \ . \tag{124}$$

This is the abstract but simple form of the laws governing the time evolution of the currents $H^{(j)}$. These equations are the "top" of the KdV theory, and form the basis of the Sato theory. It is not difficult to give them a concrete form. By using the form of the expansion (122) it is easy to show that equations (124) are equivalent to the infinite system of Riccati–type equations on the currents $H^{(j)}$:

$$\frac{\partial H^{(j)}}{\partial t_k} + H^{(j)}H^{(k)} = H^{(j+k)} + \sum_{l=1}^{j} H_l^k H^{(j-l)} + \sum_{l=1}^{k} H_l^j H^{(k-l)} \ . \tag{125}$$

It will be called the Central System (CS).

Exercise 4.2. Prove formulas (121) and (122).

4.5 The linearization process

The first reward of the previous work is the discovery of a linearization process. The equations (125) of the Central System are not directly linearizable, but they can be easily transformed into a new system of linearizable Riccati equations by a transformation in the space of currents. This idea is realized once again by a "Miura map". The novelty, however, is that this map is now operating on the space of currents rather than on the phase space of the KdV equation.

We simply give the final result. Let us consider a new family of currents $\{W^{(k)}\}_{k\geq 0}$ of the form

$$W^{(k)} = z^k + \sum_{l\geq 1} W_l^k z^{-l} \ , \tag{126}$$

and let us denote by $\mathcal{W}_+$ their linear span in $\mathcal{L}$. We define (see also [29]) a new system of equations on the currents $W^{(k)}$ by imposing the "constraints"

$$(\frac{\partial}{\partial t_k} + z^k)(\mathcal{W}_+) \subset \mathcal{W}_+ \tag{127}$$

on their linear span $\mathcal{W}_+$. It is easily seen that they take the explicit form

$$\frac{\partial W^{(k)}}{\partial t_j} + z^j W^{(k)} = W^{(j+k)} + \sum_{l=1}^{j} W_l^k W^{(j-l)} \ . \tag{128}$$

They will be called the *Sato equations* (on the "big cell of the Sato Grassmannian"). They are a system of linearizable Riccati equations. This can be seen either from the geometry of a suitable group action on the Grassmannian [17] or by means of the following more elementary considerations. We write equations (128) in the matrix form

$$\frac{\partial \mathsf{W}}{\partial t_k} + \mathsf{W} \cdot {}^T\Lambda^k - \Lambda^k \cdot \mathsf{W} = \mathsf{W}\Gamma_k\mathsf{W} \ , \tag{129}$$

where $\mathsf{W} = \left(W_l^k\right)$ is the matrix of the components of the currents $W^{(k)}$, Λ is the infinite shift matrix

$$\Lambda = \begin{bmatrix} 0 \; 1 \;\; 0 \; \cdots & & \\ 0 \; 0 \;\; 1 \;\; 0 \; \cdots & & \\ \vdots \quad \ddots \; \ddots & & \\ \vdots \qquad\quad \ddots \; \ddots & & \\ \vdots \qquad\qquad\quad \ddots & & \end{bmatrix} , \tag{130}$$

and Γ^k is the convolution matrix of level k,

$$\Gamma_k = \begin{bmatrix} 0 \cdots \quad 1 \;\; 0 \; \cdots \\ \vdots \qquad 1 \, 0 \cdots \; \cdots \\ \qquad \cdot \;\; \cdot \\ 1 \;\; 0 \\ \vdots \end{bmatrix} . \tag{131}$$

One can thus check that the matrix Riccati equation (129) is solved by the matrix

$$\mathsf{W} = \mathsf{V} \cdot \mathsf{U}^{-1} \ , \tag{132}$$

where U and V satisfy the constant coefficients linear system

$$\frac{\partial}{\partial t_k}\mathsf{U} = {}^T\Lambda^k\mathsf{U} - \Gamma_k\mathsf{V} \ , \qquad \frac{\partial}{\partial t_k}\mathsf{V} = \Lambda^k\mathsf{V} \ . \tag{133}$$

The closing remark is that the Sato equations are mapped into the Central System (125) by the following algebraic Miura map:

$$H^{(j)} = \frac{\sum_{l=0}^{j} W_{j-l}^0 W^{(l)}}{W^{(0)}} \ . \tag{134}$$

The outcome of this long chain of extensions and transformations is the following algorithm for solving the KdV equation:

i) First we solve the linear system (133), with a suitably chosen initial condition, which we do not discuss here;
ii) Then we use the projective transformation (132) and the Miura map (134) to recover the currents $H^{(j)}$;
iii) Finally, we extract the first current $H^{(1)} = h$, and we evaluate the first component h_1 of its Laurent expansion in powers of z^{-1}.

The function

$$u(x, t_3, \dots) = 2h_1|_{t_1=x} \tag{135}$$

is then a solution of the KdV equation.

4.6 The relation with the Sato approach

The equations (117) make sense for an arbitrary Laurent series h of the form (106), even if it is not a solution of the Riccati equation $h_x + h^2 = u + z^2$. Hence they define, for every j, a system of PDEs for the coefficients h_l. We will show[8] that these systems are equivalent to the celebrated KP hierarchy of the Kyoto school (see the lectures by Satsuma in these volume). The usual definition of the KP equations can be summarized as follows. Let $\Psi\mathcal{D}$ be the ring of pseudodifferential operators on the circle. It contains as a subring the space $\mathcal{D}$ of *purely differential* operators. Let us denote with $(\cdot)_+$ the natural projection from $\Psi\mathcal{D}$ onto $\mathcal{D}$. Let Q be a monic operators of degree 1,

$$Q = \partial - \sum_{j\geq 1} q_j \partial^{-j} \ . \tag{136}$$

The KP hierarchy is the set of Lax equations for Q

$$\frac{\partial}{\partial t_j} Q = [(Q^j)_+, Q] \ . \tag{137}$$

The aim of this section is to show that such a Lax representation just arises as a kind of a Euler form of the equations (117). Before stating the next result, we must observe that the relations (118) can be solved backwards, in such a way to define the Faà di Bruno elements $h^{(j)}$ for all $j \in \mathbb{Z}$.

Proposition 4.1. *Suppose the series h of the form (106) to evolve according to a conservation law,*

$$\frac{\partial h}{\partial t} = \partial_x H, \tag{138}$$

for an arbitrary H. Then the Faà di Bruno elements $h^{(j)}$, for $j \in \mathbb{Z}$, evolve according to

---
[8] See also the papers [6, 32].

$$\left(\frac{\partial}{\partial t}+H\right)h^{(j)}=\sum_{k=0}^{\infty}\binom{j}{k}(\partial_x^k H)h^{(j-k)}\ , \tag{139}$$

where

$$\binom{j}{k}=\frac{j(j-1)\cdots(j-k+1)}{k!}\ ,\qquad \binom{j}{0}=1\ .$$

Now we consider the map $\phi:\mathcal{L}\to\Psi\mathcal{D}$, from the space of Laurent series to the ring of pseudodifferential operators on the circle, which acts on the Faà di Bruno basis according to

$$\phi(h^{(j)})=\partial^j. \tag{140}$$

This map is then extended by linearity (with respect to multiplication by a *function* of x) to the whole space $\mathcal{L}$.

Definition 4.1. *We call Lax operator of the KP theory the image*

$$Q=\phi(z) \tag{141}$$

of the first element of the standard basis in $\mathcal{L}$.

If the q_j are the components of the expansion of z on the Faà di Bruno basis,

$$z=h^{(1)}-\sum_{j\geq 1}q_j h^{(-j)}\ , \tag{142}$$

then we can write

$$Q=\partial-\sum_{j\geq 1}q_j\partial^{-j} \tag{143}$$

according to the definition of the map ϕ. We note that equation (142) uniquely defines the coefficients q_j as differential polynomials of the components h_j of $h(z)$:

$$\begin{aligned} &q_1=h_1,\quad q_2=h_2,\quad q_3=h_3+h_1^2\\ &q_4=h_4+3h_1h_2-h_1h_{1x}\\ &\cdots\cdots \end{aligned} \tag{144}$$

This is an invertible relation between the h_j and the q_j, so that equation (142) may be seen as a change of coordinates in the space $\mathcal{L}$.

Proposition 4.2. *The map ϕ has the following three properties:*

i) Multiplying a vector of the Faà di Bruno basis by a power z^k of z yields

$$\phi(z^k\cdot h^{(j)})=\partial^j\cdot Q^k\ . \tag{145}$$

ii) The evolution along a conservation law of the form

$$\frac{\partial h}{\partial t} = \partial_x \left(\sum_k H_k z^k \right)$$

translates into

$$\frac{\partial}{\partial t}\left(\phi(h^{(j)})\right) = \sum_k [\partial^j, H_k] \cdot Q^k \ . \tag{146}$$

iii) If π_+ and Π_+ are respectively the projection onto the positive part $\mathcal{H}_+ \subset \mathcal{L}$ and $\mathcal{D} \subset \Psi\mathcal{D}$, then

$$\phi \circ \pi_+ = \Pi_+ \circ \phi \ . \tag{147}$$

To obtain the Sato form of the equations (117), we derive the equation

$$z = h^{(1)} - \sum_{l \geq 1} q_l h^{(-l)} \tag{148}$$

with respect to the time t_j, getting

$$\sum_{l\geq 1} \frac{\partial q_l}{\partial t_j} h^{(-l)} = \frac{\partial h^{(1)}}{\partial t_j} - \sum_{l \geq 1} q_l \frac{\partial h^{(-l)}}{\partial t_j} \ . \tag{149}$$

Applying the map ϕ to both sides of this equation we obtain

$$\sum_{l\geq 1} \frac{\partial q_l}{\partial t_j} \partial^{-l} - \sum_{k\geq 1} [\partial, H^j_k] Q \quad {}^{k} \quad \sum_{k \geq 1} q_l [\partial^{-l}, H^j_k] Q^{-k}, \tag{150}$$

or

$$\frac{\partial Q}{\partial t_j} + \sum_{k\geq 1} [Q, H^j_k] Q^{-k} = 0 \ . \tag{151}$$

Finally, we introduce the operator

$$B^{(j)} = \phi(H^{(j)}) = \phi\left(z^j + \sum_{k\geq 1} H^j_k z^{-k} \right) = Q^j + \sum_{k \geq 1} H^j_k Q^{-k} \tag{152}$$

associated with the current density $H^{(j)}$, and we note that

$$B^{(j)} = \phi(\pi_+(z^j))) = \left(\phi(z^j)\right)_+ = \left(Q^j\right)_+ \ . \tag{153}$$

Thus we can write (151) in the final form

$$\frac{\partial Q}{\partial t_j} + [Q, (Q^j)_+] = 0 \ , \tag{154}$$

which coincides with equation (137).

5 Lax representation of the reduced KdV flows

In this lecture we want to investigate more accurately the properties of the stationary KdV flows, that is, of the equations induced by the KdV hierarchy on the finite–dimensional invariant submanifolds of the singular points of any equation of the hierarchy. Examples of these reductions have already been discussed in the first lecture. In the third lecture we realized, in a couple of examples, that the reduced flows were still bi-Hamiltonian. Although not at all surprising, this property is somewhat mysterious, since it is not yet well understood how the Poisson pairs of the reductions are related to the original Poisson pairs of the KdV equation. Moreover, even if the subject is quite old and classical (see, e.g., [3, 9, 4]), it was still lacking in the literature a systematic and coordinate free proof that such reduced flows are bi-Hamiltonian (see, however, [2, 30]). In this lecture we will not provide such a proof, which is contained in [11], but we will give a sufficiently systematic algorithm to compute the reduced Poisson pair. This algorithm is based on the study of the Lax representation of the reduced equations.

5.1 Lax representation

In this section we associate a Lax matrix (polynomially depending on λ) with each element $H^{(j)}$. This matrix naturally arises from a change of basis in the linear space $\mathcal{H}_+$ attached to the point h. So far we have introduced two bases:

i) The moving frame $\{h^{(j)}\}$;
ii) The canonical basis $\{H^{(j)}\}$.

Presently we introduce a third basis by exploiting the constraint

$$\lambda(\mathcal{H}_+) \subset \mathcal{H}_+ \ , \tag{155}$$

characteristic of the KdV theory. The new basis is formed by the multiples $\{\lambda^j H^{(0)}, \lambda^j H^{(1)}\}$ of the first two currents. Formally we define

iii) the Lax basis: $(\lambda^j, \lambda^j h)$.

The use of this basis leads to a new representation of the currents $H^{(j)}$, where each current is written as a linear combination of the first two, $H^{(0)} = 1$ and $H^{(1)} = h$, with coefficients that are polynomials in λ. Let us consider a few examples:

$$\begin{aligned}
H^{(0)} &= 1 + 0 \cdot h \\
H^{(1)} &= 0 \cdot 1 + 1 \cdot h \\
H^{(2)} &= \lambda \cdot 1 + 0 \cdot h \\
H^{(3)} &= -h_2 \cdot 1 + (\lambda - h_1) \cdot h \\
H^{(4)} &= \lambda^2 \cdot 1 \\
H^{(5)} &= (-\lambda h_2 + h_1 h_2 - h_4) \cdot 1 + (\lambda^2 - \lambda h_1 + h_1^2 - h_3) \cdot h \ .
\end{aligned} \tag{156}$$

This new representation also affects our way of writing the action of the operators $(\frac{\partial}{\partial t_j} + H^{(j)})$. Let these operators act on $H^{(0)}$ and $H^{(1)}$. For the basic invariance condition (124), we get an element in $\mathcal{H}_+$ which can be represented on the Lax basis. As a result we can write

$$\left(\frac{\partial}{\partial t_j} + H^{(j)}\right)\begin{bmatrix} 1 \\ h \end{bmatrix} = L^{(j)}(\lambda)\begin{bmatrix} 1 \\ h \end{bmatrix}, \tag{157}$$

where $L^{(j)}(\lambda)$ is the *Lax matrix* associated with the current $H^{(j)}$. We shall see below the explicit form of some of these matrices.

It becomes now very easy to rewrite the Central System in the form of equations on the Lax matrices $L^{(j)}(\lambda)$. We simply have to notice that the equations (125) entail the "exactness condition"

$$\frac{\partial H^{(j)}}{\partial t_k} = \frac{\partial H^{(k)}}{\partial t_j}, \tag{158}$$

from which it follows that the operators $(\frac{\partial}{\partial t_j} + H^{(j)})$ and $(\frac{\partial}{\partial t_k} + H^{(k)})$ commute:

$$\left[\frac{\partial}{\partial t_j} + H^{(j)}, \frac{\partial}{\partial t_k} + H^{(k)}\right] = 0 . \tag{159}$$

It is now sufficient to evaluate this condition on $(H^{(0)}, H^{(1)})$ and to expand on the Lax basis to find the "zero curvature representation" of the KdV hierarchy:

$$\frac{\partial L^{(j)}}{\partial t_k} - \frac{\partial L^{(k)}}{\partial t_j} + \left[L^{(j)}, L^{(k)}\right] = 0 . \tag{160}$$

Suppose now that we are on the invariant submanifold formed by the singular points of the j–th member of the KdV hierarchy. On this submanifold

$$\frac{\partial L^{(k)}}{\partial t_j} = 0 \qquad \forall k, \tag{161}$$

and the zero curvature representation becomes the Lax representation

$$\frac{\partial L^{(j)}}{\partial t_k} = \left[L^{(k)}, L^{(j)}\right] . \tag{162}$$

We have thus shown that all the stationary reductions of the KdV hierarchy admit a Lax representation. As a matter of fact, this Lax representation coincides [11] with the Lax representation of the GZ systems on Lie–Poisson manifolds studied in Sect. 3. The latter are bi-Hamiltonian systems. Therefore, we end up stating that the stationary reductions of the KdV theory are bi-Hamiltonian, and we can construct the associated Poisson pairs. We shall now see a couple of examples.

5.2 First example

We study anew the simplest invariant submanifold of the KdV hierarchy, defined by the equation

$$u_{xxx} - 6uu_x = 0 \; . \tag{163}$$

In this example we consider the constraint from the point of view of the Central System. Since the constraint is the stationarity of the time t_3, we have to consider only the first three Lax matrices. As for the matrix $L^{(1)}$, the following computation,

$$\begin{aligned} &(\frac{\partial}{\partial t_1} + H^{(1)})1 = 0 \cdot 1 + 1 \cdot h \\ &(\frac{\partial}{\partial t_1} + H^{(1)})H^{(1)} \stackrel{(eq:125)}{=} H^{(2)} + 2h_1 \stackrel{(156)}{=} (\lambda + 2h_1) \cdot 1 + 0 \cdot h \; , \end{aligned} \tag{164}$$

shows that

$$L^{(1)} = \begin{pmatrix} 0 & 1 \\ \lambda + 2\,h_1 & 0 \end{pmatrix} . \tag{165}$$

Similarly, the computation

$$\begin{aligned} (\frac{\partial}{\partial t_3} + H^{(3)})1 &\stackrel{(156)}{=} -h_2 \cdot 1 + (\lambda - h_1) \cdot h \\ (\frac{\partial}{\partial t_3} + H^{(3)})H^{(1)} &\stackrel{(125)}{=} H^{(4)} + h_1 H^{(2)} + h_2 H^{(1)} + h_3 + H_1^3 \\ &\stackrel{(156)}{=} (\lambda^2 + \lambda h_1 + 2h_3 - h_1^2) \cdot 1 + h_2 \cdot h \end{aligned} \tag{166}$$

yields

$$L^{(3)} = \begin{pmatrix} -h_2 & \lambda - h_1 \\ \lambda^2 + h_1\lambda + 2h_3 - {h_1}^2 & h_2 \end{pmatrix} . \tag{167}$$

On the submanifold M_3 defined by equation (163) this matrix verifies the Lax equation

$$\frac{\partial L^{(3)}}{\partial t_1} = \left[L^{(1)}, L^{(3)}\right] \; . \tag{168}$$

This equation completely defines the time evolution of the first three components (h_1, h_2, h_3) of the current $H^{(1)} = h$. These components play the role of coordinates on M_3. We get

$$\begin{aligned} \frac{\partial h_1}{\partial t_1} &= -2h_2 \\ \frac{\partial h_2}{\partial t_1} &= -2h_3 - {h_1}^2 \\ \frac{\partial h_3}{\partial t_1} &= -4h_1 h_2 \end{aligned} \tag{169}$$

By the change of coordinates

$$h_1 = \frac{1}{2}u\ , \qquad h_2 = -\frac{1}{4}u_x\ , \qquad h_3 = \frac{1}{8}(u_{xx} - u^2)\ ,$$

coming from the inversion (107) of the Miura map, these equations take the form

$$\frac{\partial u}{\partial t_1} = u_x, \quad \frac{\partial u_x}{\partial t_1} = u_{xx}, \quad \frac{\partial u_{xx}}{\partial t_1} = 6uu_x\ , \tag{170}$$

already encountered in Lecture 1. This shows explicitly the connection between the two points of view.

To find the connection between these equations and the GZ equations dealt with in the first example of Lecture 3, we compare the Lax matrix

$$L^{(3)}(\lambda) = \lambda^2 \begin{pmatrix} 0 & 0 \\ 1 & 0 \end{pmatrix} + \lambda \begin{pmatrix} 0 & 1 \\ h_1 & 0 \end{pmatrix} + \begin{pmatrix} -h_2 & -h_1 \\ 2h_3 - h_1^2 & h_2 \end{pmatrix}$$

with the Lax matrix

$$S(\lambda) = \lambda^2 \begin{pmatrix} 0 & 0 \\ 1 & 0 \end{pmatrix} + \lambda \begin{pmatrix} p_1 & 1 \\ q_1 & -p_1 \end{pmatrix} + \begin{pmatrix} p_0 & -(q_1 + p_1^2) \\ q_0 & -p_0 \end{pmatrix}$$

associated with the points of the symplectic leaf defined by (75). We easily identify $L^{(3)}$ with the restriction of $S(\lambda)$ to $p_1 = 0$ upon setting

$$p_0 = -h_2, \quad q_1 = h_1, \quad q_0 = 2h_3 - h_1^2\ . \tag{171}$$

By comparing these equations with the projection (78), which allows to pass from the symplectic leaf S to the quotient space $N = S/E$, we obtain the change of coordinates

$$u_1 = h_1, \quad u_2 = -h_2, \quad u_3 = 2h_3 - h_1^2, \tag{172}$$

connecting the reduction (169) of the Central System to the GZ system (86) dealt with in the third Lecture. The latter was, by construction, a bi-Hamiltonian system. We argue that also the reduction of the Central System herewith considered is a bi-Hamiltonian vector field, and that its Poisson pair is obtained by geometric reduction. Basic for this identification is the property of the Lax matrix $L^{(3)}$ of being a *section* of the fiber bundle $\pi : S \to S/E$ appearing in the geometric reduction. It is this property which allows to set an invertible relation among the coordinates (u_1, u_2, u_3), coming from the geometric reduction, and the coordinates (h_1, h_2, h_3) coming from the reduction of the Central System.

5.3 The generic stationary submanifold

It is now not hard to give the general form of the matrices $L^{(j)}$ for an arbitrary odd integer $2j+1$. First we observe that

$$(\frac{\partial}{\partial t_{2j+1}} + H^{(2j+1)})1 = H^{(2j+1)} \stackrel{(116)}{=} -\frac{1}{2}\alpha_x^{(j)} + \alpha^{(j)}h \ . \tag{173}$$

Then we notice that

$$\begin{aligned}(\frac{\partial}{\partial t_j} + H^{(j)})h &= H^{(j+1)} + \sum_{l=1}^{j} h_l H^{(j-l)} + H_1^j \\ &= -\frac{1}{2}(\alpha_x^{(j+1)} + \sum_{l=1}^{j} h_l \alpha_x^{(j-1)}) + H_1^j + (\alpha^{(j+1)} + \sum_{l=1}^{j} h_l \alpha^{(j-1)})h \ .\end{aligned} \tag{174}$$

Therefore

$$L^{(j)} = \begin{pmatrix} -\frac{1}{2}\alpha_x^{(j)} & \alpha^{(j)} \\ L_{21}^{(j)} & L_{22}^{(j)} \end{pmatrix}, \tag{175}$$

with

$$\begin{aligned} L_{21}^{(j)} &= -\frac{1}{2}(\alpha_x^{(j+1)} + \sum_{l=1}^{j} h_l \alpha_x^{(j-1)}) + H_1^j \\ L_{22}^{(j)} &= \alpha^{(j+1)} + \sum_{l=1}^{j} h_l \alpha^{(j-1)} \ . \end{aligned}$$

By using the definition (111) of the Lenard series $\alpha(z)$ of which the polynomials $\alpha^{(j)}$ are the partial sums, it is easy to prove that $L^{(j)}$ is a traceless matrix.

We leave to the reader to specialize the matrix $L^{(5)}$, and to write explicitly the Lax equations

$$\frac{\partial L^{(5)}}{\partial t_1} = \left[L^{(1)}, L^{(5)}\right] \ , \qquad \frac{\partial L^{(5)}}{\partial t_3} = \left[L^{(3)}, L^{(5)}\right] \ . \tag{176}$$

They should be compared with the reduced KdV equations (31) and (32) on the invariant submanifold defined by the constraint

$$u_{xxxxx} - 10uu_{xxx} - 20u_x u_{xx} + 30u^2 u_x = 0 \ . \tag{177}$$

They should also be compared with the GZ equations obtained via the geometric reduction process applied to the Lie–Poisson pairs defined on three copies of $\mathfrak{sl}(2)$ by equations (89) and (90). We have not displayed explicitly these equations yet. We will give their form in the next lecture.

5.4 What more?

There is nothing "sacred" with the KdV theory. As we know, it is related with the constraint

$$z^2(\mathcal{H}_+) \subset (\mathcal{H}_+) , \tag{178}$$

which defines an invariant submanifold of the Central System. Many other constraints can be considered. For instance, the constraint

$$z^3(\mathcal{H}_+) \subset (\mathcal{H}_+) \tag{179}$$

leads to the so–called Boussinesq theory, and is studied in [12]. What is remarkable is that the change of constraint does not affect the algorithm for the study of the reduced equations. All the previous reasonings are valid without almost no change. The only difference resides in the fact that the computations become more involved. This remark allows to better appreciate the meaning of the process leading from the KdV equation to the Central System. We have not only given a new formulation to known equations. We have actually found a much bigger hierarchy, possessing remarkable properties, which coincides with the KdV hierarchy on a (small) proper invariant subset. The integrability properties belong to the bigger hierarchy, and hold outside the KdV submanifold. Many other interesting equations can be found by other processes of reduction. There is some evidence that a very large class of evolution equations possessing some integrability properties can be eventually recovered as a suitable reduction of the Central System, or of strictly related systems. However, we shall not pursue this point of view further, since it would lead us too far away from our next topic, the *separability* of the reduced KdV flows.

6 Darboux–Nijenhuis coordinates and separability

In this lecture we shall consider the reduced KdV flows from a different point of view. Our aim is to probe the study of the geometry of the Poisson pair which, as realized in the third and fifth lectures, is associated with these flows. The final goal is to show the existence of a suitable set of coordinates defined by and adapted to the Poisson pair. They are called *Darboux–Nijenhuis* coordinates. We shall prove that they are separation coordinates for the Hamilton–Jacobi equations associated with the reduced flows.

To keep the presentation within a reasonable size, we shall mainly deal with a particular example, and we shall not discuss thoroughly the theoretical background, referring to [11] for more details. We shall use the example to display the characteristic features of the geometry of the reduced manifolds. The reader is asked to believe that all that will be shown is general inside the class of the reduced stationary KdV manifolds, whose Poisson pencils are of maximal rank. A certain care must be used in trying to extend these conclusions to other examples like the Boussinesq stationary reductions, whose Poisson pencils are not of maximal rank. They will not be covered in these lecture notes. The example worked out is the reduction of the first and the third KdV equations on the invariant submanifold defined by the equation

$$u_{xxxxx} - 10uu_{xxx} - 20u_x u_{xx} + 30u^2 u_x = 0 \ , \tag{180}$$

a problem addressed at the end of Sect. 5.3.

6.1 The Poisson pair

As we mentioned several times, the invariant submanifold M_5 defined by equation (180) has dimension five. From the standpoint of the Central System, it is characterized by the two equations

$$z^2(\mathcal{H}_+) \subset (\mathcal{H}_+) \ , \qquad H^{(5)}h = \lambda^3 + \sum_{l=1}^{5} h_l H^{(5-l)} + H_1^5 \ . \tag{181}$$

We recall that the first constraint means that, inside the big cell of the Sato Grassmannian, we are working on the special submanifold corresponding to the KdV theory. The second constraint means that, inside the phase space of the KdV theory, we are working on the set (180) of singular points of the fifth flow. The two constraints play the following roles. The first constraint sets up a relation among the currents $H^{(j)}$: All the currents are expressed as linear combinations (with polynomial coefficients) of the first two currents $H^{(0)} = 1$ and $H^{(1)} = h$. So this constraint drastically reduces the number of the unknowns H_l^j to the coefficients h_l of h. The second constraint then further cuts the degrees of freedom to a finite number, by setting relations among the coefficients h_l. It can be shown that only the first five coefficients $(h_1, h_2, h_3, h_4, h_5)$ survive as free parameters. All the other coefficients can be expressed as polynomial functions of the previous ones. By a process of elimination of the exceeding coordinate, one proves that the restriction of the first and third flows of the KdV hierarchy are represented by the following differential equations:

$$\begin{aligned}
\frac{\partial h_1}{\partial t_1} &= -2h_2 \\
\frac{\partial h_2}{\partial t_1} &= -2h_3 - {h_1}^2 \\
\frac{\partial h_3}{\partial t_1} &= -2h_1 h_2 - 2h_4 \\
\frac{\partial h_4}{\partial t_1} &= -2h_5 - {h_2}^2 - 2h_1 h_3 \\
\frac{\partial h_5}{\partial t_1} &= -4h_3 h_2 + 2{h_1}^2 h_2 - 4h_1 h_4 \ ,
\end{aligned} \tag{182}$$

and

$$\begin{aligned}
\frac{\partial h_1}{\partial t_3} &= -2h_4 + 2h_1h_2 \\
\frac{\partial h_2}{\partial t_3} &= -2h_5 + {h_2}^2 + {h_1}^3 \\
\frac{\partial h_3}{\partial t_3} &= -2h_1h_4 + 4{h_1}^2h_2 - 2h_3h_2 \\
\frac{\partial h_4}{\partial t_3} &= -2{h_3}^2 - 2h_2h_4 + 2h_1{h_2}^2 + {h_1}^4 + {h_1}^2h_3 \\
\frac{\partial h_5}{\partial t_3} &= 2{h_1}^2h_4 - 4h_3h_4 + 2{h_1}^3h_2 \,.
\end{aligned} \tag{183}$$

They can also be seen as the Lax equations (176). However, for our purposes, it is more important to recognize that the above equations are the GZ equations of the Poisson pencil defined on M_5. This pencil can be computed according to the reduction procedure explained in the third lecture. The final outcome is that the reduced Poisson bivector is given by

$$\begin{aligned}
\dot{h}_1 &= 2\frac{\partial H}{\partial h_2} + 2(h_1-\lambda)\frac{\partial H}{\partial h_4} + 2h_2\frac{\partial H}{\partial h_5} \\
\dot{h}_2 &= -2\frac{\partial H}{\partial h_1} + 2(\lambda-2h_1)\frac{\partial H}{\partial h_3} - 2h_2\frac{\partial H}{\partial h_4} + (4\lambda h_1 - 2h_3 - h_1^2)\frac{\partial H}{\partial h_5} \\
\dot{h}_3 &= 2(2h_1-\lambda)\frac{\partial H}{\partial h_2} + (2h_3 + 2h_1^2 - 4\lambda h_1)\frac{\partial H}{\partial h_4} + 2(h_4 + h_1h_2)\frac{\partial H}{\partial h_5} \\
\dot{h}_4 &= 2(\lambda-h_1)\frac{\partial H}{\partial h_1} + 2h_2\frac{\partial H}{\partial h_2} - (2h_3 + 2h_1^2 - 4\lambda h_1)\frac{\partial H}{\partial h_3} \\
&\quad + (2h_5 - 6h_1h_3 + h_2^2 + 2h_1^3 + 4\lambda h_3 + 2\lambda h_1^2)\frac{\partial H}{\partial h_5} \\
\dot{h}_5 &= -2h_2\frac{\partial H}{\partial h_1} + (2h_3 + h_1^2 - 4\lambda h_1)\frac{\partial H}{\partial h_2} - 2(h_4 + h_1h_2)\frac{\partial H}{\partial h_3} \\
&\quad - (2h_5 - 6h_1h_3 + h_2^2 + 2h_1^3 + 4\lambda h_3 + 2\lambda h_1^2)\frac{\partial H}{\partial h_4} \,.
\end{aligned} \tag{184}$$

The Casimir function of this pencil is a quadratic polynomial,

$$C(\lambda) = C_0\lambda^2 + C_1\lambda + C_2 \,, \tag{185}$$

and the coefficients are

$$\begin{aligned}
C_0 &= {h_1}^3 - 2h_1h_3 + h_5 \\
C_1 &= h_2h_4 - h_1h_5 + \tfrac{3}{2}{h_1}^2h_3 - \tfrac{1}{2}h_1{h_2}^2 - \tfrac{1}{2}{h_3}^2 - \tfrac{1}{2}{h_1}^4 \\
C_2 &= \tfrac{1}{2}h_3{h_2}^2 - h_3h_5 + \tfrac{1}{2}{h_1}^5 + h_1{h_3}^2 - h_1h_2h_4 \\
&\quad -\tfrac{3}{2}{h_1}^3h_3 + {h_1}^2h_5 + \tfrac{1}{2}{h_4}^2
\end{aligned} \tag{186}$$

The Lenard chain is

$$\begin{aligned} P_0 dC_0 &= 0 \\ P_0 dC_1 &= P_1 dC_0 = \frac{\partial \mathbf{h}}{\partial t_1} \\ P_0 dC_2 &= P_1 dC_1 = \frac{\partial \mathbf{h}}{\partial t_3} \\ P_1 dC_2 &= 0 \,, \end{aligned} \tag{187}$$

where $\mathbf{h}$ is the vector $(h_1, h_2, h_3, h_4, h_5)$. It shows that the reduced flows are bi-Hamiltonian. Finally, if one uses the coordinate change (107) from the coordinates $(h_1, h_2, h_3, h_4, h_5)$ to the coordinates $(u, u_x, u_{xx}, u_{xxx}, u_{xxxx})$, one can put the equations (182) and (183) in the form (31) and (32) considered in the first lecture.

6.2 Passing to a symplectic leaf

We aim to solve equations (182) and (183) by the Hamilton–Jacobi method. This requires to set the study of such equations on a symplectic manifold. This can be easily accomplished by noticing that these vector fields are already tangent to the submanifold S_4 defined by the equation

$$C_0 = E \,, \tag{188}$$

for a constant E. We know that this submanifold is symplectic since C_0 is the Casimir of P_0. The dimension of S_4 is four, and the variables (h_1, h_2, h_3, h_4) play the role of coordinates on it.

For our purposes it is crucial to remark an additional property of S_4: It is a bi-Hamiltonian manifold. This means that also the second bivector P_1 induces, by a process of reduction, a Poisson structure on S_4 compatible with the natural restriction of P_0. This is not a general situation. It holds as a consequence of a peculiarity of the Poisson pencil (184). The property we are mentioning concerns the vector field

$$Z = \frac{\partial}{\partial h_5}. \tag{189}$$

One can easily check that:

i) Z is transversal to the symplectic leaf S_4.
ii) The functions which are invariant along Z form a Poisson subalgebra with respect to the pencil.

In simpler terms, the Poisson bracket of functions which are independent of h_5 is independent on h_5 as well. Since they coincide with the functions on S_4 (by the transversality condition), this property allows us to define a pair of Poisson brackets also on S_4. The first bracket is associated with the symplectic 2–form ω_0 on S_4. It can be easily checked that

$$\omega_0 = h_1 dh_1 \wedge dh_2 + \frac{1}{2}(dh_2 \wedge dh_3 + dh_4 \wedge dh_1) \ . \tag{190}$$

The second Poisson bracket can be represented in the form

$$\{f, g\}_1 = \omega_0(NX_f, X_g) \ , \tag{191}$$

where X_f and X_g are the Hamiltonian vector fields associated with the functions f and g by the symplectic 2–form ω_0, and N is a $(1,1)$–tensor field on S_4, called the Nijenhuis tensor associated with the pencil (see, e.g., [15]). In our example one obtains

$$\begin{aligned} N =& \Big(- h_1 \frac{\partial}{\partial h_1} - h_2 \frac{\partial}{\partial h_2} + (h_3 - 3h_1^2) \frac{\partial}{\partial h_3} - 2h_1 h_2 \frac{\partial}{\partial h_4} \Big) \otimes dh_1 \\ & + (h_3 - h_1^2) \frac{\partial}{\partial h_4} \otimes dh_2 + \Big(\frac{\partial}{\partial h_1} + 2h_1 \frac{\partial}{\partial h_3} + h_2 \frac{\partial}{\partial h_4} \Big) \otimes dh_3 \\ & + \Big(\frac{\partial}{\partial h_2} + h_1 \frac{\partial}{\partial h_4} \Big) \otimes dh_4 \ . \end{aligned} \tag{192}$$

Thus we arrive at the following picture of the GZ hierarchy considered in this lecture. It is formed by a pair of vector fields, X_1 and X_3, defined by (182) and (183). They are tangent to the symplectic leaf (S_4, ω_0) defined by equations (188) and (190). This symplectic manifold is still bi-Hamiltonian, and therefore there exists a Nijenhuis tensor field N, defined by equation (191). The vector fields X_1 and X_3 span a Lagrangian subspace which is invariant with respect to N. One finds that they obey the following "modified Lenard recursion relations"

$$\begin{aligned} NX_1 &= X_3 + \quad (\tfrac{1}{2}\mathrm{Tr}\ N)X_1 \\ NX_3 &= \qquad + (-\sqrt{\det N})X_1 \ . \end{aligned} \tag{193}$$

From them we can extract the matrix

$$\mathsf{F} = \begin{pmatrix} \frac{1}{2}\mathrm{Tr}\ N & 1 \\ -\sqrt{\det N} & 0 \end{pmatrix} \tag{194}$$

whose transpose represents the action of N on the abovementioned Lagrangian subspace. It will play a fundamental role in the upcoming discussion of the separability of the vector fields.

Exercise 6.1. Compute the expression of the reduced pencil on S_4 and check the form of the Nijenhuis tensor, as well as the modified Lenard recursion relations (193).

6.3 Darboux–Nijenhuis coordinates

We are now in a position to introduce the basic tool of the theory of separability in the bi-Hamiltonian framework: The concept of Darboux–Nijenhuis coordinates on a symplectic bi-Hamiltonian manifold, like S_4.

Given a symplectic 2–form ω_0 and a Nijenhuis tensor N coming from a Poisson pencil defined on a $2n$–dimensional manifold $\mathcal{M}$, under the assumption that the eigenvalues of N are real and functionally independent, one proves [18] the existence of a system of coordinates $(\lambda_1, \ldots, \lambda_n; \mu_1, \ldots, \mu_n)$ which are canonical for ω_0,

$$\omega_0 = \sum_{i=1}^{n} d\mu_i \wedge d\lambda_i \ , \tag{195}$$

and which allows to put N^* (the adjoint of N) in diagonal form:

$$N^* d\lambda_i = \lambda_i d\lambda_i \ , \qquad N^* d\mu_i = \lambda_i d\mu_i \ . \tag{196}$$

The coordinates λ_i are the eigenvalues of N^*, and therefore can be computed as the *zeroes* of the minimal polynomial of N:

$$\lambda^n + c_1 \lambda^{n-1} + \cdots + c_n = 0 \ . \tag{197}$$

The coordinates μ_j can be computed as the *values* that a conjugate polynomial

$$\mu = f_1 \lambda^{n-1} + \cdots + f_n \tag{198}$$

assumes on the eigenvalues λ_j, that is,

$$\mu_j = f_1 \lambda_j^{n-1} + \cdots + f_n, \quad j = 1, \ldots, n \ . \tag{199}$$

The determination of this polynomial, which is not uniquely defined by the geometric structures present in the theory, requires a certain care. Although there is presently a sufficiently developed theory on the Darboux–Nijenhuis coordinates and on their computation, for the sake of brevity we shall not tackle this problem, but rather limit ourselves to display these polynomials in the example at hand. They are

$$\begin{aligned} &\lambda^2 - h_1 \lambda + (h_1^2 - h_3) = 0 \\ &\mu - h_2 \lambda + (h_1 h_2 - h_4) = 0 \end{aligned} \tag{200}$$

The important idea emerging from the previous discussion is that the GZ equations are often coupled with a special system of coordinates related with the Poisson pair.

Exercise 6.2. Check that the polynomials (200) define a system of Darboux–Nijenhuis coordinates for the pair (ω_0, N) considered above.

6.4 Separation of variables

We start from the classical Stäckel theorem on the separability, in orthogonal coordinates, of the Hamilton–Jacobi equation associated with the natural Hamiltonian

$$H(q,p) = \frac{1}{2}\sum g^{ii}(q)p_i^2 + V(q_1,\ldots,q_n) \tag{201}$$

on the cotangent bundle of the configuration space. According to Stäckel, this Hamiltonian is separable if and only if there exists an invertible matrix $S(q_1,\ldots,q_n)$ and a vector $U(q_1,\ldots,q_n)$ such that H is among the solutions $(H_1,\ldots,H_n)$ of the linear system

$$\frac{1}{2}p_i^2 = U_i(q) + \sum_{j=1}^{n} S_{ij}(q)H_j \ , \tag{202}$$

and S and U verify the Stäckel condition:

The rows of S and U depend only on the corresponding coordinate.

This means for instance that the elements S_{1j} and U_1 depend only on the first coordinate q_1, and so on. Such a matrix S is called a Stäckel matrix (and U a Stäckel vector).

The strategy we shall follow to prove the separability of the Hamilton–Jacobi equations associated with the GZ vector fields X_1 and X_3 on the manifold S_4 considered above, is to show that the Darboux–Nijenhuis coordinates allow to define a Stäckel matrix for the corresponding Hamiltonians.

The construction of the Stäckel matrix starts from the matrix F which relates the vector field X_1 and X_3 to the Nijenhuis tensor N (see equation (194)). One can prove that this matrix satisfies the remarkable identity

$$N^*d\mathsf{F} = \mathsf{F}d\mathsf{F} \ . \tag{203}$$

This is a matrix equation which must be interpreted as follows: $d\mathsf{F}$ is a matrix of 1–forms, and N^* acts separately on each entry of this matrix; $\mathsf{F}d\mathsf{F}$ denotes the matrix multiplication of the matrices F and $d\mathsf{F}$, which amounts to linearly combine the 1–forms appearing in $d\mathsf{F}$. In our example, equation (203) becomes

$$\begin{aligned} N^*d(\tfrac{1}{2}\mathrm{Tr}N) &= -d(\sqrt{\det N}) + (\tfrac{1}{2}\mathrm{Tr}N)\, d(\tfrac{1}{2}\mathrm{Tr}N) \\ N^*d(\sqrt{\det N}) &= \qquad\qquad\quad + (\sqrt{\det N})\, d(\tfrac{1}{2}\mathrm{Tr}N) \end{aligned} \tag{204}$$

Exercise 6.3. Check that this equations are verified by the Nijenhuis tensor (192).

We leave for a moment the particular case we are dealing with, and we suppose that, on a symplectic bi-Hamiltonian manifold fulfilling the conditions of Sect. 6.3, a family of n vector fields $(X_1, X_3, \ldots, X_{2n-1})$ is given. We assume that they are Hamiltonian with respect to P_0, say, $X_{2i-1} = P_0 dC_i$, and that there exists a matrix F such that

$$NX_{2i-1} = \sum_{j=1}^{n} \mathsf{F}_i^j X_{2j-1} \qquad \text{for all } i \ . \tag{205}$$

Finally, we suppose that F satisfies condition (203). Then, from the matrix F we build up the matrix T whose rows are the left–eigenvectors of F. In other words, we construct a matrix T such that

$$\mathsf{F} = \mathsf{T}^{-1} \Lambda \mathsf{T} \ , \tag{206}$$

where $\Lambda = \operatorname{diag}(\lambda_1, \ldots, \lambda_n)$ is the diagonal matrix of the eigenvalues of F, coinciding with the eigenvalues of N. The matrix T is normalized by imposing that in each row there is a constant component. A suitable normalization criterion, for instance, is to set the entries in the last column equal to 1.

Theorem 6.1. *If the matrix F verifies condition (203) (as it is always true in our class of examples), then the matrix T is a (generalized) Stäckel matrix in the Darboux–Nijenhuis coordinates.*

This theorem means that the rows of the matrix T verify the following generalized Stäckel condition: The entries of the first row of T depend only on the canonical pair (λ_1, μ_1), those of the second row on (λ_2, μ_2), and so on. With respect to the classical case recalled at the beginning of this lecture, we notice that by generalizing the class of Hamiltonians considered, we have been obliged to extend a little bit the notion of Stäckel matrix. However, this extension does not affect the theorem of separability. Indeed, as a consequence of the fact that the matrix F is defined by the vector fields $(X_1, X_3, \ldots, X_{2n-1})$ themselves through equation (205), one can prove that T is a Stäckel matrix for the corresponding Hamiltonians $(C_1, \ldots, C_n)$.

Theorem 6.2. *The column vector*

$$\mathbf{U} = \mathsf{T}\mathbf{C} \ , \tag{207}$$

where $\mathbf{C}$ is the column vector of the Hamiltonians $(C_1, \ldots, C_n)$, verifies the (generalized) Stäckel condition in the Darboux–Nijenhuis coordinates. This means that the first component of $\mathbf{U}$ depends only on the pair (λ_1, μ_1), the second on (λ_2, μ_2), and so on.

We shall not prove these two theorems here, preferring to see them "at work" in the example at hand. First we consider the matrix T. Due to the form (194) of the matrix F, it is easily proved that

$$\mathsf{T} = \begin{pmatrix} \lambda_1 & 1 \\ \\ \lambda_2 & 1 \end{pmatrix} . \tag{208}$$

Indeed, the equation $\mathsf{TF} = \Lambda\mathsf{T}$ follows directly from the characteristic equation for the tensor N. It should be noted that the matrix T has been computed without computing explicitly the eigenvalues λ_1 and λ_2. It is enough to use the first of equations (200), defining the Darboux–Nijenhuis coordinates. The matrix T clearly possess the Stäckel property (even in the classical, restricted sense).

The vector $\mathbf{U}$ can be computed as well without computing explicitly the coordinates (λ_j, μ_j). It is sufficient, once again, to use the equations (200). We now pass to prove that equation (207), in our example, has the particular form

$$\begin{aligned} \frac{1}{2}\mu_1^2 - \frac{1}{2}\lambda_1^2 - E\lambda_1^2 &= \lambda_1 C_1 + C_2 \\ \frac{1}{2}\mu_2^2 - \frac{1}{2}\lambda_2^2 - E\lambda_2^2 &= \lambda_2 C_1 + C_2 \ . \end{aligned} \tag{209}$$

We notice that proving this statement is tantamount to proving that the following equality between polynomials,

$$\mu(\lambda)^2 - \lambda^5 = 2C(\lambda) \ , \tag{210}$$

is verified in correspondence of the eigenvalues of N. This can be done as follows. Let us write the polynomials defining the Darboux–Nijenhuis coordinates in the symbolic form

$$\begin{aligned} \lambda^2 &= e_1\lambda + e_2 \\ \mu &= f_1\lambda + f_2 \ . \end{aligned} \tag{211}$$

The coefficients (e_j, f_j) of these polynomials must be regarded as known functions of the coordinates on the manifold. By squaring the second polynomial and by eliminating λ^2 by means of the first equation, we get

$$\begin{aligned} \mu^2 &= f_1^2(e_1\lambda + e_2) + 2f_1 f_2\lambda + f_1 f_2 \\ &= (f_1^2 e_1 + 2f_1 f_2)\lambda + (f_1^2 e_2 + f_2^2) \ . \end{aligned} \tag{212}$$

In the same way we obtain

$$\begin{aligned} \lambda^5 = \lambda \cdot \lambda^4 &= \lambda[(e_1^3 + 2e_1e_2)\lambda + (e_1^2 e_2 + e_2^2)] \\ &= (e_1^4 + 3e_1^2 e_2 + e_2^2)\lambda + (e_1^3 e_2 + 2e_1 e_2^2) \ . \end{aligned} \tag{213}$$

Finally,

$$C(\lambda) = C_0\lambda^2 + C_1\lambda + C_2 = (C_0e_1 + C_1)\lambda + (C_0e_2 + C_2) \ . \tag{214}$$

By inserting these expressions into equation (210), we see that the resulting equation splits into two parts, according to the "surviving" powers of λ:

$$\begin{aligned} \lambda : &\quad (f_1^2 e_1 + 2f_1 f_2) - (e_1^4 + 3e_1^2 e_2 + e_2^2) = 2(C_0e_1 + C_1) \\ 1 : &\quad (e_1^2 + e_2 + e_2^2) - (e_1^3 e_2 + 2e_1 e_2^2) = 2(C_0e_2 + C_2) \ . \end{aligned} \tag{215}$$

This method allows to reduce the proof of the separability of the Hamilton–Jacobi equation(s) to the procedure of checking that explicitly known functions identically coincide on the manifold.

We end our discussion of the separability at this point. Our aim was simply to introduce the method of Poisson pairs, and to show by means of concrete examples how it can be profitably used to define and solve special classes of integrable Hamiltonian equations. We hope that the examples discussed in these lectures might be successful in giving at least a feeling of the nature and the potentialities of this method.

Acknowledgments

We thank B. Dubrovin, P. Casati, G. Tondo, and J.P. Zubelli for useful discussions. We are also grateful to F. Musso for a careful reading of the manuscript.

References

1. S.I. Alber, *On stationary problems for equations of Korteweg–de Vries type.* Comm. Pure Appl. Math. **34** (1981), 259–272.
2. M. Antonowicz, A.P. Fordy, S. Wojciechowski, *Integrable stationary flows, Miura maps and bi–Hamiltonian structures.* Phys. Lett. A **124** (1987), 455–462.
3. O.I. Bogoyavlensky, S.P. Novikov, *The relationship between Hamiltonian formalism of stationary and non-stationary problems.* Funct. Anal. Appl. **10** (1976), 8–11.
4. V.M. Buchstaber, V.Z. Enol'skii, D.V. Leykin, *Hyperelliptic Kleinian functions and applications.* In: *Solitons, Geometry and Topology: On the Crossroad*, AMS Trans. Ser. 2, Advances in Math. Sciences (V.M. Buchstaber and S.P. Novikov, eds.), Vol. **179**, 1997, pp. 1–34.
5. P. Casati, F. Magri, M. Pedroni, *Bi-Hamiltonian Manifolds and τ–function.* In: Mathematical Aspects of Classical Field Theory 1991 (M. J. Gotay et al. eds.), Contemporary Mathematics vol. **132**, American Mathematical Society, Providence, R. I., 1992, pp. 213-234.
6. I.V. Cherednik, *Differential equations for the Baker-Akhiezer functions of Algebraic curves.* Funct. Anal. Appl. **10** (1978), 195–203.
7. E. Date, M. Jimbo, M. Kashiwara, T. Miwa, *Transformation Groups for Soliton Equations.* Proceedings of R.I.M.S. Symposium on Nonlinear IntegrableSystems–Classical Theory and Quantum Theory (M. Jimbo, T. Miwa, eds.), World Scientific, Singapore, 1983, pp. 39–119.
8. L.A. Dickey, *Soliton Equations and Hamiltonian Systems.* Adv. Series in Math. Phys. Vol. **12**, World Scientific, Singapore, 1991.
9. B.A. Dubrovin, I.M. Krichever, S.P. Novikov, *Integrable Systems I.* Enciclopædya of Mathematical Sciences, vol. **4** (Dynamical Systems IV), V.I. Arnol'd and S.P. Novikov, eds., Springer Verlag, Berlin, 1990.

10. G. Falqui, F. Magri, M. Pedroni, *Bihamiltonian Geometry, Darboux Coverings, and Linearization of the KP Hierarchy.* Commun. Math. Phys. **197** (1998), 303-324.
11. G. Falqui, F. Magri, M. Pedroni, J. P. Zubelli, *A Bi-Hamiltonian theory for stationary KdV flows and their separability.* Regular and Chaotic Dynamics **5** (2000), 33–52.
12. G. Falqui, F. Magri, G. Tondo, *Bihamiltonian systems and separation of variables: an example from the Boussinesq hierarchy.* Theor. Math. Phys. **122** (2000), 212-230.
13. I.M. Gel'fand, I. Zakharevich, *On the local geometry of a bi-Hamiltonian structure.* In The Gel'fand Mathematical Seminars 1990-1992 (L. Corwin et al., eds.), Birkhäuser, Boston, 1993, pp. 51–112.
14. I.M. Gel'fand, I. Zakharevich, *Webs, Lenard schemes, and the local geometry of bi-Hamiltonian Toda and Lax structures.* Selecta Math. (N.S.) **6** (2000), 131–183.
15. Y. Kosmann–Schwarzbach, F. Magri, *Poisson–Nijenhuis structures.* Ann. Inst. Poincaré (Phys. Théor.) **53** (1990), 35–81.
16. B.A. Kupershmidt, *On the nature of the Gardner transformation.* Jour. Math. Phys. **22** (1981), 449–451.
17. P. Libermann, C.M. Marle, *Symplectic Geometry and Analytical Mechanics.* Reidel, Dordrecht, 1987.
18. F. Magri, *Geometry and soliton equations.* In: *La Mécanique Analytique de Lagrange et son héritage*, Atti Acc. Sci. Torino Suppl. **124** (1990), 181–209.
19. F. Magri, *Eight lectures on Integrable Systems.* In: Integrability of Nonlinear Systems (Y. Kosmann-Schwarzbach et al. eds.), Lecture Notes in Physics **495**, Springer Verlag, Berlin-Heidelberg, 1997, pp. 256–296.
20. G. Magnano, F. Magri, *Poisson–Nijenhuis Structures and Sato Hierarchy.* Rev. Math. Phys. **3** (1991), 403-466.
21. J.E. Marsden, T. Ratiu, *Reduction of Poisson Manifolds.* Lett. Math. Phys. **11** (1986), 161–169.
22. V.B. Matveev, M.A. Salle, *Darboux transformations and solitons.* Springer Verlag, New York, 1991.
23. R.M. Miura, C.S. Gardner, M.D. Kruskal, *Korteweg-de Vries equation and generalizations. II. Existence of conservation laws and constants of motion.* Jour. Math. Phys. **9** (1968), 1204–1209.
24. A. C. Newell, *Solitons in Mathematics and Physics.* S.I.A.M., Philadelphia, 1985.
25. A.G. Reyman, M.A. Semenov-Tian-Shansky, *Compatible Poisson structures for Lax equations: an r-matrix approach.* Phys. Lett. A **130** (1988), 456–460.
26. E. Sklyanin, *Separations of variables: new trends.* Progr. Theor. Phys. Suppl. **118** (1995), 35–60.
27. M. Sato, Y. Sato, *Soliton equations as dynamical systems on infinite–dimensional Grassmann manifold.* Nonlinear PDEs in Applied Sciences (US-Japan Seminar, Tokyo), P. Lax and H. Fujita, eds., North-Holland, Amsterdam, 1982, pp. 259–271.
28. G. Segal, G. Wilson, *Loop Groups and equations of the KdV type.* Publ. Math. IHES **61** (1985), 5–65.
29. K. Takasaki, *Geometry of Universal Grassmann Manifold from Algebraic point of view.* Rev. Math. Phys. **1** (1989), 1–46.

30. G. Tondo, *On the integrability of stationary and restricted flows of the KdV hierarchy.* J. Phys. A **28** (1995), 5097–5115.
31. I. Vaisman, *Lectures on the Geometry of Poisson Manifolds.* Progress in Math., Birkhäuser, 1994.
32. G. Wilson, *On two Constructions of Conservation Laws for Lax equations.* Quart. J. Math. Oxford **32** (1981), 491–512.

Nonlinear superposition formulae of integrable partial differential equations by the singular manifold method

Micheline Musette

Dienst Theoretische Natuurkunde, Vrije Universiteit Brussel, Pleinlaan 2, B-1050 Brussels, Belgium; *mmusette@vub.ac.be*

Summary. We study by the singular manifold method a few $1+1$-dimensional partial differential equations which possess N-soliton solutions for arbitrary N, i. e. classes of solutions particularly stable under the nonlinear interaction. The existence of such solutions represents one of the different aspects of the property of *integrability*, and it can be connected to the existence of a *Bäcklund transformation* from which a *nonlinear superposition formula* can be established.

1 Introduction

This report which corresponds to the content of two seminars given during the CIME session may be viewed as a complement to the Conte lecture [8] and to a course delivered in Cargèse [37] on the direct approach of nonlinear partial differential equations (PDEs) by singularity analysis.

The existence of N-soliton solutions for some classes of nonlinear evolution equations is among the conditions which characterize the integrability of a system with an infinite number of degrees of freedom. One way of proceeding to construct those solutions is to establish from the Bäcklund transformation (BT) a nonlinear superposition formula (NLSF) which links four different solutions of the equation. The NLSF is based on the validity for the BT of the permutability theorem which for the first time has been proved by Darboux [14] in infinitesimal geometry for the sine-Gordon equation. However there exists no general proof of this theorem which must be reconsidered for each integrable PDE. In the particular case of the sine-Gordon equation, the NLSF has the nice feature of being purely algebraic

$$\tan\left(\frac{u_{12}-u}{4}\right) = \frac{\lambda_2+\lambda_1}{\lambda_2-\lambda_1}\tan\frac{u_2-u_1}{4}, \quad u_{xt}=\sin u \tag{1}$$

but in general it is not the case. For example, it reads for the KdV or Boussinesq equation as

$$f_{12}f = f_{1,x}f_2 - f_{2,x}f_1, \quad u = 2\partial_x^2 \log f \tag{2}$$

and it was shown that the N-soliton solution is related to a Wronskian [61, 52, 45].

M. Musette, Nonlinear superposition formulae of integrable partial differential equations by the singular manifold method, Lect. Notes Phys. **632**, 137–170 (2003)
http://www.springerlink.com/

The six first sections provide to the reader the concepts which are needed for the understanding of our main topic in the framework of the Painlevé analysis. Giving the definition of the NLSF in Sect. 8, we also give in Sect. 9 the details for obtaining the NLSF of KdV, modified KdV and sine-Gordon equations. Then in Sect. 10, we first present the results [43] for two fifth order integrable evolution equations, the Sawada-Kotera (SK) and Kaup-Kupershmidt (KK) equations, which are associated with two exactly solvable cases of the Hénon-Heiles Hamiltonian system [17, 60]. We establish their NLSF from their respective Bäcklund transformation obtained by singularity analysis and show the link of their N-soliton solution with a Pfaffian for SK and a Grammian for KK.

The third example is the classical Tzitzéica equation [56] which is a PDE of hyperbolic type like sine-Gordon but which possesses a different BT [9]. From the permutability theorem established by Schief [55] by iteration of the Moutard transformation [36] we obtain the expression of the NLSF which has the same form as the one for SK. Therefore its N-soliton solution is also linked to a determinant of Pfaffian type.

2 Integrability by the singularity approach

The problem we intend to solve is the following one: if a nonlinear PDE is a candidate for integrability, find its auto-BT by singularity analysis *only*, in order to prove consistently its integrability. Then, determine the associated nonlinear superposition formula to build some classes of solutions of the PDE under study.

The method adopted to achieve this goal is the *singular manifold method* [65], which consists of two parts:

1. **the Painlevé test**, based on a local analysis of the equation in the neigbourhood of a noncharacteristic singular manifold $\varphi(x,t) = 0$. The expansion of the general solution of the nonlinear PDE as a Laurent series of φ provides a set of necessary conditions for the Painlevé property. This analysis also yields the *singular part operator* denoted $\mathcal{D}$ defining the *Darboux transformation* (DT)
$$u = U + \mathcal{D} \log \tau, \ \ E(u) = 0, \tag{3}$$
with u solution of the PDE under consideration, τ some expression related to the ratio of two "entire" functions, and U a function which is *a priori* unconstrained.
2. **the truncation method**, in which the Laurent series is truncated at some level of the expansion and produces an overdetermined set of equations to be solved. If their general solution is found, it must yield the *Lax pair*, consisting in a set of two linear differential operators L and M which commute iff $E(U) = 0$. These two linear operators in association with the DT yield by using some elimination procedure the BT.

3 Bäcklund transformation: definition and example

For simplicity, but this is not a restriction, we give the basic definitions for a PDE defined as a single scalar equation for one dependent variable u and two independent variables (x, t).

Definition 3.1. *(Refs. [7] vol. III chap. XII, [51]) A* ***Bäcklund transformation*** *(BT) between two given PDEs*

$$E_1(u, x, t) = 0, \;\; E_2(U, X, T) = 0 \tag{4}$$

is a pair of relations

$$F_j(u, x, t, U, X, T) = 0, \;\; j = 1, 2, \tag{5}$$

with some transformation between (x, t) and (X, T), in which F_j depends on the derivatives of $u(x, t)$ and $U(X, T)$, such that the elimination of u (resp. U) between (F_1, F_2) implies $E_2(U, X, T) = 0$ (resp. $E_1(u, x, t) = 0$). In case the two PDEs are the same, the BT is called the ***auto-BT****.*

Example 3.1. Given u and U, two different solutions of the sine-Gordon equation

$$E_1 \equiv u_{xt} - \sin u = 0, \;\; E_2 \equiv U_{xt} - \sin U = 0, \tag{6}$$

this equation admits the auto-BT [14, 4]

$$F_1(u, U; \lambda) \equiv (u + U)_x - 2\lambda \sin \frac{u - U}{2} = 0, \tag{7}$$

$$F_2(u, U; \lambda) \equiv (u - U)_t - \frac{2}{\lambda} \sin \frac{u + U}{2} = 0, \tag{8}$$

in which λ is an arbitrary complex constant, called the *Bäcklund parameter.*

The elimination of u or U between F_1 and F_2 gives respectively:

$$F_{1,t} + F_{2,x} + (1/\lambda)\cos((u + U)/2)F_1 + \lambda \cos((u - U)/2)F_2 \equiv 2E_1, \tag{9}$$

$$F_{1,t} - F_{2,x} - (1/\lambda)\cos((u + U)/2)F_1 + \lambda \cos((u - U)/2)F_2 \equiv 2E_2. \tag{10}$$

4 Singularity analysis of nonlinear differential equations

4.1 Nonlinear ordinary differential equations

Definition 4.1. ***Painlevé property.*** *A nonlinear ODE possesses the Painlevé property (PP) if the general solution has no movable, critical singularities.*

Definition 4.2. *Group of invariance*. *Given the differential equation $E(u,x)=0$, the PP is preserved under the group of homographic transformation (H), defined as*

$$(u,x) \to (\tilde{u},\tilde{x}): \quad u = \frac{\alpha(x)\tilde{u}+\beta(x)}{\gamma(x)\tilde{u}+\delta(x)}, \ \tilde{x}=\xi(x), \ \alpha\delta-\beta\gamma \neq 0. \tag{11}$$

A larger group which also preserves the PP is the group of birational transformations, see (23) for an example.

Definition 4.3. Painlevé test. *The test consists of performing a pole-like expansion of the general solution u in the neighbourhood of each movable singularity $x-x_0=0$. This kind of analysis was considered for the first time by Sophie Kovalewski [28] in her study of the spinning top.*

Definition 4.4. Classification of Painlevé and Gambier. *Using a systematic method, Painlevé [48] and Gambier [20] have classified all second order and first degree nonlinear ODEs of the form:*

$$y'' = f(y,y';x), \tag{12}$$

rational in y', algebraic in y, analytic in x. The result of their classification yields 53 equations having the PP, among them 50 are rational in y' and y. More particularly:

- *six of them define the new transcendents P1, P2... P6,*
- *some are algebraic transforms of the elliptic equation $\wp'^2 = 4\wp^3 - g_2\wp - g_3$,*
- *some are linearizable into differential equations of order 2,3 or 4.*

In the case of 1 + 1-dimensional equations considered in this report, one of the two equations defining the BT to be found is an ODE. This nonlinear ODE for $Y = u - U$, with coefficients depending on U and an arbitrary constant λ, has two properties. Firstly, it is linearizable since it results from the Lax pair, a linear system, and the DT [34] by an elimination process [5]. Secondly, it has the Painlevé property (PP) since it is linearizable. Therefore, if its order is small (at most three), it belongs to the finite list established by the Painlevé school.

These very special nonlinear ODEs provide a link to both the Lax pair, *via* their linearizing transformation, and the Darboux transformation, *via* an involution which leaves them invariant.

The only ODE of first order and first degree with the PP is the *Riccati equation*, linearizable into a second order linear equation and defining a unique choice (94) for describing scattering problems which have order two.

Next, we consider the ODEs of order two and degree one, inequivalent under the homographic group of transformations, but linearizable into a third order equation, defined as

$$\psi_{xxx} - a\psi_x - b\psi = 0, \tag{13}$$

where a and b are functions of x. Introducing the two components

$$Y_1 = \frac{\psi_x}{\psi},\ Y_2 = \frac{\psi_{xx}}{\psi} + K\left(\frac{\psi_x}{\psi}\right)^2, \tag{14}$$

where K is a constant, (13) is equivalent to

$$Y_{1,x} = Y_2 - (K+1)Y_1^2, \tag{15}$$
$$Y_{2,x} = aY_1 + (2K-1)Y_1Y_2 - K(2K+1)Y_1^3 + b. \tag{16}$$

This first order system defines a birational transformation between Y_1 and Y_2 iff

$$K(2K+1) = 0, \tag{17}$$

a constraint which is assumed from now on. The elimination of Y_2 yields the second order ODE for Y_1

$$Y_{1,xx} + 3Y_1Y_{1,x} + Y_1^3 - aY_1 - b = 0, \tag{18}$$

which corresponds to the G5 equation of the Gambier classification, while the elimination of Y_1 gives:

$$W_{xx} - \left(1 + \frac{K+1}{2K-1}\right)W^{-1}W_x^2 + \left(1 - \frac{2(K+1)}{2K-1}\right)\left(\frac{a_x}{2K-1} + b\right)WW_x$$
$$-\frac{K+1}{2K-1}\left(\frac{a_x}{2K-1} + b\right)^2 W^3 + \left(\frac{a_{xx}}{2K-1} + b_x\right)W^2 - aW + 2K - 1 = 0,$$
$$W = \left(Y_2 + \frac{a}{2K-1}\right)^{-1}. \tag{19}$$

In the case $K = 0$, the equations (15)–(16) represent a coupled Riccati system of projective type [2], therefore the equation for W is necessarily homographically equivalent to the G5 equation.

In the second case, $K = -1/2$, making the transformation $W = \varphi Z$ in (19) the equation for Z possesses the PP iff

$$\varphi^{-1} = -\frac{a_x}{2} + b \equiv \lambda(t) \neq 0, \tag{20}$$

which allows us to identify it with the G25 equation of the Gambier classification

$$Z_{xx} - \frac{3}{4}Z^{-1}Z_x^2 + \frac{3}{2}ZZ_x + \frac{1}{4}Z^3 - aZ - 2\lambda(t) = 0, \tag{21}$$

$$Z = \frac{\lambda(t)\psi^2}{\psi_{xx}\psi - (1/2)\psi_x^2 - (a/2)\psi^2} \equiv \partial_x \log\left(\psi_{xx}\psi - \frac{1}{2}\psi_x^2 - \frac{a}{2}\psi^2\right), \tag{22}$$

taking account of the definition (14) for Y_2, the linear equation (13) and the constraint (20).

In conclusion, the only two generic choices among the ODEs of second order and first degree for describing scattering problems of third order like (13) are the equations (18) and (21), which are equivalent under the birational transformation

$$Z = \frac{\lambda(t)}{Y_{1,x} + \frac{1}{2}Y_1^2 - \frac{a}{2}},\ Y_1 = \frac{Z_x + Z^2}{2Z}. \tag{23}$$

The involutions on (18) and (21) given by

$$(Y_1, b, a) \to (-Y_1, -b, a - 6Y_{1,x}), \tag{24}$$

$$(Z, \lambda, a) \to (-Z, -\lambda, a - 3Z_x), \tag{25}$$

can be seen as a simplified definition of the DT, see the full definition in next section.

4.2 Nonlinear partial differential equations

The extension of the Painlevé analysis to partial differential equations was performed in 1983 by Weiss, Tabor and Carnevale [65]. This consists in two parts:

1. the **Painlevé test**, which is completely algorithmic. Consider the PDE $E(u; x, t) = 0$, polynomial in u and its partial derivatives Du, where u is assumed to be a single valued function of the singular manifold variable $\varphi(x,t)$. One considers the local expansion of the solution u and the equation $E(u)$ in the neighbourhood of $\varphi = 0$ and builds up step by step the Laurent series

$$u = \varphi^{-p} \sum_{j=0}^{\infty} u_j(x,t)\varphi^j,\ E(u) = \varphi^{-q} \sum_{j=0}^{\infty} E_j(x,t)\varphi^j, \tag{26}$$

   where p and q are respectively the order of the pole of u and $E(u)$. If for every family each index (or resonance) $j = r$ is compatible, all the coefficients u_r are arbitrary functions and the corresponding coefficients E_r of the expansion of the equation are identically equal to zero. One then concludes that the Painlevé test is satisfied.
2. the **Weiss truncation**. When the Painlevé test is satisfied, one of the different possibilities to prove the Painlevé property is to determine the Bäcklund transformation associated with the equation under study. To achieve this goal, the method introduced by Weiss [63] consists in truncating the Laurent series for u

$$u_T = \varphi^{-p} \sum_{j=0}^{p} u_j(x,t)\varphi^j \equiv \mathcal{D}\log\varphi + u_p, \tag{27}$$

then solving an overdetermined set of equations $\{E_j = 0\}$ depending on the derivatives of φ, which are the coefficients of the expansion

$$E_T = E(u) = \varphi^{-q} \sum_{j=0}^{q} E_j(x,t)\varphi^j, \tag{28}$$

in order to obtain the Lax pair and the Darboux transformation, which we now define.

5 Lax Pair and Darboux transformation

Definition 5.1. Lax Pair *[30]. The Lax pair of a nonlinear PDE $E(u) = 0$ is a set of two linear differential operators L, M such that*

$$([L, M] = 0) \Leftrightarrow (E(u) = 0). \tag{29}$$

For example, in the case of KdV equation

$$E(u) \equiv u_t + u_{xxx} + 6uu_x = 0, \tag{30}$$

the operators L and M are

$$L \equiv \partial_x^2 + u - \lambda, \quad \lambda \text{ constant}, \tag{31}$$

$$M \equiv \partial_t + 4\partial_x^3 + 3\partial_x u + 3u\partial_x, \tag{32}$$

and the scattering operator L is of second order.
For the p-Boussinesq equation

$$E(V) \equiv V_{xxxx} + 3\left(V_x^2\right)_x - 3\alpha^2 V_{tt} = 0, \quad \alpha^2 = \pm 1 \tag{33}$$

the operators L and M are

$$L \equiv -\lambda + 4\partial_x^3 + 3\partial_x V_x + 3V_x\partial_x - 3\alpha V_t, \quad \lambda \text{ constant}, \tag{34}$$

$$M \equiv \alpha\partial_t + \partial_x^2 + V_x, \tag{35}$$

and in this case the scattering operator L is of third order. However the PDE's (30) and (33) are two different reductions of the $2+1$ dimensional Kadomtsev-Petviashvili (KP) equation [26, 34]

$$E(u) \equiv \left(u_t + u_{xxx} + 3\left(u^2\right)_x\right)_x - 3\alpha^2 u_{yy} = 0, \tag{36}$$

which possesses the Lax pair

$$L \equiv \alpha\partial_y + \partial_x^2 + u, \tag{37}$$
$$M \equiv \partial_t + 4\partial_x^3 + 3\partial_x u + 3u\partial_x - 3\alpha w, \ (w_x = u_y). \tag{38}$$

The operators L and M are acting on the wave function $\psi(x, y, t)$ for which one may consider either the reduction

$$\psi(x, y, t) = \mathrm{e}^{-\lambda y}\tilde{\psi}(x, t) \text{ and } u_y = 0, \tag{39}$$

which yields the Lax pair (31)-(32) for KdV, or the reduction

$$\psi(x, y, t) = \mathrm{e}^{-\lambda t}\tilde{\psi}(x, y) \text{ and } u_t = 0, \ u = V_x, \tag{40}$$

which yields, after the substitution of y by t and the interchanging of L and M, the Lax pair (34)-(35) for the p-Boussinesq equation. We shall see that the NLSF for KdV and Boussinesq are related to the same second order operator (respectively $L+\lambda I$ for KdV, with L given by (31) and $M-\alpha\partial_t$ for Boussinesq, with M given by (35)). It is the reason why in the introduction we may associate the unique formula (2) with both equations.

Definition 5.2. Darboux transformation *[13]. This is a transformation on both the wave function ψ and the potential u expressing the covariance of a given Lax pair.*

Let us give three examples.

5.1 Second order scalar scattering problem

The second order spectral problem

$$L^{(2)}(\psi; u, \lambda) \equiv \psi_{xx} + (u - \lambda)\psi = 0 \tag{41}$$

is covariant under the transformation $(\psi, u, \lambda) \mapsto (\tilde{\psi}, \tilde{u}, \lambda)$

$$\tilde{\psi} = (\partial_x - \frac{\theta_x}{\theta})\psi, \quad L^{(2)}(\theta; u, \mu) = 0, \tag{42}$$
$$\tilde{u} = u + 2\partial_x^2 \log\theta, \tag{43}$$

which for historical reasons is called *classical* Darboux transformation. The DT defined by the gauge operator $G = \theta\partial_x\theta^{-1}$ also guarantees the covariance of (32) such that if $L\psi = M\psi = 0$, one has that $\tilde{L}(G(\psi)) = \tilde{M}(G(\psi)) = 0$ where $\tilde{L}$ and $\tilde{M}$ are obtained from L, M by replacing u by $\tilde{u}$. Setting $w = \int u \, \mathrm{d}x$, $\tilde{w} = \int \tilde{u} \, \mathrm{d}x$, the BT is a set of two nonlinear equations in $(w \pm \tilde{w})$ obtained from the elimination of θ between (43) and the Lax operators (31) and (32) acting on θ.

Iterating the transformation (42) N times, one obtains the function

$$\tilde{\psi} = \frac{W(\psi_1, \psi_2, \ldots, \psi_N, \psi)}{W(\psi_1, \psi_2, \ldots, \psi_N)}, \tag{44}$$

where $\psi_1, \psi_2, \dots \psi_N$ are eigenfunctions of (41) associated with parameters $\lambda_1, \lambda_2, \dots, \lambda_N$ and the symbol W represents the Wronskian determinant. This wave function solves the equation (41) for the potential

$$\tilde{u} = u + 2\partial_x^2 \log W(\psi_1, \psi_2, \dots, \psi_N). \tag{45}$$

This transformation introduced by Crum [12] is a key in the theory of nonlinear integrable evolution equations for building soliton solutions and understanding their "asymptotically linear" superposition rules. The Wronskian formula (45) was used by Satsuma [52] for building the N-soliton solution of KdV from N copies of solutions ψ_i $(i = 1, \cdots N)$ of the Schrödinger equation with potential $u = 0$ (vacuum) and spectral parameter λ_i.

The next sections deal with scalar spectral problems of order greater than two of the Gelfand-Dikii type [21], or with matrix spectral problems of matrix order $N > 2$. This has been investigated by many authors [32, 34, 47, 44, 31].

5.2 Third order scalar scattering problem

Let first consider two particular reductions of the third order scalar problem

$$L^{(3)}(\psi; u, \lambda) \equiv \psi_{xxx} + u\psi_x + (v - \lambda)\psi = 0, \tag{46}$$

corresponding to $v = 0$ and $v = u_x/2$.

These two cases can be associated with the scattering problem of two fifth order integrable evolution equations, respectively the Sawada-Kotera [54] and Kaup-Kupershmidt [27] equations (see the expressions (157) and (161) for the operators M defining the t-part of the Lax pair). The transformation which guarantees the covariance of the Lax pair is

$$\tilde{\psi} = \psi - \frac{2\mu\theta\delta(\theta, \psi)}{(\lambda + \mu)\delta(\theta, \theta)}, \quad L^{(3)}(\theta; u, \mu) = 0, \tag{47}$$

$$\tilde{u} = u + 3\partial_x^2 \log \delta(\theta, \theta), \tag{48}$$

where the binary forms δ are respectively

- $v = 0$ (SK case):

$$\delta(\theta, \psi) = \mu\theta\psi - \theta_{xx}\psi_x + \theta_x\psi_{xx}, \tag{49}$$

$$\delta(\theta, \theta) = \mu\theta^2. \tag{50}$$

- $v = u_x/2$ (KK case):

$$\delta(\theta, \psi) = \theta_{xx}\psi + \theta\psi_{xx} - \theta_x\psi_x + u\theta\psi \equiv (\lambda + \mu)\int^x \theta\psi, \tag{51}$$

$$\delta(\theta, \theta) = 2\theta_{xx}\theta - \theta_x^2 + u\theta^2 \equiv 2\mu \int^x \theta^2. \tag{52}$$

As opposed to the second order scattering problem, the Darboux covariance requires the introduction of the product of two gauge operators [47, 44]. The DT is for this reason called a *binary Darboux transformation.*

5.3 A third order matrix scattering problem

The third order matrix Lax pair [35]

$$L = \partial_x I + \begin{pmatrix} u_x/2 & 0 & i\lambda^{-1}\mathrm{e}^{u/2} \\ \mathrm{e}^{-u} & -u_x/2 & 0 \\ 0 & -i\lambda \mathrm{e}^{u/2} & 0 \end{pmatrix}, \tag{53}$$

$$M = \partial_t I + \begin{pmatrix} -u_t/2 & \mathrm{e}^{-u} & 0 \\ 0 & u_t/2 & i\lambda^{-1}\mathrm{e}^{u/2} \\ -i\lambda \mathrm{e}^{u/2} & 0 & 0 \end{pmatrix}, \tag{54}$$

$$[L, M] = \begin{pmatrix} -1 & 0 & 0 \\ 0 & 1 & 0 \\ 0 & 0 & 0 \end{pmatrix} E(u), \tag{55}$$

$$E(u) \equiv u_{xt} - \mathrm{e}^{u} + \mathrm{e}^{-2u} = 0, \tag{56}$$

is equivalent to the triad [57]

$$\partial_x^2 \psi = u_x \partial_x \psi + \lambda \mathrm{e}^{-u} \partial_t \psi, \tag{57}$$
$$\partial_t^2 \psi = u_t \partial_t \psi + \lambda^{-1} \mathrm{e}^{-u} \partial_x \psi, \tag{58}$$
$$\partial_x \partial_t \psi = \mathrm{e}^{u} \psi, \tag{59}$$

and to the nonlinear coupled system

$$Y_{1,x} = -Y_1^2 + u_x Y_1 + \lambda \mathrm{e}^{-u} Y_2, \tag{60}$$
$$Y_{2,x} = -Y_1 Y_2 + \mathrm{e}^{u}, \tag{61}$$
$$Y_{1,t} = -Y_1 Y_2 + \mathrm{e}^{u}, \tag{62}$$
$$Y_{2,t} = -Y_2^2 + u_t Y_2 + \lambda^{-1} \mathrm{e}^{-u} Y_1, \tag{63}$$
$$\text{with} \quad Y_1 = \frac{\psi_x}{\psi}, \quad Y_2 = \frac{\psi_t}{\psi}. \tag{64}$$

The nonlinear equation (56) was originally found by Georges Tzitzéica [56], who looked for surfaces $\Sigma = \{\overrightarrow{\mathbf{r}}(x,t)\}$ on which the total curvature is proportional to the fourth power of the distance from a fixed point to the tangent plane of Σ. Each of the coordinates of the point P on Σ reported to its asymptotic lines (x, t) satisfies the linear system (57)–(59) which is compatible if u is a solution of (56).
Tzitzéica also built a second family of surfaces $\tilde{\Sigma}$ with the same properties via the transformation [58]

$$\tilde{\psi} = \psi - \frac{2e^{-u}}{(\lambda - \mu)\theta}(\lambda\theta_x\psi_t - \mu\theta_t\psi_x), \tag{65}$$

where θ is a solution of the system (57)–(58)–(59) with parameter μ. Therefore, the quantity

$$e^{\tilde{u}} = e^u - 2\partial_x\partial_t \log\theta \tag{66}$$

is a second solution of the Tzitzéica equation.
Note that by eliminating Y_2 between (60) and (61 or between (62) and (63) one obtains the nonlinear equations

$$Y_{1,xx} + 3Y_1Y_{1,x} + Y_1^3 - (u_x^2 + u_{xx})Y_1 = \lambda \tag{67}$$
$$Y_{1,tt} + 3Y_1Y_{1,t} + Y_1^3 - (u_t^2 + u_{tt})Y_1 = \lambda^{-1} \tag{68}$$

which respectively linearize into

$$\psi_{xxx} - (u_x^2 + u_{xx})\psi_x = \lambda\psi, \tag{69}$$
$$\psi_{ttt} - (u_t^2 + u_{tt})\psi_t = \lambda^{-1}\psi. \tag{70}$$

Those two equations are defined by Gaffet [19] as a Lax pair for equation (56).
One can also remark that (65) is equivalent to (47)-(49)-(50) if, using the linear equation (57) for ψ and the analogue for θ, one eliminates in (49)-(50) the second derivative of ψ and θ.

6 Different truncations in Painlevé analysis

The result of the Painlevé test (necessary conditions) is independent of the explicit expression for the expansion variable χ in a Laurent series of u and $E(u)$ but some particular choices are better than others as well as the level at which one truncates the series for obtaining the Lax pair and the DT.

Let us recall the different choices for the expansion variable χ and the associated truncations which have succeeded for recovering the BT of some PDEs in $1 + 1$ dimensions.

1. **WTC truncation** [65]
   The variable of the Laurent expansion and the truncated series are:

$$\chi = \varphi, \tag{71}$$
$$u = \sum_{j=0}^{p} u_j\varphi^{j-p}, \; E = \sum_{j=0}^{q} E_j\varphi^{j-q}, \tag{72}$$

   where the coefficients (u_j, E_j) are rational in the derivatives $D\varphi$ of φ, u and E are polynomial in χ^{-1}. In terms of the linear singular part operator $\mathcal{D}$, this truncation becomes:

$$u = \mathcal{D}\log\varphi + u_p. \tag{73}$$

2. **WTC truncation versus invariant Painlevé analysis** [6]
The variable χ is related to φ as follows

$$\chi = \frac{\varphi}{\varphi_x - \varphi_{xx}\varphi/(2\varphi_x)} \sim_{\varphi\to 0} \frac{\varphi}{\varphi_x}, \tag{74}$$

$$u_j = u_j(S,C),\ E_j = E_j(S,C), \tag{75}$$

$$S = \{\varphi; x\} = \left(\frac{\varphi_{xx}}{\varphi_x}\right)_x - (1/2)\left(\frac{\varphi_{xx}}{\varphi_x}\right)^2,\ C = -\frac{\varphi_t}{\varphi_x}, \tag{76}$$

$$\chi_x = 1 + \frac{S}{2}\chi^2,\ \chi_t = -C + C_x\chi - \frac{1}{2}(CS + C_{xx})\chi^2, \tag{77}$$

$$X = 2\left[(\chi_t^{-1})_x - (\chi_x^{-1})_t\right] \equiv S_t + C_{xxx} + 2C_xS + CS_x = 0. \tag{78}$$

In terms of the singular part operator, one has

$$u = \mathcal{D}\log\psi + U, \tag{79}$$

and u and E are polynomial in $\chi^{-1} = \psi_x/\psi$. The linearization of system (77) of the Riccati type by the transformation $\chi = \psi/\psi_x$ allows one to explain why the Weiss truncation is successful for integrable PDEs with one family of poles, such as the KdV equation, which possesses a Lax pair the x-part of which is a second order linear operator of the Schrödinger type. However, for equations possessing two families of poles with opposite residues, such as the modified KdV (MKdV) and sine-Gordon equations, the truncation must include positive powers of the Laurent series to reproduce the AKNS [1] Lax pair.
3. **First extension: two-opposite families truncation** [40, 49]
For equation admitting two families of poles with opposite residues, one introduces the expansion variable Y related to χ by the following homographic transformation

$$Y = \frac{B^{-1}\chi}{1 + A\chi},\quad B \neq 0. \tag{80}$$

It satisfies the Riccati system

$$Y_x = B^{-1} - (2A + B^{-1}B_x)Y + B\left(-A_x + A^2 + \frac{S}{2}\right)Y^2, \tag{81}$$

$$Y_t = -CB^{-1} + (2AC + C_x - B^{-1}B_t)Y$$
$$-B\left(CA^2 + AC_x + \frac{CS + C_{xx}}{2} + A_t\right)Y^2. \tag{82}$$

The two-family truncation is defined as

$$u = \mathcal{D}\log Y + U, \tag{83}$$

and, taking account that Y satisfies (81) and (82), the truncated expansions

$$u_T = \sum_{j=0}^{2p} u_j Y^{j-p}, \; E_T = \sum_{j=0}^{2q} E_j Y^{j-q} \tag{84}$$

involve positive and negative powers of Y.

4. **Second extension: Lax pair of order greater than two**
In the case of the Boussinesq, Sawada-Kotera, Kaup-Kupershmidt or Tzitzéica equations, their singularity structure does not require extension of the truncation to the positive powers of the Laurent series, but it requires the introduction of more than one component in the series expansion, otherwise the Painlevé analysis would yield restricted results such as a Miura transformation, particular solutions, etc, but not the Lax pair. If the order of the Lax pair is assumed to be three, with the x-part given by the scalar equation (13), equivalently represented by a projective Riccati system in the two components

$$Y_1 = \frac{\psi_x}{\psi}, \; Y_2 = \frac{\psi_{xx}}{\psi}, \tag{85}$$

the one-family truncation is defined as

$$u_T = \mathcal{D} \log \psi + U, \tag{86}$$

$$E_T = \sum_{k \ge 0} \sum_{l \ge 0} E_{k,l} Y_1^k Y_2^l. \tag{87}$$

7 Method for a one-family equation

Let us explain the route to follow, starting from the Painlevé analysis of a PDE with only one family of movable singularities and obtaining the BT of this equation. A preliminary step consists in making a local analysis for checking if the equation passes the Painlevé test (necessary condition of integrability). This analysis also yields the *singular part operator* $\mathcal{D}$. Next,

First step. Assume the existence of the transformation (3), with solution u of the PDE under consideration, τ some expression related to the ratio of two "entire" functions and U a function *a priori* unconstrained.

Second step. Choose the order two, then three, then ..., for the unknown scalar Lax pair and represent this Lax pair by an equivalent, multicomponent, Riccati pseudopotential $\mathbf{Y} = (Y_1, \dots)$. A second order scalar Lax pair in canonical form, written here in the case of two independent variables,

$$L\psi \equiv \psi_{xx} + \frac{S}{2}\psi = 0, \tag{88}$$

$$M\psi \equiv \psi_t + C\psi_x - \frac{C_x}{2}\psi = 0, \tag{89}$$

$$2[L, M] \equiv S_t + C_{xxx} + CS_x + 2C_x S = 0, \tag{90}$$

is equivalently represented by two Riccati equations in $\mathbf{Y} = \psi_x/\psi = \chi^{-1}$.

A third order scalar Lax pair in canonical form

$$L\psi \equiv \psi_{xxx} - a\psi_x - b\psi = 0, \tag{91}$$

$$M\psi \equiv \psi_t - c\psi_{xx} - d\psi_x - e\psi = 0, \tag{92}$$

$$[L, M] \equiv X_0 + X_1\partial_x + X_2\partial_x^2, \tag{93}$$

where X_0, X_1, X_2 depend on a, b, c, d, e and their derivatives is similarly represented by a two-component Riccati pseudopotential $\mathbf{Y} = (Y_1, Y_2)$ of projective type [2, 38, 39] as defined in (15)–(16) when $K = 0$.

Third step. Choose an explicit link

$$\mathcal{D}\log\tau = f(\psi), \tag{94}$$

between the function τ and the solution ψ of a scalar Lax pair. For most PDEs, this is simply $\tau = \psi$, but, in the case of a third order Lax pair, there also exists another choice corresponding to the transformation (22) which linearizes G25.

Fourth step. With the assumptions (3) for a DT, (94) for a link between τ and ψ, define the truncation

$$E(u) = \sum_{j=0}^{q} E_j(S, C, U)\chi^{j-q} \text{ (for second order)}, \tag{95}$$

$$E(u) = \sum_{k\geq 0}\sum_{l\geq 0} E_{k,l}(a, b, c, d, e, U)Y_1^k Y_2^l \text{ (for third order)}, \tag{96}$$

and solve the set of *determining equations*

$$\forall j \quad E_j(S, C, U) = 0 \text{ (for second order)}, \tag{97}$$

$$\forall k\ \forall l \quad E_{k,l}(a, b, c, d, e, U) = 0 \text{ (for third order)}, \tag{98}$$

for the unknown coefficients (S, C) or (a, b, c, d, e) as functions of U.

The second, third and fourth steps must be repeated until a success occurs. The process is successful iff

1. U comes out unconstrained, apart from being a solution of the PDE,
2. the vanishing of the commutator $[L, M]$ is equivalent to $E(U) = 0$,
3. in the $1+1$-dimensional case only, the coefficients depend on an arbitrary constant λ, the spectral or Bäcklund parameter.

Fifth step. Obtain the two equations for the BT by eliminating ψ [5] between the DT and the scalar Lax pair. This operation may become easier by eliminating the appropriate component of the multicomponent pseudopotential $\mathbf{Y}$ rather than ψ, and this is the only reason for introducing $\mathbf{Y}$ in the second step.

8 Nonlinear superposition formula

Definition 8.1. *The nonlinear superposition formula establishes a link between four different solutions of an integrable PDE. It requires both the existence of a BT with spectral parameter λ relating two solutions v and $\tilde{v}$, schematically represented by*

$$BT\ : v \xrightarrow{\lambda} \tilde{v}, \tag{99}$$

and the validity of the permutability theorem for the BT. This theorem states that, if $v_n, \tilde{v}_n$ are the transforms of v_{n-1} under the BT with the respective parameters λ_n and λ_{n+1}, then the solution v_{n+1} is at the same time the transform of v_n and $\tilde{v}_n$ with respective parameters λ_{n+1} and λ_n. The content of this theorem is schematically represented by the Bianchi diagram, Fig. 1.

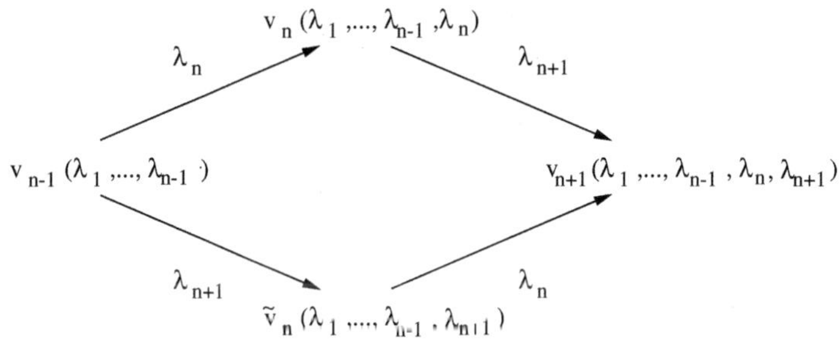

Fig. 1. Bianchi diagram

9 Results for PDEs possessing a second order Lax pair

Let us first consider PDEs which possess one family of movable poles and in particular the KdV equation.

9.1 First example: KdV equation

The conservative form of the Korteweg-de Vries equation is (30). First, we will briefly summarize the results of the Painlevé test and the truncation in the invariant formalism, referring for more details to the Conte lecture [6].

- **Painlevé test**
  Considering the Laurent series expansions for u and $E(u)$ in the variable χ (74), the algorithmic results of the Painlevé test for equation (30) are

$$p = 2, \ q = 5, \ u_0 = -2, \ \text{indices} : 4, 6 \ \text{compatible}, \ \mathcal{D} = 2\partial_x^2 \quad (100)$$

$$u = -2\chi^{-2} + (C - 4S)/6 + O(\chi) \equiv 2\partial_x^2 \log \psi + U, \ \chi = \psi/\psi_x . \quad (101)$$

- **One-family truncation**
  Solving the equations of the truncation $E_3 = 0, E_5 = 0$ ($E_4 \equiv 0$ because the index $j = 4$ is compatible), one obtains the singular manifold equation

$$C - S - 6\lambda = 0, \ \lambda \ \text{arbitrary constant}, \quad (102)$$

  which, associated with the compatility condition (78), yields the parametrization

$$S = 2(U - \lambda), \ C = 2(U + 2\lambda), \ \text{KdV}(U) = 0. \quad (103)$$

- **Lax pair**
  The parametrization (103) of S and C provides by linearizing the Riccati equations (77) the second order Lax pair

$$\psi_{xx} + (U - \lambda)\psi = 0, \quad (104)$$

$$\psi_t + 2(U + 2\lambda)\psi_x - U_x\psi = 0, \quad (105)$$

  satisfying the cross-derivative condition $\psi_{xxt} - \psi_{txx} \equiv 2 \ \text{KdV}(U)\psi = 0$. It is easy to show that, taking account of (104) to eliminate λ in (105) the Lax pair provided here coincides with the two operators (31) and (32).
- **Darboux transformation**
  The link between two solutions u_T and U of KdV coming from the truncation is

$$u_T = 2(\log \psi)_{xx} + (C + 2S)/6 \equiv 2(\log \psi)_{xx} + U. \quad (106)$$

  Thus, the one-family truncation yields both the Lax pair (104)–(105) and the DT (106) of the KdV equation.
- **Bäcklund transformation**
  The Bäcklund transformation results from the elimination of ψ between the DT and the Lax pair in the following way. Defining in the DT the transformation $u_T = w_x, U = W_x$, and integrating with respect to x, one performs in the Lax pair, written in Riccati form, the substitution $\psi_x/\psi \equiv Y = (w - W)/2$ and obtains

$$x\text{-BT}(w, W; \lambda) \equiv (w + W)_x - 2\lambda + (w - W)^2/2 = 0, \quad (107)$$

$$t\text{-BT}(w, W; \lambda) \equiv (w + W)_t + 2(w_x^2 + w_x W_x + W_x^2) + (w - W)(w - W)_{xx} = 0, \quad (108)$$

  where λ is the Bäcklund parameter.
  The one-soliton is built from the vacuum solution $U = 0$ by setting $\lambda = k^2/4$ in the Lax pair, which possesses as general solution $\psi_0 = A \exp[(kx - k^3 t)/2] + B \exp[-(kx - k^3 t)/2]$, where A and B are arbitrary constants.

- **Nonlinear superposition formula**
Considering four copies of the BT (107), respectively x-BT$(w_n, w_{n-1}; \lambda_n)$, x-BT$(\tilde{w}_n, w_{n-1}; \lambda_{n+1})$, x-BT$(w_{n+1}, w_n; \lambda_{n+1})$, x-BT$(w_{n+1}, \tilde{w}_n; \lambda_n)$, and eliminating the four first derivatives $w_{n+1,x}$, $w_{n,x}$, $\tilde{w}_{n,x}$ and $w_{n-1,x}$ between these four equations, one obtains the nonlinear superposition formula [62, 29]

$$(w_{n+1} - w_{n-1})(\tilde{w}_n - w_n) = 4(\lambda_{n+1} - \lambda_n). \tag{109}$$

The two-soliton solution u_{12} corresponding to $n = 1$ is built from the vacuum $w_0 \equiv 0$. With the notation $w_{n+1} \equiv w_{12}$, $w_n \equiv w_1$, $\tilde{w}_n \equiv w_2$, $w_{n-1} \equiv w_0$ and

$$w_{12} = 4\frac{(\lambda_2 - \lambda_1)}{w_2 - w_1}, \tag{110}$$

$$w_i \equiv 2\partial_x \log \psi_{0,i} = k_i \tanh \frac{1}{2}(k_i x - k_i^3 t + \eta_i),\ k_i^2 = 4\lambda_i,\ i = 1,2 \tag{111}$$

this two-soliton is therefore

$$u_{12} \equiv \partial_x w_{12} = 2\partial_x^2 \log \left[1 + \mathrm{e}^{\theta_1} + \mathrm{e}^{\theta_2} + \kappa_{12}\mathrm{e}^{\theta_1+\theta_2}\right], \tag{112}$$

$$\theta_i = k_i x - k_i^3 t + \delta_i,\ \kappa_{12} = \left(\frac{k_1 - k_2}{k_1 + k_2}\right)^2. \tag{113}$$

The equivalence of (110) with (2) is easily shown in taking account that

$$w_{12} = 2\partial_x \log f_{12},\ w_i = 2\partial_x \log f_i,\ \lambda_i = f_{i,xx}/f_i. \tag{114}$$

As to the p-Boussinesq equation (33), the NLSF (2) can be derived by considering four copies of the BT

$$(v + V)_{xx} - (v - V)_t + (v - V)(v - V)_x = 0, \tag{115}$$

given in [9] (formula (53)) or in this volume (Conte lecture, formula (234)), respectively for the couples (v_0, v_1), (v_0, v_2), (v_1, v_{12}), (v_2, v_{12}) and by eliminating between those four equations the t-derivative of v_0,v_1,v_2,v_{12}. Note that in this case f_i is solution of the third order linear equation

$$f_{i,xxx} - \lambda_i f_i = 0, \qquad i = 1, 2 \tag{116}$$

and $f_0 = 1$.

9.2 Second example: MKdV and sine-Gordon equations

- **PDEs with two opposite families**
The Weiss truncation does not always produce the BT when the PDE $E(u)$ possesses several families of poles. This number of families is determined by the degree of the algebraic equation for the first coefficient

u_0 of the Laurent expansion for u. In particular, when the equation possesses two opposite families of poles, the following extension [40, 49] has been proposed for recovering the BT. We will handle the MKdV and sine-Gordon equations.
Let us first illustrate the principle of this truncation on the MKdV

$$E(u) \equiv u_t + u_{xxx} - 2a^{-2}(u^3)_x = 0. \tag{117}$$

This equation passes the Painlevé test, which yields the results

$$u_0 = \pm a, \ \mathcal{D} = u_0 \partial_x. \tag{118}$$

As the one-family truncation does not provide the Lax pair of the MKdV (see the details in [40, 37]), in order to take account of the two opposite families we consider the following extension

$$u = (a\partial_x \log \psi_1 - a\partial_x \log \psi_2) + U \equiv a\partial_x \log Y + U, \ Y = \psi_1/\psi_2, \tag{119}$$

where Y is the ratio of two linear independent solutions of a second order linear equation. Then, one assumes that

$$Y^{-1} = B(\chi^{-1} + A), \ B \neq 0, \tag{120}$$

where χ satisfies the normalized Riccati system (77) with compatibility condition (78). This extension of the Weiss truncation produces the BT of the MKdV and sine-Gordon equations [40, 49].
Here we choose to analyze in detail the sine-Gordon equation

$$E(u) \equiv u_{xt} - \sin u = 0, \tag{121}$$

allowing one to recover the BT (7)–(8) and the nonlinear superposition formula [4].

- **From Painlevé analysis to BT**
  To apply the Painlevé test we first transform the sine-Gordon equation into a polynomial form. Taking account of the parity in u of (121), it writes

$$E(v) \equiv 2vv_{xt} - 2v_x v_t - v^3 + v = 0, \ v \equiv v_\pm = \mathrm{e}^{\pm \mathrm{i}u}. \tag{122}$$

The equation passes the Painlevé test, which yields the results

$$p = 2, \ q = 6, \ \text{index:2 compatible}, \ \mathcal{D} = -4\partial_x\partial_t. \tag{123}$$

The two-family truncation for v is defined as

$$v_+ - v_- \equiv 2\mathrm{i}\sin u = -4\partial_x\partial_t \log Y + 2\mathrm{i}V, \tag{124}$$

where Y given by (120) satisfies the Riccati system (81)–(82). Taking account of (121), one can integrate twice equation (124), which provides the Darboux-like transformation

$$u = 2\mathrm{i} \log Y + U, \text{ with } U_{xt} \equiv V. \tag{125}$$

The truncation of $E(v)$

$$E_T(v) \equiv Y^{-4} \sum_{j=0}^{4} E_j(U, A, B, S, C) Y^{j-2} = 0 \tag{126}$$

generates six determining equations $\{E_j = 0\}$ and (78), whose resolution is explained in the Conte lecture [8]. We only mention the result:

$$E_0 : \ B^2 \mathrm{e}^{\mathrm{i}U} = -4C, \tag{127}$$

$$E_1 : \ A = -\frac{1}{2} (\log C)_x, \tag{128}$$

$$E_2 \equiv 0, \tag{129}$$

$$E_3 : \ S = -F(x) + \frac{C_x^2}{2C^2} - \frac{C_{xx}}{C}, \tag{130}$$

$$E_4 : \ CC_{xt} - C_x C_t + F(x) C^3 - F(x)^{-1} C = 0, \tag{131}$$

$$X : \ F'(x) = 0, \tag{132}$$

in which F is a function of integration. Therefore, $F(x)$ must be a constant

$$F(x) = 2\lambda^2. \tag{133}$$

and $\log C$ is proportional to a second solution $\tilde{U}$ of the PDE

$$C = -\frac{1}{4} \lambda^{-2} \mathrm{e}^{\mathrm{i}\tilde{U}}, \ E(\tilde{U}) = 0. \tag{134}$$

From (125) and (127), one obtains the Darboux transformation

$$u = 2\mathrm{i} \log y + \tilde{U}, \ y = \lambda B Y, \tag{135}$$

in which y satisfies the Riccati system

$$y_x = \lambda + \mathrm{i}\tilde{U}_x y - \lambda y^2, \tag{136}$$

$$y_t = -\frac{1}{4} \lambda^{-1} \mathrm{e}^{\mathrm{i}\tilde{U}} - \frac{1}{4} \lambda^{-1} \mathrm{e}^{-\mathrm{i}\tilde{U}} y^2, \tag{137}$$

$$y_{xt} - y_{tx} = E(\tilde{U}) y. \tag{138}$$

The BT (7)–(8) is obtained by setting in (136) and (137)

$$\log y = \frac{1}{2\mathrm{i}} (u - \tilde{U}). \tag{139}$$

– **Nonlinear superposition formula**
Let us consider four copies of the x-part of the BT, i. e. $F_1(u_n, u_{n-1}; \lambda_n)$, $F_1(\tilde{u}_n, u_{n-1}; \lambda_{n+1})$, $F_1(u_{n+1}, u_n; \lambda_{n+1})$, $F_1(u_{n+1}, \tilde{u}_n; \lambda_n)$, and eliminate the first derivatives of u_{n+1}, u_n,$\tilde{u}_n$,u_{n-1} between these four relations for obtaining the algebraic relation

$$\begin{aligned}\lambda_{n+1}\Big(\sin\frac{u_{n+1}-u_n}{2} + \sin\frac{\tilde{u}_n - u_{n-1}}{2}\Big)\\ = \lambda_n\Big(\sin\frac{u_{n+1}-\tilde{u}_n}{2} + \sin\frac{u_n - u_{n-1}}{2}\Big).\end{aligned} \tag{140}$$

This equation can be solved for u_{n+1}, using the trigonometric relation

$$\begin{aligned}(\lambda_{n+1}-\lambda_n)\sin\frac{u_{n+1}-u_{n-1}}{4}\cos\frac{\tilde{u}_n-u_n}{4}\\ = -(\lambda_{n+1}+\lambda_n)\cos\frac{u_{n+1}-u_{n-1}}{4}\sin\frac{\tilde{u}_n-u_n}{4},\end{aligned} \tag{141}$$

and dividing this last equation by $\cos\frac{u_{n+1}-u_{n-1}}{4}\cos\frac{\tilde{u}_n-u_n}{4}$. One then obtains the well known relation [4]

$$u_{n+1}-u_{n-1} = 4\arctan\left[\frac{\lambda_{n+1}+\lambda_n}{\lambda_{n+1}-\lambda_n}\tan\frac{u_n-\tilde{u}_n}{4}\right], \tag{142}$$

which may be used to build the two-soliton solution from the vacuum $u_{0,x} \equiv u_{n-1,x} = 0$.

10 PDEs possessing a third order Lax pair

10.1 Sawada-Kotera, KdV$_5$, Kaup-Kupershmidt equations

The fifth order nonlinear partial differential equation

$$\beta u_t = (u_{xxxx} + (8a-2b)uu_{xx} - 2(a+b)u_x^2 - (20/3)abu^3)_x \tag{143}$$

is known to be integrable for only three values of the ratio $b/a=(-1,-6,-16)$, corresponding respectively to the Sawada-Kotera (SK), KdV$_5$, and Kaup-Kupershmidt (KK) equations, which possess N-soliton solutions. They are related by the reduction $(x,t) \to \xi = x - ct, u(x,t) = q_1(\xi)$ to the Hénon-Heiles system [17]:

$$q_1'' = bq_1^2 - aq_2^2 + \frac{c}{4a},\ q_2'' = -2aq_1q_2 - \frac{\Lambda^2}{4a^2}q_2^{-3},\ \Lambda \text{ arbitrary constant}, \tag{144}$$

which is integrable for the only same three values of the ratio b/a. The SK and KK equations are dual equations related to each other by a Bäcklund transformation [18] (BT) and they are both associated with a scattering problem of third order [27]. However, the auto-BT for SK has been known since 1977 [16] and can be easily recovered [39] by singularity approach [65], the one previously obtained for KK [50] did not contain any Bäcklund parameter, whose introduction is crucial for determining the nonlinear superposition formula. Last but not least, there still remains a challenge for the Hirota method to provide for KK a bilinear BT which can be linearized in the appropriate way.

The correct result for the auto-BT was in the end obtained [41] in the framework of the Painlevé analysis. We have seen in Sect. 4.1 that the association of a third order spectral problem of the Gel'fand-Dikii type with a two-component Riccati system led us to consider two particular equations belonging to the Gambier classification, the G5 and G25 equations, equivalent under the group of birational transformations. In this section we explain how the solution of the Weiss truncation implies that the x-part of the BT is respectively related to the G5 equation for SK and to the G25 equation for KK. Then, we derive for both evolution equations the nonlinear superposition formula [42, 43, 59], recovering for SK a result obtained by Hu and Li [25] in the framework of the bilinear Hirota formalism, and for KK a result derived by Loris [33] in the context of symmetry reductions of the CKP hierarchy.

10.2 Painlevé test

Because of their duality [27, 64], the pSK and pKK equations defined as

$$\mathrm{pSK}(v) \equiv v_t + v_{xxxxx} + \frac{30}{\alpha} v_x v_{xxx} + \frac{60}{\alpha^2} v_x^3 = 0, \tag{145}$$

$$\mathrm{pKK}(v) \equiv v_t + v_{xxxxx} + \frac{30}{\alpha} v_x v_{xxx} + \frac{45}{2\alpha} v_{xx}^2 + \frac{60}{\alpha^2} v_x^3 = 0, \tag{146}$$

are handled at the same time, with the conservative field u related to v by $u = v_x$.

Both equations have the PP [64]. Each of them has two families of movable singularities, whose leading order $v \sim v_0 \chi^p$ and Fuchs indices are the following [64]

$$\mathrm{pSK, F1}: p = -1,\ v_0 = \alpha,\ \text{indices}\ -1, 1, 2, 3, 10, \tag{147}$$

$$\mathrm{pSK, F2}: p = -1,\ v_0 = 2\alpha,\ \text{indices}\ -2, -1, 1, 5, 12, \tag{148}$$

$$\mathrm{pKK, F1}: p = -1,\ v_0 = \alpha/2,\ \text{indices}\ -1, 1, 3, 5, 7, \tag{149}$$

$$\mathrm{pKK, F2}: p = -1,\ v_0 = 4\alpha,\ \text{indices}\ -7, -1, 1, 10, 12. \tag{150}$$

The singular part operator $\mathcal{D}$ attached to a given family is $\mathcal{D} = v_0 \partial_x$.

The two families have residues which are not opposite, which makes inapplicable the two-singular manifold method [40]. Fortunately, each potential equation possesses in its hierarchy a "minus-one" equation [66]

$$\mathrm{pSK}_{-1}: v_{xxt} + \frac{6}{\alpha} v_x v_t = 0, \tag{151}$$

$$\mathrm{pKK}_{-1}: v_t v_{xxt} - \frac{3}{4} v_{xt}^2 + \frac{6}{\alpha} v_x v_t^2 = 0, \tag{152}$$

which has only one family. The equation (151), initially written by Hirota and Satsuma [24], has already been processed successfully [38, 39] by the one-family method recalled in Sect. 7. This is a strong indication that the same method should also succeed for the KK_{-1} and the KK equations, by restricting to their unique common family F1 (149).

10.3 Truncation with a second order Lax pair

The one-family truncation for pSK and pKK with the assumption

$$v = v_0 \partial_x \log \psi + V, \tag{153}$$

where V is unconstrained and ψ is a solution of the second order Lax pair (88)–(89), provides in both cases the BT (or Miura transformation) linking the solutions of SK and KK equations:

$$\alpha(w + W/2)_x + (w - W)^2 = 0, \tag{154}$$

$$\left[(w - W)(72W_x^2/\alpha^2 + 6W_{xxx}/\alpha) - 72w_x W_{xx}/\alpha - 3W_{xxxx}\right]_x$$
$$+2(w - W)_t = 0,\ w = v_{SK},\ W = v_{KK}, \tag{155}$$

a result previously obtained by Hirota [22]. As this truncation only determines the link between the solutions of two different PDEs and does not provide any arbitrary constant, we repeat the procedure with a third order Lax pair.

10.4 Truncation with a third order Lax pair

With the order three for the Lax pair and the link $\tau = \psi$ associated with the linearizing transformation of the G5 ODE, the process is successful for SK but not for KK. It provides the Lax pair [64, 7] :

$$L = \partial_x^3 + 6\frac{U}{\alpha}\partial_x - \lambda, \tag{156}$$

$$M = \partial_t + (18\frac{U_x}{\alpha} - 9\lambda)\partial_x^2 + (36\frac{U^2}{\alpha^2} - 6\frac{U_{xx}}{\alpha})\partial_x - 36\lambda\frac{U}{\alpha}, \tag{157}$$

$$[L, M] = (6/\alpha)\mathrm{SK}(U), \tag{158}$$

previously obtained in the bilinear formalism by Satsuma and Kaup [53].

We then perform for KK the truncation in the basis $(\psi_x/\psi, \psi_{xx}/\psi - ((\psi_x/\psi)^2 - a)/2)$, associated with the G25 equation (21) and obtain the result $\lambda_t = 0,\ a = -6U/\alpha\ (U = V_x)$ and [41]

$$\frac{\tau_x}{\tau} = \frac{\lambda}{\psi_{xx}/\psi - (1/2)(\psi_x/\psi)^2 + 3(U/\alpha)}, \tag{159}$$

$$L = \partial_x^3 + 6\frac{U}{\alpha}\partial_x + 3\frac{U_x}{\alpha} - \lambda, \tag{160}$$

$$M = \partial_t - 9\lambda\partial_x^2 + (3\frac{U_{xx}}{\alpha} + 36\frac{U^2}{\alpha^2})\partial_x - 3\frac{U_{xxx}}{\alpha} - 72\frac{UU_x}{\alpha^2} - 36\lambda\frac{U}{\alpha}, \tag{161}$$

$$[L, M] = (6/\alpha)\mathrm{KK}(U)\partial_x + (3/\alpha)\mathrm{KK}(U)_x. \tag{162}$$

This is the Lax pair given in [27] and the DT given in [32] with :

$$\tau = \psi\psi_{xx} - (1/2)\psi_x^2 + 3(U/\alpha)\psi^2, \ \tau_x = \lambda\psi^2. \tag{163}$$

10.5 Bäcklund transformation

The two equations which define the BT result from the elimination of ψ between the DT and the scalar Lax pair.

In order to perform this elimination easily, it is convenient to choose one of the two components of the pseudopotential $\mathbf{Y}$ so as to characterize the DT

$$\mathrm{SK} : \frac{v - V}{\alpha} = \frac{\psi_x}{\psi} = Y_1, \tag{164}$$

$$\mathrm{KK} : \frac{2(v - V)}{\alpha} = \frac{\tau_x}{\tau} = Z. \tag{165}$$

The chosen equivalent systems are the canonical projective Riccati system in $(Y_1 = \psi_x/\psi, Y_2 = \psi_{xx}/\psi)$ associated for SK with the third order Lax pair (156)–(157) acting on ψ, and for KK the following system in $(Y_1 = \psi_x/\psi, Z = \tau_x/\tau)$

$$Y_{1,x} = -Y_1^2/2 + \lambda Z^{-1} - 3U/\alpha, \tag{166}$$

$$Z_x = 2Y_1 Z - Z^2, \tag{167}$$

as well as the corresponding system for the t-derivatives of Y_1 and Z explicitly given in [41]. The x-BT of SK results from the elimination of Y_2, which provides the equation for $Y_1 \equiv Y$

$$Y_{xx} + 3YY_x + Y^3 + 6(U/\alpha)Y - \lambda = 0, \tag{168}$$

followed by the substitution $Y = (v - V)/\alpha$

$$x\text{-BT}(v, V; \lambda) \equiv (v - V)_{xx}/\alpha + 3(v - V)(v + V)_x/\alpha^2 + (v - V)^3/\alpha^3 - \lambda = 0. \tag{169}$$

See again in [41] the corresponding expression for the t-part.

The x-BT of KK arises from the elimination of Y_1 between (166) and (167) which writes for $Z \equiv Y$

$$Y_{xx} - (3/4)Y_x^2/Y + 3YY_x/2 + Y^3/4 + 6(U/\alpha)Y - 2\lambda = 0, \tag{170}$$

followed by the substitution $Y = 2(v - V)/\alpha$

$$x\text{-BT}(v, V; \lambda) \equiv (v - V)_{xx}/\alpha - (3/4)(v - V)_x^2/(\alpha(v - V)) \\ +3(v - V)(v + V)_x/\alpha^2 + (v - V)^3/\alpha^3 - \lambda = 0. \quad (171)$$

An analogous procedure for obtaining the t-part of the BT is described in [41].

10.6 Nonlinear superposition formula for Sawada-Kotera

We first derive the SK nonlinear superposition formula previously obtained by Hu and Li [25] with the Hirota bilinear formalism.

Let v_n and $\tilde{v}_n$ be solutions of (145) generated by application of the BT to a known solution v_{n-1} (depending of $n-1$ parameters), with the respective Bäcklund parameters λ_n and λ_{n+1}. Assuming that the permutability theorem [4] is valid, we denote by v_{n+1} the solution generated by application of the BT to v_n with parameter λ_{n+1}, or by application of the BT to $\tilde{v}_n$ with parameter λ_n. This can be represented by the Bianchi diagram Fig. 1, and the first step for obtaining the nonlinear superposition formula is to eliminate the four second derivatives of $v_{n+1}, v_n, \tilde{v}_n, v_{n-1}$ between the four equations equal to the x-part of the BT (169). This yields for v_{n+1} an equation of Riccati type with coefficients depending on $v_n, \tilde{v}_n$ and v_{n-1}:

$$(v_{n+1} - v_{n-1})_x + \frac{1}{\alpha}(v_{n+1}^2 - v_{n-1}^2) \\ +(v_{n+1} - v_{n-1}) \left(\frac{v_{n,x} - \tilde{v}_{n,x}}{\tilde{v}_n - v_n} - \frac{1}{\alpha}(v_n + \tilde{v}_n) \right) = 0. \quad (172)$$

In the tau-function representation

$$v = \alpha \partial_x \log f, \quad (173)$$

the equation (172) linearizes into the second order ODE for f_{n+1}

$$(f_{n-1} f_{n+1,xx} - f_{n+1} f_{n-1,xx})(\tilde{f}_n f_{n,x} - \tilde{f}_{n,x} f_n) \\ +(\tilde{f}_n f_{n,xx} - \tilde{f}_{n,xx} f_n)(f_{n-1} f_{n+1,x} - f_{n+1} f_{n-1,x}) = 0. \quad (174)$$

Introducing the bilinear operator D_x, this last equation is best written as:

$$\partial_x \left(\log D_x(f_{n+1} \cdot f_{n-1}) - \log D_x(f_n \cdot \tilde{f}_n) \right) = 0, \quad (175)$$

which can be integrated twice to yield the general solution:

$$f_{n+1} = K_1(t) f_{n-1} \int^x \frac{D_x(f_n \cdot \tilde{f}_n)}{f_{n-1}^2} \mathrm{d}x + K_2(t) f_{n-1} . \quad (176)$$

where $K_1(t)$ and $K_2(t)$ are two arbitrary functions. The permutability theorem implies that $K_2(t) \equiv 0$.

For $n = 1$, switching to the usual notation

$$f_0 \equiv f_{n-1},\ f_1 \equiv f_n,\ f_2 \equiv \tilde{f}_n,\ f_{12} \equiv f_{n+1}, \tag{177}$$

and starting from $f_0 = 1$, the 2-parameter tau-function $f_{12}(\lambda_1, \lambda_2)$ is obtained from the one-parameter functions $f_1(\lambda_1), f_2(\lambda_2), \lambda_1 \neq \lambda_2$ as:

$$f_{12} = \int^x (f_{1,x} f_2 - f_{2,x} f_1) \mathrm{d}x, \tag{178}$$

setting in (176) $K_1 = 1$, $K_2 = 0$.

The construction of the two-soliton solution then follows. Setting in (169) $V = 0$, and identifying v with v_i, λ with λ_i, for $(i = 1, 2)$, it results from the DT (153) with $\psi \equiv f_i$ solution of the third order Lax pair (156)–(157) for $U \equiv V_x = 0$ that:

$$f_{i,xxx} - \lambda_i f_i = 0, \quad f_{i,t} - 9\lambda_i f_{i,xx} = 0, \tag{179}$$

with the general solution

$$f_i = A_i \mathrm{e}^{p_i x + 9p_i^5 t} + B_i \mathrm{e}^{r_i x + 9r_i^5 t} + C_i \mathrm{e}^{s_i x + 9s_i^5 t}, \tag{180}$$

where A_i, B_i, C_i are arbitrary constants, p_i, r_i, s_i are the three different cubic roots of λ_i subjected to the constraint:

$$p_i^2 + r_i^2 + p_i r_i = 0 \tag{181}$$

as well as two other relations deduced from (181) by cyclic permutation. Defining $k_i = r_i - p_i$, one may easily check, taking account of (181), that $9(r_i^5 - p_i^5) = -k_i^5$. Then, setting $C_i = 0$ in (180), one obtains for the two-soliton solution the well known expressions

$$u_{12} = \alpha \partial_x^2 \log \left[1 + \mathrm{e}^{\theta_1} + \mathrm{e}^{\theta_2} + A_{12}^{\mathrm{SK}} \mathrm{e}^{\theta_1 + \theta_2} \right], \tag{182}$$

$$A_{12}^{\mathrm{SK}} = \frac{(k_1 - k_2)^2 (k_1^2 + k_2^2 - k_1 k_2)}{(k_1 + k_2)^2 (k_1^2 + k_2^2 + k_1 k_2)}, \tag{183}$$

$$\theta_i = k_i x - k_i^5 t + \delta_i, \tag{184}$$

$$\delta_i = \log \frac{B_i (r_i - p_j)(p_i + p_j)}{A_i (r_i + p_j)(p_i - p_j)}, \quad (i, j = 1, 2), i \neq j. \tag{185}$$

10.7 Nonlinear superposition formula for Kaup-Kupershmidt

In the case of the pKK equation (146), the similar elimination of the second derivatives of $v_{n+1}, v_n, \tilde{v}_n, v_{n-1}$ between the four equations equal to the x-BT (171) yields a differential equation of first order and second degree in v_{n+1}:

$$(v_{n+1,x}+\frac{2}{\alpha}v_{n+1}^2-Av_{n+1}+\alpha B)^2=\frac{\alpha^2C^2(v_n-v_{n+1})(\tilde{v}_n-v_{n+1})}{(v_n-v_{n-1})(\tilde{v}_n-v_{n-1})}, \quad (186)$$

with coefficients

$$A=\frac{2}{\alpha}(v_n+\tilde{v}_n)+\frac{\tilde{v}_{n,x}-v_{n,x}}{\tilde{v}_n-v_n},\ B=\frac{2}{\alpha^2}v_n\tilde{v}_n+\frac{\tilde{v}_{n,x}v_n-v_{n,x}\tilde{v}_n}{\alpha(\tilde{v}_n-v_n)}, \quad (187)$$

$$C=B+\frac{1}{\alpha}\left(v_{n-1,x}+\frac{2}{\alpha}v_{n-1}^2-Av_{n-1}\right). \quad (188)$$

Under the transformation $v_{n+1}=(\alpha/2)\partial_x\log f_{n+1}$, equation (186) becomes

$$(f_{n+1,xx}-Af_{n+1,x}+2Bf_{n+1})^2$$
$$=C^2\frac{\alpha^2f_{n+1,x}^2-2\alpha(\tilde{v}_n+v_n)f_{n+1,x}f_{n+1}+4v_n\tilde{v}_nf_{n+1}^2}{(v_n-v_{n-1})(\tilde{v}_n-v_{n-1})}. \quad (189)$$

This equation of the Appell type [3, 10, 11] has the general solution

$$f_{n+1}=c_1^2f_{n+1}^{(1)}+c_1c_2f_{n+1}^{(2)}+c_2^2f_{n+1}^{(3)}, \quad (190)$$

in which c_1,c_2 are arbitrary, and $f_{n+1}^{(j)}, j=1,2,3$, are three independent solutions of the linear third order equation

$$f_{n+1,xxx}+p_1(x)f_{n+1,xx}+p_2(x)f_{n+1,x}+p_3(x)f_{n+1}=0. \quad (191)$$

The coefficients p_1,p_2,p_3 are functions of $f_{n-1},f_n,\tilde{f}_n$, with

$$v_{n-1}=\frac{\alpha}{2}\partial_x\log f_{n-1},\ v_n=\frac{\alpha}{2}\partial_x\log f_n,\ \tilde{v}_n=\frac{\alpha}{2}\partial_x\log\tilde{f}_n. \quad (192)$$

One can show [43] that a particular solution of (191) is:

$$f_{n+1}=\frac{f_n\tilde{f}_n}{f_{n-1}}-f_{n-1}(R_{n+1})^2, \quad (193)$$

$$\text{with}\quad \partial_xR_{n+1}=\sqrt{\left(\frac{f_n}{f_{n-1}}\right)_{,x}\left(\frac{\tilde{f}_n}{f_{n-1}}\right)_{,x}}, \quad (194)$$

which corresponds to the permutability theorem for KK equation.

For $n=1$, switching to the notation (177), and starting from $f_0=1$, equation (191) becomes

$$f_{12,xxx}-\left(B\frac{(f_1f_2)_x}{f_{1,x}f_{2,x}}+\frac{f_{1,x}f_{2,xxx}-f_{2,x}f_{1,xxx}}{f_{1,x}f_{2,xx}-f_{2,x}f_{1,xx}}\right)f_{12,xx}$$
$$+\left(B\frac{f_1f_{2,xx}+f_2f_{1,xx}}{f_{1,x}f_{2,x}}+\frac{f_{1,xx}f_{2,xxx}-f_{2,xx}f_{1,xxx}}{f_{1,x}f_{2,xx}-f_{2,x}f_{1,xx}}\right)f_{12,x}=0, \quad (195)$$

$$B=(f_{2,xx}f_{1,x}-f_{1,xx}f_{2,x})/(2(f_{2,x}f_1-f_{1,x}f_2)). \quad (196)$$

Its general solution in terms of $f_1(\lambda_1), f_2(\lambda_2)$ is

$$f_{12} = K_1(t) + K_2(t) \int^x \sqrt{f_{1,x} f_{2,x}} \\ +K_3(t) \left(f_1 f_2 - \left(\int^x \sqrt{f_{1,x} f_{2,x}} \right)^2 \right). \tag{197}$$

The construction of the two-soliton solution is done as follows. After solving the Lax equations (160)–(161) with $U = 0$ for the vacuum wave functions $\psi_i(\lambda_i)$ $(i = 1, 2)$, the function f_i is given by

$$f_i = \lambda_i \int^x \psi_i^2 \mathrm{d}x, \tag{198}$$

where ψ_i satisfies the linear superposition (180). Choosing $C_i = 0$ in (180), the expression (198) becomes

$$f_i = \lambda_i \left(\frac{A_i^2}{2p_i} \mathrm{e}^{2p_i x + 18 p_i^5 t} + \frac{B_i^2}{2r_i} \mathrm{e}^{2r_i x + 18 r_i^5 t} + \frac{2A_i B_i}{p_i + r_i} \mathrm{e}^{(p_i + r_i)x + 9(p_i^5 + r_i^5)t} \right), \tag{199}$$

and yields the one-soliton solution

$$u_i^{(1)} = \frac{\alpha}{2} \partial_x^2 \log \left(1 + 4\mathrm{e}^{(k_i x - k_i^5 t + \delta_i)} + \mathrm{e}^{2(k_i x - k_i^5 t + \delta_i)} \right), \tag{200}$$

$$\delta_i = \log \left(\frac{B_i p_i}{A_i (p_i + r_i)} \right). \tag{201}$$

Then, taking account of (198), the expression (197) can equivalently be written as

$$K_1 = K_2 = 0, \; K_3 = 1, \; f_{12} = \lambda_1 \lambda_2 \begin{vmatrix} \int^x \psi_1^2 & \int^x \psi_1 \psi_2 \\ \int^x \psi_1 \psi_2 & \int^x \psi_2^2 \end{vmatrix}. \tag{202}$$

Choosing again $C_i = 0$ in (198), the two-soliton tau-function becomes

$$\begin{aligned} f_{12} = {} & A_1^2 A_2^2 \frac{(p_1 - p_2)^2}{4 p_1 p_2 (p_1 + p_2)^2} \mathrm{e}^{2(p_1 + p_2)x + 18(p_1^5 + p_2^5)t} \\ & \times \Big[1 + 4(\mathrm{e}^{\theta_1} + \mathrm{e}^{\theta_2}) + \mathrm{e}^{2\theta_1} + \mathrm{e}^{2\theta_2} \\ & + 8 \frac{2k_1^4 - k_1^2 k_2^2 + 2k_2^4}{(k_1 + k_2)^2 (k_1^2 + k_1 k_2 + k_2^2)} \mathrm{e}^{\theta_1 + \theta_2} \\ & + 4 A_{12}^{SK} (\mathrm{e}^{\theta_1 + 2\theta_2} + \mathrm{e}^{\theta_2 + 2\theta_1}) + \left(A_{12}^{SK} \right)^2 \mathrm{e}^{2(\theta_1 + \theta_2)} \Big], \end{aligned} \tag{203}$$

$$\theta_i = k_i x - k_i^5 t + \Delta_i, \tag{204}$$

$$\Delta_i = \log \left(\frac{B_i p_i (r_i - p_j)(p_i + p_j)}{A_i (p_i + r_i)(r_i + p_j)(p_i - p_j)} \right), \; i \neq j, \tag{205}$$

and the expression for the two-soliton solution of KK is

$$u_{12} = \frac{\alpha}{2}\partial_x^2 \log \Big(1 + 4(e^{\theta_1} + e^{\theta_2}) + e^{2\theta_1} + e^{2\theta_2} + 4A_{12}^{SK}(e^{\theta_1+2\theta_2} + e^{\theta_2+2\theta_1})$$
$$+8\frac{2k_1^4 - k_1^2k_2^2 + 2k_2^4}{(k_1+k_2)^2(k_1^2+k_1k_2+k_2^2)}e^{\theta_1+\theta_2} + \left(A_{12}^{SK}\right)^2 e^{2(\theta_1+\theta_2)}\Big). \tag{206}$$

Now, the expression (202) suggests that the N-parameter tau-function can be written as

$$\tau^{(N)} = \det\left[\int^x \psi_i\psi_j \mathrm{d}x\right]_{1\le i,j\le N}, \tag{207}$$

which indeed coincides with the result obtained in [33] by symmetry reduction of the CKP hierarchy.

The N-soliton solution is obtained by computing the second logarithmic derivative of (207), setting $C_i = 0, 1 \le i \le N$, in the vacuum wave function ψ_i.

The result (202) and the corresponding expression for u_{12} can be directly obtained by iteration of the binary DT associated with the KK equation. Let $(\theta_1; u, \lambda_1)$ and $(\theta_2; u, \lambda_2)$ be solutions of the third order scattering problem

$$\theta_{i,xxx} + 6u\theta_{i,x} + (3u_x - \lambda_i)\theta_i = 0,\ i = 1, 2; \tag{208}$$

then, from (47) and (48), one has

$$\theta_{12} = \theta_2 - \frac{2\lambda_1\theta_1\delta(\theta_1,\theta_2)}{(\lambda_1+\lambda_2)\delta(\theta_1,\theta_1)}, \tag{209}$$

$$u_{12} = u_1 + 3\partial_x^2 \log \delta(\theta_{12},\theta_{12}). \tag{210}$$

Taking account of (51) and (52), one can evaluate the square of θ_{12} in terms of θ_1 and θ_2

$$\theta_{12}^2 = \left(\theta_2 - \frac{\theta_1\int^x\theta_1\theta_2}{\int^x\theta_1^2}\right)^2$$
$$= \theta_2^2 - \partial_x\frac{(\int^x\theta_1\theta_2)^2}{\int^x\theta_1^2}, \tag{211}$$

such that the δ-form for θ_{12} becomes

$$\delta(\theta_{12},\theta_{12}) = 2\lambda_2\left(\int\theta_2^2 - \frac{(\int\theta_1\theta_2)^2}{\int\theta_1^2}\right), \tag{212}$$

and the expression for u_{12} is

$$u_{12} = \frac{1}{2}\partial_x^2 \log\left[\delta(\theta_{12},\theta_{12})\delta(\theta_1,\theta_1)\right]$$
$$= \frac{1}{2}\partial_x^2 \log\left[\int\theta_1^2\int\theta_2^2 - (\int\theta_1\theta_2)^2\right] \tag{213}$$

if one begins the iteration with $u = 0$.

10.8 Tzitzéica equation

For the hyperbolic equation (56), the transformation (66), which is a particular case of the Moutard transformation [36], has been iterated by Schief [55] in a way similar to that of previous section. Let $(\theta_1; u, \lambda_1)$ and $(\theta_2; u, \lambda_2)$ be solutions of the linear triad (57)–(59) with the respective parameters λ_1 and λ_2. These two functions relate the solutions $\tilde{u}_i\,(i = 1, 2)$ and u by

$$e^{\tilde{u}_i} = e^u - 2\partial_x\partial_t \log \theta_i, \;\; i = 1, 2. \tag{214}$$

This transformation can be iterated as follows: building a new solution of the linear triad corresponding to the potential $\tilde{u}_1$ and parameter λ_2

$$\theta_{12} = \theta_2 - \frac{2e^{-u}}{(\lambda_2 - \lambda_1)\theta_1}(\lambda_2\theta_{1,x}\theta_{2,t} - \lambda_1\theta_{1,t}\theta_{2,x}), \tag{215}$$

one obtains, besides $u, \tilde{u}_1, \tilde{u}_2$, a fourth solution of the Tzitzéica equation

$$e^{u_{12}} = e^{\tilde{u}_1} - 2\partial_x\partial_t \log \theta_{12}, \tag{216}$$

or, explicitly

$$e^{u_{12}} = e^u - 2\partial_x\partial_t \log \Big(\theta_1\theta_2 - \frac{2e^{-u}}{(\lambda_2 - \lambda_1)}(\lambda_2\theta_{1,x}\theta_{2,t} - \lambda_1\theta_{1,t}\theta_{2,x})\Big). \tag{217}$$

For obtaining a permutability theorem for the Tzitzéica equation, one must eliminate the eigenfunctions θ_i in favour of $\tilde{u}_i$.
Switching to the tau-representation

$$e^u = 1 - 2\partial_x\partial_t \log f_0, \; e^{\tilde{u}_i} = 1 - 2\partial_x\partial_t \log f_i, \; e^{u_{12}} = 1 - 2\partial_x\partial_t \log f_{12}, \tag{218}$$

and solving (214) for θ_i, one obtains

$$\theta_i = \frac{f_i}{f_0}, \tag{219}$$

where the functions of integration are set equal to zero. Thus (217) becomes

$$\begin{aligned}(\lambda_2 - \lambda_1)(f_{0,xt}f_0 - f_{0,x}f_{0,t} - f_0^2/2)(f_{12}f_0 - f_1f_2) \\ = \lambda_2(f_{1,x}f_0 - f_{0,x}f_1)(f_{2,t}f_0 - f_2f_{0,t}) \\ -\lambda_1(f_{2,x}f_0 - f_{0,x}f_2)(f_{1,t}f_0 - f_1f_{0,t})\end{aligned} \tag{220}$$

an expression solvable for f_{12}. The vacuum for building the two-soliton solution corresponds to $f_0 = 1$. In this case, the functions f_i are solutions of the linear triad (59) for $u = 0$ and $\lambda \equiv \lambda_i$ $i = 1, 2$. Therefore, computing the x-derivative of f_{12} one obtains

$$f_{12,x} = \frac{\lambda_1 + \lambda_2}{\lambda_2 - \lambda_1}(f_{2,x}f_1 - f_{1,x}f_2) \tag{221}$$

the integral of which is identical to (178). Therefore the 2-soliton solution reads

$$e^{u_{12}} - 1 = -2\partial_x\partial_t \log\left[1 + e^{\theta_1} + e^{\theta_2} + A_{12}e^{\theta_1+\theta_2}\right], \tag{222}$$

$$\theta_i = k_i x + 3k_i^{-1}t + \delta_i,\ A_{12} = \frac{(k_1 - k_2)^2(k_1^2 + k_2^2 - k_1 k_2)}{(k_1 + k_2)^2(k_1^2 + k_2^2 + k_1 k_2)}, i = 1, 2 \tag{223}$$

In [9] we consider two possibilities of writing a BT from the third order matrix Lax pair [35] of the Tzitzéica equation. Making the transformation

$$e^U = 1 + W_t, \qquad e^u = 1 + w_t \tag{224}$$

where U and u are two solutions of (56), it follows that W and w satisfy the equation

$$E_{pHS}(w) \equiv w_{txx} - 3(1 + w_t)w_x = 0 \tag{225}$$

the relation between the equation E_{TZI} (56) and E_{pHS} (225) given by [23]

$$\left(E_{TZI}(w)(1 + w_t)^2\right)_x = (1 + w_t)^2 \left(\frac{E_{pHS}(w)}{1 + w_t}\right)_t. \tag{226}$$

The x-BT of (225) writes

$$Y_{xx} + 3YY_x + Y^3 - 3W_xY - \lambda = 0, \quad Y = w - W \tag{227}$$

which is associated, like the x-BT of SK, to the fifth Gambier equation.

Equation (225) which is a model equation for a shallow water wave, has been considered by Hirota and Satsuma [24] which have proved in the framework of the bilinear method that the coupling factor of the 2-soliton solution is similar to the one obtained for the SK equation. Moreover, Nimmo and Willox [46] have shown that the Tzitzéica equation considered as a reduction of a two-dimensional Toda lattice possesses solutions of Pfaffian type.

Acknowledgments

The author would like to thank Prof. Greco for his invitation to this session, the CIME Fundazione for the financial support, and the IUAP Contract No. P4/08 funded by the Belgian government.

References

1. M. J. Ablowitz, D. J. Kaup, A. C. Newell and H. Segur, The inverse scattering transform-Fourier analysis for nonlinear problems, Stud. Appl. Math. **53** (1974) 249–315.

2. R. Anderson, J. Harnad, and P. Winternitz, Systems of ordinary differential equations with nonlinear superposition principles, Physica D **4** (1982) 164–182.
3. P. Appell, Sur les équations différentielles algébriques et homogènes par rapport à la fonction inconnue et à ses dérivées, C. R. Acad. Sc. Paris **104** (1887) 1776–1779.
4. L. Bianchi, Sulla trasformazione di Bäcklund per le superficie pseudosferiche, Rend. Accad. Naz. Lincei **1** (1892) 3–12; *Luigi Bianchi, Opere*, vol. V, 163–173 (Edizioni Cremonese (Unione Matematica Italiana), Roma, 1957).
5. H. H. Chen, General derivation of Bäcklund transformations from inverse scattering problems, Phys. Rev. Lett. **33** (1974) 925–928.
6. R. Conte, Invariant Painlevé analysis of partial differential equations, Phys. Lett. A **140** (1989) 383–390.
7. R. Conte, Towards the equivalence between integrability and Painlevé test for partial differential equations, *Nonlinear and chaotic phenomena in plasmas, solids and fluids*, 94–101, eds. W. Rozmus and J. A. Tuszyński (World Scientific, Singapore, 1991).
8. R. Conte, Chapter 1, this volume.
9. R. Conte M. Musette and A. M. Grundland, Bäcklund transformation of partial differential equations obtained from the Painlevé-Gambier classification. II. Tzitzéica equation, J. Math. Phys. **40** (1999) 2092–2106.
10. C. M. Cosgrove, A new formulation of the field equations for the stationary axisymmetric vacuum gravitational field I. General theory, J. Phys. A **11** (1978) 2389–2404.
11. C. M. Cosgrove and G. Scoufis, Painlevé classification of a class of differential equations of the second order and second degree, Stud. Appl. Math. **88** (1993) 25–87.
12. M. M. Crum, Associated Sturm-Liouville systems, Quart. J. Math. Oxford **6** (1955) 121–127.
13. G. Darboux, Sur une proposition relative aux équations linéaires, C. R. Acad. Sc. Paris **94** (1882) 1456–1459.
14. G. Darboux, Sur les surfaces dont la courbure totale est constante, C. R. Acad. Sc. Paris **97** (1883) 848–850.
15. G. Darboux, *Leçons sur la théorie générale des surfaces et les applications géométriques du calcul infinitésimal*, vol. III (Gauthier-Villars, Paris, 1894). Reprinted, *Théorie générale des surfaces* (Chelsea, New York, 1972). Reprinted (Gabay, Paris, 1993).
16. R. K. Dodd and J. D. Gibbon, The prolongation structure of a higher order Korteweg-de Vries equation, Proc. Roy. Soc. London A **358** (1977) 287–296.
17. A. P. Fordy, The Hénon-Heiles system revisited, Physica D **52** (1991) 204–210.
18. A. P. Fordy and J. Gibbons, Some remarkable nonlinear transformations, Phys. Lett. A **75** (1980) 325–325.
19. B. Gaffet,
20. B. Gambier, Sur les équations différentielles du second ordre et du premier degré dont l'intégrale générale est à points critiques fixes, Thèse, Paris (1909); Acta Math. **33** (1910) 1–55.
21. I. M. Gel'fand and L. A. Dikii, The resolvent and Hamiltonian systems, Funktsional'nyi Analiz i Ego Prilozheniya **11** (1977) 11–27 [English: Funct. Anal. Appl. **11** (1977) 93–105].

22. R. Hirota, Fundamental properties of the binary operators in soliton theory and their generalization, *Dynamical problems in soliton systems*, 42–49, ed. S. Takeno (Springer, Berlin, 1985).
23. R. Hirota and A. Ramani, The Miura transformations of Kaup's equation and of Mikhailov's equation, Phys. Lett. A **76** (1980) 95–96.
24. R. Hirota and J. Satsuma, N-soliton solutions of model equations for shallow water waves, J. Phys. Soc. Japan Letters **40** (1976) 611–612.
25. Hu Xing-biao and Li Yong, Some resuts on the Caudrey-Dodd-Gibbon-Kotera-Sawada equation, J. Phys. A **24** (1991) 3205–3212.
26. B. B. Kadomtsev and V. I. Petviashvili, Dokl. Akad. Nauk SSSR **192** (1970) 753–756 [English: Sov. Phys. Dokl. **15** (1970) 539–?].
27. D. J. Kaup, On the inverse scattering problem for cubic eigenvalue problems of the class $\psi_{xxx} + 6Q\psi_x + 6R\psi = \lambda\psi$, Stud. Appl. Math. **62** (1980) 189–216.
28. Sophie Kowalevski, Sur le problème de la rotation d'un corps solide autour d'un point fixe, Acta Math. **12** (1889) 177–232.
29. G. L. Lamb Jr, Bäcklund transformations for certain nonlinear evolution equations, J. Math. Phys. **15** (1974) 2157–2165.
30. P. D. Lax, Integrals of nonlinear equations of evolution and solitary waves, Comm. Pure Appl. Math. **21** (1968) 467–490.
31. S. B. Leble and N. V. Ustinov, Third order spectral problems: reductions and Darboux transformations, Inverse Problems **10** (1994) 617–633.
32. D. Levi and O. Ragnisco, Non-isospectral deformations and Darboux transformations for the third-order spectral problem, Inverse Problems **4** (1988) 815–828.
33. I. Loris, On reduced CKP equations, Inverse Problems **15** (1999) 1099–1109.
34. V. B. Matveev and M. A. Salle, *Darboux transformations and solitons* (Springer-Verlag, Berlin, 1991).
35. A. V. Mikhailov, Integrability of a two-dimensional generalization of the Toda chain, Pis'ma Zh. Eksp. Teor. Fiz. **30** (1979) 443–448 [English : Soviet Physics JETP Letters **30** (1979) 414–418].
36. Moutard, Sur la construction des équations de la forme $\frac{1}{z}\mathrm{d}^2 z/\mathrm{d}x\mathrm{d}y = \lambda(x,y)$, qui admettent une intégrale générale explicite, Journal de l'École Polytechnique **45** (1878) 1–11.
37. M. Musette, Painlevé analysis for nonlinear partial differential equations, *The Painlevé property, one century later*, 517–572, ed. R. Conte, CRM series in mathematical physics (Springer, Berlin, 1999) (Cargèse school, June 1996).
38. M. Musette and R. Conte, Algorithmic method for deriving Lax pairs from the invariant Painlevé analysis of nonlinear partial differential equations, J. Math. Phys. **32** (1991) 1450–1457.
39. M. Musette and R. Conte, Solitary waves and Lax pairs from polynomial expansions of nonlinear differential equations, *Nonlinear evolution equations and dynamical systems*, 161–170, eds. M. Boiti, L. Martina, and F. Pempinelli (World Scientific, Singapore, 1992),
40. M. Musette and R. Conte, The two–singular manifold method, I. Modified KdV and sine-Gordon equations, J. Phys. A **27** (1994) 3895–3913.
41. M. Musette and R. Conte, Bäcklund transformation of partial differential equations obtained from the Painlevé-Gambier classification, I. Kaup-Kupershmidt equation, J. Math. Phys. **39** (1998) 5617–5630.

42. M. Musette, R. Conte, and C. Verhoeven, Bäcklund transformation and nonlinear superposition formula of the Kaup-Kupershmidt and Tzitzéica equations, CRM Proceedings and Lecture Notes **29** (2001) 1–18.
43. M. Musette and C. Verhoeven, Nonlinear superposition formula for the Kaup-Kupershmidt partial differential equation, Physica D **144** (2000) 211–220.
44. J. J. C. Nimmo, Darboux transformations from reductions of the KP hierarchy, *Nonlinear evolution equations and dynamical systems*, 168–177, eds. V. G. Makhankov, A. R. Bishop, and D. D. Holm (World Scientific, Singapore, 1995).
45. J. J. C. Nimmo and N. C. Freeman, A method of obtaining the N-soliton solution of the Boussinesq equation in terms of a Wronskian, Phys. Lett.A **95** (1983) 4–6.
46. J. J. C. Nimmo and R. Willox, Darboux transformations for the two-dimensional Toda system, Proc. R. Soc. Lond. A **453** (1997) 2497–2525.
47. W. Oevel and W. Schief, Darboux theorems and the KP hierarchy, *Applications of analytic and geometric methods to nonlinear differential equations*, 193–206, ed. P. A. Clarkson (Plenum, New York, 1993).
48. P. Painlevé, Mémoire sur les équations différentielles dont l'intégrale générale est uniforme, Bull. Soc. Math. France **28** (1900) 201–261.
49. A. Pickering, The singular manifold method revisited, J. Math. Phys. **37** (1996) 1894–1927.
50. C. Rogers and S. Carillo, On reciprocal properties of the Caudrey-Dodd-Gibbon and Kaup-Kupershmidt hierarchies, Physica Scripta **36** (1987) 865–869.
51. C. Rogers and W. F. Shadwick, *Bäcklund transformations and their applications* (Academic press, New York, 1982).
52. J. Satsuma, A Wronskian representation of $N-$soliton solutions of nonlinear evolution equations, J. Phys. Soc. Japan Letters **46** (1979) 359–360.
53. J. Satsuma and D. J. Kaup, A Bäcklund transformation for a higher order Korteweg-de Vries equation, J. Phys. Soc. Japan **43** (1977) 692–697.
54. K. Sawada and T. Kotera, A method for finding N-soliton solutions of the K.d.V. equation and K. d. V.-like equation, Prog. Theor. Phys. **51** (1974) 1355–1367.
55. W. Schief, Self-dual Einstein spaces via a permutability theorem for the Tzitzéica equation, Phys. Lett. A **223** (1996) 55–62.
56. G. Tzitzéica, Sur une nouvelle classe de surfaces, C. R. Acad. Sc. Paris **144** (1907) 1257–1259.
57. G. Tzitzéica, Sur une nouvelle classe de surfaces, Rendiconti del Circolo Matematico di Palermo **25** (1908) 180–187.
58. G. Tzitzéica, Sur une nouvelle classe de surfaces, C. R. Acad. Sc. Paris **150** (1910) 955–956.
59. C. Verhoeven and M. Musette, Extended soliton solutions for the Kaup-Kupershmidt equation, J. Phys. A **34** (2001) 2515–2523.
60. C. Verhoeven, M. Musette and R. Conte, Integration of a generalized Hénon-Heiles Hamiltonian, J. Math. Phys. **43** (2002) 1906–1915.
61. H. D. Wahlquist, Bäcklund transformation of potentials of the Korteweg-de Vries equation and the interaction of solitons with cnoidal waves, in *Bäcklund transformations* Lecture Notes in Mathematics **515** (1976), ed. R. M. Miura, pp.162–183.
62. H. D. Wahlquist and F. B. Estabrook, Bäcklund transformation for solutions of the Korteweg-de Vries equation, Phys. Rev. Lett. **31** (1973) 1386–1390.

63. J. Weiss, The Painlevé property for partial differential equations. II: Bäcklund transformation, Lax pairs, and the Schwarzian derivative, J. Math. Phys. **24** (1983) 1405–1413.
64. J. Weiss, On classes of integrable systems and the Painlevé property, J. Math. Phys. **25** (1984) 13–24.
65. J. Weiss, M. Tabor and G. Carnevale, The Painlevé property for partial differential equations, J. Math. Phys. **24** (1983) 522–526.
66. R. Willox, private communication (1995).

Hirota bilinear method for nonlinear evolution equations

Junkichi Satsuma

Graduate School of Mathematical Sciences, The University of Tokyo, 3–8–1 Komaba, Meguro-ku, Tokyo 153–8914, Japan; *Satsuma@Poisson.ms.u-tokyo.ac.jp*

Summary. The bilinear method introduced by Hirota to obtain exact solutions for nonlinear evolution equations is discussed. Firstly, several examples including the Korteweg-deVries, nonlinear Schrödinger and Toda equations are given to show how solutions are derived. Then after considering multi-dimensional systems such as the Kadomtsev-Petviashvili, two dimensional Toda and Hirota-Miwa equations, the algebraic structure of such nonlinear evolution systems is explained. Finally, extensions of the method including q-analogue, ultra-discrete systems and trilinear forms are also presented.

1 Introduction

The bilinear method, which originates with Hirota, now almost 30 years ago, has played a crucial role in the study of integrable nonlinear systems. The formalism is perfectly suitable for obtaining not only multi-soliton solutions but also several types of special solutions of many nonlinear evolution equations. Moreover, it has been used for the investigation of the algebraic structure of integrable evolution equations as well as for obtaining extensions of such systems.

In this article, we attempt to present a brief survey of the bilinear formalism. At the same time however we shall also try to present several recent developments. The main focus will always lie with obtaining explicit solutions of various classes of nonlinear evolution equations. Section 2 is devoted to an explanation of the use of the bilinear procedure for obtaining soliton solutions. Several examples including the Korteweg-deVries (KdV) equation, the nonlinear Schrödinger (NLS) equation and the Toda equation are given. In the bilinear method, the notion of a dependent variable transformation is crucial and the transformed variable turns out to be a key function. We shall call it the τ function. For multi-soliton solutions, it is written in the form of a polynomial in exponential functions.

In Sect. 3, we consider multi-dimensional soliton equations. By using the fact that τ functions can be expressed in terms of Wronski determinants or Casorati determinants, we show that the τ function of soliton equations satisfies special algebraic identities in bilinear form. This result reflects the richness of the algebraic structure which is common to most, if not all, soliton

J. Satsuma, Hirota bilinear method for nonlinear evolution equations, Lect. Notes Phys. **632**, 171–222 (2003)
http://www.springerlink.com/

equations. Actually, the so called Kyoto school has developed a grand theory explaining this algebraic structure: it is commonly referred to as *Sato theory.* In Sect. 4, we first survey Sato theory, thereby revealing the importance of the τ function. Afterwards we introduce the so-called Fermion analysis based on the connection between Sato theory and infinite dimensional Lie algebras.

Finally in Sect. 4, we discuss a number of extensions of the bilinear formalism. The first one is the q-analogue of soliton equations. It will be shown that q-soliton equations naturally arise in the bilinear formalism. Main focus will be on the q-Toda equations. The second one is special function solutions for soliton equations. It will be shown that the class of solutions satisfy nonautonomous equations which are obtained by reduction of the discrete Kadomtsev-Petviashvili (dKP) equation. The third is the extension to ultra-discrete systems. It will be shown that the ideas of the bilinear formalism are also applicable to cellular automata, i.e. to (time) evolution systems in which all the variables are discrete. Finally we shall discuss the trilinear formalism which gives a multi-dimensional extension of the soliton equations.

2 Soliton solutions

The key point of the bilinear method lies in finding a suitable dependent variable transformation. Let us first illustrate this on a simple case.

2.1 The Burgers equation

The Burgers equation,

$$u_t + uu_x = u_{xx}, \tag{1}$$

under the Cole-Hopf transformation $u = -2(\log f)_x$, is mapped to

$$f_t = f_{xx} + c(t)f,$$

in which $c(t)$ is an arbitrary integration function. The change of variable $f = e^{\int^t c(t)dt}F$ converts it to the linear equation $F_t = F_{xx}$, which admits the particular solutions $F(x,t) = \sum_p e^{-px+p^2t}$. Among them the simplest nontrivial one is

$$F = 1 + e^{-px+p^2t},$$

or in the original variable:

$$u = 2p\frac{e^\eta}{1+e^\eta}, \quad \eta = -px + p^2t.$$

This is a wave travelling towards $+\infty$ for $p \geq 0$. The superposition $F = 1 + e^{-px+p^2t} + e^{-qx+q^2t}$ gives a nonstationary solution showing fusion of two travelling waves.

We want to stress that in the case of the Burgers equation, the Cole-Hopf is a *linearizing* transformation.

2.2 The Korteweg-de Vries equation

Let us apply the Cole-Hopf-type transformation $u = 2(\log f)_{xx}$ to the KdV equation,

$$u_t + 6uu_x + u_{xxx} = 0. \tag{2}$$

After one integration, we have

$$f_{xt}f - f_x f_t + f_{xxxx}f - 4f_{xxx}f_x + 3f_{xx}^2 = 0, \tag{3}$$

which is not linear but quadratic. This bilinear equation can be written in a "nicer" form if we use the following operators.

Definition 2.1. *The Hirota bilinear operator D_x is defined by*

$$D_x^n D_t^m a \cdot b = \left(\frac{\partial}{\partial_x} - \frac{\partial}{\partial_{x'}}\right)^n \left(\frac{\partial}{\partial_t} - \frac{\partial}{\partial_{t'}}\right)^m a(x,t)b(x',t')|_{x'=x,t'=t}. \tag{4}$$

For example:

$$\begin{aligned}
D_x a \cdot b &= a_x b - ab_x, \\
D_x^2 a \cdot b &= a_{xx}b - 2a_x b_x + ab_{xx}, \\
D_x^3 a \cdot b &= a_{xxx}b - 3a_{xx}b_x + 3a_x b_{xx} - ab_{xxx}.
\end{aligned}$$

Equation (3) then becomes

$$(D_x D_t + D_x^4) f \cdot f = 0, \tag{5}$$

which we call the bilinear form of the KdV equation.

To obtain particular solutions for the KdV equation, Hirota [1] used the formal perturbation,

$$f = 1 + \varepsilon f_1 + \varepsilon^2 f_2 + \ldots,$$

$$(D_x D_t + D_x^4)\left((1 + \varepsilon f_1 + \varepsilon^2 f_2 + \ldots) \cdot (1 + \varepsilon f_1 + \varepsilon^2 f_2 + \ldots)\right) = 0.$$

At various powers of ε one finds:

$$\begin{aligned}
O(\varepsilon) &: \mathcal{L} f_1 \equiv 2(\partial_x \partial_t + \partial_x^4) f_1 = 0, \\
O(\varepsilon^2) &: \mathcal{L} f_2 = (D_x D_t + D_x^4) f_1 \cdot f_1 = 0, \\
O(\varepsilon^3) &: \mathcal{L} f_3 = -2(D_x D_t + D_x^4) f_1 \cdot f_2 = 0, \\
&\vdots
\end{aligned}$$

One particular solution for the equation in $O(\varepsilon)$ is given by

$$f_1 = \sum_{j=1}^{N} \exp \eta_j, \quad \eta_j = p_j(x - p_j^2 t) + \eta_j^{(0)}.$$

In the case $N = 1$ we have $f_1 = 1 + e^{\eta_1}$ in which case we can take $f_2 = f_3 = \cdots = 0$ to obtain

$$u = 2(\log f)_{xx} = \frac{1}{2} p_1^2 \operatorname{sech}^2 \frac{1}{2} \{p_1(x - p_1^2 t) + \eta_1^{(0)}\}.$$

This is nothing but the one soliton solution of the KdV equation.
For $N = 2$, one takes

$$f_1 = e^{\eta_1} + e^{\eta_2}, \quad f_2 = e^{\eta_1 + \eta_2 + A_{12}},$$

with $e^{A_{12}} = [(p_1 - p_2)/(p_1 + p_2)]^2$. Then we can take $f_j = 0, \ j \geq 3$ to obtain the two-soliton solution showing a nonlinear interaction of two solitons. Similarly for arbitrary N, we obtain an exact solution exhibiting a multiple collision of solitons, which we call the N-soliton solution. Note that a useful formula in the above procedure is

$$D_x^n \left(e^{\alpha x} \cdot e^{\beta x} \right) = (\alpha - \beta)^n e^{(\alpha + \beta) x}.$$

2.3 The nonlinear Schrödinger equation

The bilinear method is applicable to a much wider class of equations than the above. For the NLS equation,

$$i\psi_t + \psi_{xx} + 2|\psi|^2 \psi = 0, \tag{6}$$

we substitute $\psi = \dfrac{g}{f}$, $\overline{\psi} = \dfrac{\overline{g}}{f}$, where f is a real function and $\overline{g}$ is the complex conjugate of g [2]. Then we obtain

$$\frac{i}{f^2}(g_t f - g f_t) + \frac{1}{f^2}(g_{xx} f - g f_{xx}) - \frac{2f_x(g_x f - g f_x)}{f^3} + 2\frac{g^2 \overline{g}}{f^3} = 0,$$

i.e.

$$i(g_t f - g f_t) + g_{xx} f - 2g_x f_x + g f_{xx} - 2\frac{g}{f}(f_{xx} f - f_x^2 - g\overline{g}) = 0.$$

Since we have introduced three fields $f, g, \overline{g}$, we need to decouple these two (real) equations into three (hopefully bilinear) equations. This results in

$$\begin{aligned} &(iD_t + D_x^2) g \cdot f = 0, \\ &(-iD_t + D_x^2) \overline{g} \cdot f = 0, \\ &D_x^2 f \cdot f = 2g\overline{g}. \end{aligned}$$

When it comes to constructing soliton solutions, an argument similar to the one used in the case of the KdV equation, together with parity considerations, yields the Ansatz

$$\begin{aligned} f &= 1 + \varepsilon^2 f_2 + \varepsilon^4 f_4 + \dots, \\ g &= \varepsilon g_1 + \varepsilon^3 g_3 + \dots, \\ g_1 &= \sum_{j=1}^{N} e^{\eta_j}, \quad \eta_j = P_j x + i P_j^2 t + \eta_j^{(0)}, \quad P_j,\ \eta_j^{(0)} \in \mathcal{C}. \end{aligned}$$

The choice

$$g = e^{\eta}, \quad f = 1 + \frac{1}{(P + \overline{P})^2} e^{\eta + \overline{\eta}}, \quad \text{with } P = p + ik$$

gives the (bright) one-soliton solution,

$$\psi = p \operatorname{sech} p(x - 2kt) e^{i(kx - (k^2 - p^2)t)}.$$

2.4 The Toda equation

The Toda equation,

$$\ddot{y}_n = e^{-(y_n - y_{n-1})} - e^{-(y_{n+1} - y_n)} \tag{7}$$

is a famous differential-difference system which models a lattice with non-linear interactions between the nodes. Let us introduce the new variable $r_n = y_n - y_{n-1}$. Then eq.(7) is written as

$$\ddot{r}_n = -e^{-r_{n-1}} + 2e^{-r_n} - e^{-r_{n+1}}.$$

If we further introduce $V_n = e^{-r_n} - 1$, then we have

$$\frac{\mathrm{d}^2}{\mathrm{d}t^2} \log(1 + V_n) = V_{n-1} - 2V_n + V_{n+1}. \tag{8}$$

The substitution $V_n = \frac{\mathrm{d}^2}{\mathrm{d}t^2} \log \tau_n$ with the boundary condition $\lim_{|n|\to\infty} V_n = 0$ yields after two integrations in t:

$$\log(1 + \frac{\mathrm{d}^2}{\mathrm{d}t^2} \log \tau_n) = \log \frac{\tau_{n+1}\tau_{n-1}}{\tau_n^2},$$

or

$$\ddot{\tau}_n \tau_n - \dot{\tau_n}^2 \equiv \frac{1}{2} D_t^2 \tau_n \cdot \tau_n = \tau_{n+1}\tau_{n-1} - \tau_n^2. \tag{9}$$

This is the bilinear form of the Toda equation [3].

Let us introduce the difference operator (or translation operator)

$$e^{\varepsilon\partial_x} f(x) = f(x + \varepsilon) \quad \text{or} \quad e^{\partial_n} f_n = f_{n+1}$$

and its bilinear extension

$$e^{D_n} f_n \cdot f_n = e^{\partial_n - \partial_{n'}} (f_n f_{n'})|_{n'=n} = f_{n+1} f_{n-1}.$$

Then eq.(9) is written as

$$\left(D_t^2 - 4\sinh^2 \frac{D_n}{2} \right) \tau_n \cdot \tau_n = 0.$$

Just as for the KdV equation, we can obtain soliton solutions to this equation by applying a formal perturbation procedure. The one-soliton solution is given by

$$\tau_n = 1 + e^{2\eta}, \quad \eta = pn - \Omega t + \eta^{(0)}, \quad \Omega^2 = \sinh p^2, \quad V_n = \Omega^2 \operatorname{sech}^2 \eta,$$

where p is an arbitrary parameter.

2.5 Painlevé equations

Other important ODE's can be bilinearized as well. Consider for example the second Painlevé equation P_2

$$\frac{\mathrm{d}^2 W}{\mathrm{d}x^2} = 2W^3 - 2xW + \alpha, \tag{10}$$

for $\alpha = 2N+1$, with N integer. The transformation

$$W = \left(\log \frac{\tau_{N+1}}{\tau_N}\right)_x$$

converts it to the bilinear system

$$D_x^2\, \tau_{N+1} \cdot \tau_N = 0,$$
$$(D_x^3 + 2xD_x - \alpha)\, \tau_{N+1} \cdot \tau_N = 0.$$

We obtain special solutions of the Painlevé equation through this bilinear system. It is remarked that a similar idea applies to the discrete Painlevé equations [4].

2.6 Difference vs differential

When using the bilinear procedure for obtaining soliton solutions, one rapidly notices that difference systems are much more tractable than differential systems. It is a universal fact that the algebraic structure in the former type is easier to understand than in the latter one. Here we show some simple examples of the correspondence between difference and differential systems.

The difference equation,

$$u(t+\Delta t) - u(t) = \alpha \Delta t\, u(t), \quad u(0) = u_0, \tag{11}$$

is a simple model system describing population dynamics called Malthus' law. The solution is easily obtained algebraically as

$$u(\Delta t) = (1+\alpha\Delta t)u_0,$$
$$\ddots,$$
$$u(n\Delta t) = (1+\alpha\Delta t)^n u_0.$$

If we introduce $\tau = n\Delta t$ and take the limit, $n \to \infty,\ \Delta t \to 0,\ \tau$ finite $\neq 0$, then this solution is reduced to the exponential function, i.e.:

$$u(\tau) = \{(1+\frac{\alpha}{n}\tau)^{n/\alpha\tau}\}^{\alpha\tau} u_0 \to e^{\alpha\tau} u_0.$$

The exponential function is the solution of the differential equation,

$$\frac{du(t)}{dt} = \alpha u(t),$$

which is the continuum analogue of eq.(11).

Let us move to the multi-variable case. The partial difference equation,

$$u(x, t + \Delta t) = \frac{1}{2}\{u(x - \Delta x, t) + u(x + \Delta x, t)\} \tag{12}$$

describes a random walk in one dimension. If we start with the initial condition,

$$u(0, 0) = 1, \quad u(x, 0) = 0 \quad \text{for} \quad x \neq 0, \tag{13}$$

then the solution diffuses obeying a binomial distribution. It is noted that the propagation speed is finite in this discrete model.

By the Taylor expansion of eq.(12), we have

$$\frac{\partial u}{\partial t} = \frac{(\Delta x)^2}{2\Delta t}\frac{\partial^2 u}{\partial x^2} + O\left(\Delta t, \frac{(\Delta x)^4}{\Delta t}\right).$$

If one assumes the coefficient $\frac{(\Delta x)^2}{2\Delta t}$ to be constant then the limit is a diffusion equation. By Fourier analysis we know that the solution with the initial condition corresponding to (13) diffuses, obeying a normal distribution. Notice that the propagating speed now becomes infinite in this continuum model.

The two-dimensional random walk is given by the difference scheme,

$$\begin{aligned} u(x, y, t + \Delta t) = \frac{1}{4}\{&u(x + \Delta x, y, t) + u(x - \Delta x, y, t) \\ &+ u(x, y + \Delta y, t) + u(x, y - \Delta y, t)\} \end{aligned} \tag{14}$$

which admits the limit

$$\frac{\partial u}{\partial t} = \frac{\partial^2 u}{\partial x^2} + \frac{\partial^2 u}{\partial y^2} = \Delta u.$$

The correspondence between the solutions of this difference system and its differential partner is essentially the same as in the one-dimensional case. If we now drop the t-dependence in the above equation, we obtain the Laplace equation, $\Delta u = 0$. The famous "maximum principle" for this equation states that, inside a domain, the solution u has no extremum and thus that any extremum must occur on the boundary of that domain. This fact is obvious on the difference equation.

The simplest imaginable (but nontrivial) partial difference scheme is:

$$u(x,t+\Delta t)=u(x-\Delta x,t). \tag{15}$$

This is a discrete analogue of the wave equation,

$$\frac{\partial u}{\partial t}=-c\frac{\partial u}{\partial x}$$

Both have the same general solution $u=f(x-ct)$, i.e. a wave propagating in + directions for positive c.

Similarly, the difference equation,

$$u(x,t+\Delta t)+u(x,t-\Delta t)=u(x+\Delta x,t)+u(x-\Delta x,t) \tag{16}$$

is a discrete analogue of the wave equation (d'Alembert equation):

$$\frac{\partial^2 u}{\partial t^2}=c^2\frac{\partial^2 u}{\partial x^2}.$$

Both systems have the general solution $u=f(x-ct)+g(x-ct)$.

So far we have seen that for linear systems, solutions of difference equations and differential equations are essentially the same, but for nonlinear systems this is no longer the case. For instance, the logistic equation

$$\frac{\mathrm{d}u}{\mathrm{d}t}=\alpha(1-\beta u)u, \tag{17}$$

with the initial condition $u(0)=u_0$, integrates as

$$u=\frac{u_0}{\beta u_0+(1-\beta u_0)e^{-\alpha t}},$$

which is analytic. It is however well known that a naive discretization of (17) as the logistic map

$$u_{n+1}=\alpha(1-\beta u_n)u_n,$$

is chaotic for $\beta=1,\ \alpha>3.57$. On the contrary, the discretization

$$u(t+\Delta t)-u(t)=\alpha\Delta t(1-\beta u(t))u(t+\Delta t),$$

under the Möbius transformation $w(t)=\dfrac{1-\beta u(t)}{u(t)}$, is mapped onto the linear equation

$$w(t+\Delta t)-w(t)=-\alpha\Delta t w(t),$$

which is integrated as

$$w(n\Delta t) = (1-\alpha\Delta t)^n w(0),$$
$$u(n\Delta t) = \frac{u_0}{\beta u_0 + (1-\beta u_0)(1-\alpha\Delta t)^n}, \quad u_0 = u(0).$$

These examples indicate that the way of discretization affects very much the structure of solutions for discrete nonlinear systems.

3 Multidimensional equations

3.1 The Kadomtsev-Petviashvili equation

The Kadomtsev-Petviashvili (KP) equation is a 2+1 dimensional soliton equation, usually written as:

$$\left(u_t - \frac{1}{4}u_{xxx} - 3uu_x\right)_x - \frac{3}{4}u_{yy} = 0. \tag{18}$$

Under the transformation $u = (\log\tau)_{xx}$, it takes on the bilinear form [5]:

$$(4D_xD_t - D_x^4 - 3D_y^2)\tau\cdot\tau = 0. \tag{19}$$

Its one-soliton solution is defined by

$$\tau = e^{\eta_1} + e^{\xi_1}, \tag{20}$$

with $\eta_j = p_jx + p_j^2y + p_j^3t + \eta_j^{(0)}$, $\xi_j = q_jx + q_j^2y + q_j^3t + \xi_j^{(0)}$, or equivalently by

$$\tau = 1 + e^{\eta_1 - \xi_1},$$

which results in a solution (to the original equation) of the form:

$$u = \frac{\partial^2}{\partial x^2}\log\{1 + e^{(q_1-p_1)x + (q_1^2-p_1^2)y + (q_1^3-p_1^3)t + \xi^{(0)}}\}.$$

The reduction $q_1 = -p_1$ yields the KdV one-soliton, in accordance with the fact that if we drop the y dependence in equation (18) (or in its bilinear form), we obtain the KdV equation (or its bilinear form).

A rather general solution to the KP equation in bilinear form (19) is given by the Wronskian determinant [6]

$$\tau = \det(\partial_x^{i-1}f^{(j)}), \quad i,j = 1,\cdots,N,$$

in which each function $f^{(j)}$ satisfies the linear system

$$\partial_y f = \partial_x^2 f, \ \partial_t f = \partial_x^3 f.$$

The N-soliton solutions are obtained by choosing the $f^{(j)}$ to be sums of exponentials such as in (20).

3.2 The two-dimensional Toda lattice equation

The two-dimensional extension of the Toda lattice (7) is given by:

$$\frac{\partial^2}{\partial x \partial y} \log(1+V_n) = V_{n-1} - 2V_n + V_{n+1}. \tag{21}$$

The transformation

$$V_n = \frac{\partial^2}{\partial x \partial y} \log \tau_n,$$

with the boundary conditions $\lim_{n\to\pm\infty} V_n = 0$, yields the bilinear form

$$D_x D_y \tau_n \cdot \tau_n = 2(\tau_{n+1}\tau_{n-1} - \tau_n^2), \tag{22}$$

which already appeared in the book of Darboux [7].

The one-soliton solution is expressed as:

$$\begin{aligned}
\tau_n &= p^n e^{px - \frac{y}{p}} + q^n e^{qx - \frac{y}{p}}, \\
V_n &= \frac{\partial^2}{\partial x \partial y} \log p^n e^{px - \frac{y}{p}} \{1 + (\frac{q}{p})^n e^{(q-p)x - (\frac{1}{q} - \frac{1}{p})y}\} \\
&= \frac{\partial^2}{\partial x \partial y} \log\{1 + (\frac{q}{p})^n e^{(q-p)x - (\frac{1}{q} - \frac{1}{p})y}\}.
\end{aligned}$$

The reduction $q = e^P, p = e^{-P}$, which eliminates the $x - y$ dependence, yields

$$V_n = \frac{\partial^2}{\partial x \partial y} \log\{1 + e^{2Pn + 2\sinh P(x+y)}\},$$

which is the (one-dimensional) Toda soliton in the variable $t = x + y$ (the reduction to the one-dimensional Toda equation (7) can therefore be thought of as setting $x = y$ in the 2D Toda equation).

The N-soliton solution is given by a Casorati determinant (a discrete analogue of the Wronskian determinant) [8]

$$\tau_n = \det(f_{n+j}^{(i)}), \; i = 1, \cdots, N, \; j = 0, \cdots, N-1, \tag{23}$$

in which the functions satisfy the linear system

$$\frac{\partial}{\partial x} f_m^{(i)} = f_{m+1}^{(i)}, \quad \frac{\partial}{\partial y} f_m^{(i)} = -f_{m-1}^{(i)}.$$

We shall now explain how to prove this fact. A similar proof (but then for the existence of Wronskian solutions for the KP equation) will be discussed in the section on Sato theory.

For $N = 1, \tau_n = f_n^{(1)}$, the lhs of the (bilinear) 2D Toda equation (22) evaluates to

$$\begin{aligned} &D_x D_y \tau_n \cdot \tau_n \\ &= 2(f_{n,xy}^{(1)} f_n^{(1)} - f_{n,x}^{(1)} f_{n,y}^{(1)}) \\ &= -2(f_n^{(1)} f_n^{(1)} - f_{n+1}^{(1)} f_{n-1}^{(1)}) = 2(\tau_{n+1}\tau_{n-1} - \tau_n^2), \end{aligned}$$

which indeed equals the rhs.

For $N = 2$, one first notices the identity

$$0 \equiv \begin{vmatrix} a_0 & a_1 & a_2 & a_3 \\ b_0 & b_1 & b_2 & b_3 \\ 0 & a_1 & a_2 & a_3 \\ 0 & b_1 & b_2 & b_3 \end{vmatrix}$$

$$= \begin{vmatrix} a_0 & a_1 \\ b_0 & b_1 \end{vmatrix} \begin{vmatrix} a_2 & a_3 \\ b_2 & b_3 \end{vmatrix} - \begin{vmatrix} a_0 & a_2 \\ b_0 & b_2 \end{vmatrix} \begin{vmatrix} a_1 & a_3 \\ b_1 & b_3 \end{vmatrix} + \begin{vmatrix} a_0 & a_3 \\ b_0 & b_3 \end{vmatrix} \begin{vmatrix} a_1 & a_2 \\ b_1 & b_2 \end{vmatrix},$$

obtained by Laplace expansion of the original 4×4 determinant (which is of rank 2). Let us (schematically) rewrite this Laplace expansion in a form which, with hindsight, can be identified as a Plücker relation:

$$0 = \begin{vmatrix} 0 & 1 \end{vmatrix} \begin{vmatrix} 2 & 3 \end{vmatrix} - \begin{vmatrix} 0 & 2 \end{vmatrix} \begin{vmatrix} 1 & 3 \end{vmatrix} + \begin{vmatrix} 0 & 3 \end{vmatrix} \begin{vmatrix} 1 & 2 \end{vmatrix},$$

or, introducing the bilinear notation of Sato (the so called Maya-diagram),

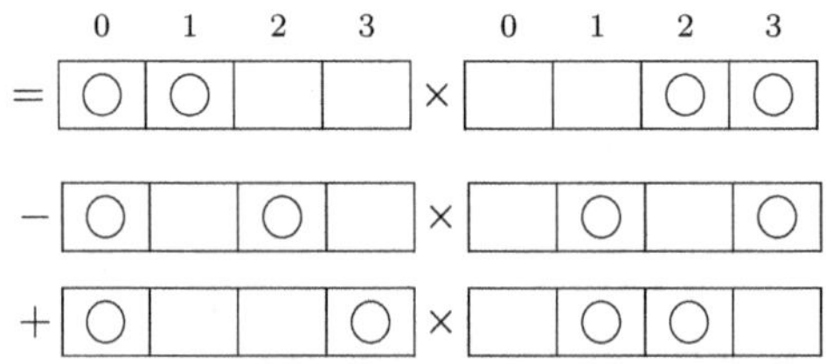

For the two-soliton, let us denote τ_n as

$$\tau_n = \begin{vmatrix} f_n^{(1)} & f_{n+1}^{(1)} \\ f_n^{(2)} & f_{n+1}^{(2)} \end{vmatrix} = \text{notation} : \begin{vmatrix} 0 & 1 \end{vmatrix},$$

and consequently:

$$\tau_{n+1} = \begin{vmatrix} 1\ 2 \end{vmatrix}, \qquad \tau_{n-1} = \begin{vmatrix} -1\ 0 \end{vmatrix}, \qquad \tau_{n,x} = \begin{vmatrix} 0\ 2 \end{vmatrix},$$
$$\tau_{n,y} = -\begin{vmatrix} -1\ 1 \end{vmatrix}, \qquad \tau_{n,xy} = -\begin{vmatrix} -1\ 2 \end{vmatrix} - \begin{vmatrix} 0\ 1 \end{vmatrix}.$$

In this notation, the 2D Toda equation becomes

$$\begin{vmatrix} -1\ 2 \end{vmatrix}\begin{vmatrix} 0\ 1 \end{vmatrix} - \begin{vmatrix} 0\ 2 \end{vmatrix}\begin{vmatrix} -1\ 1 \end{vmatrix} + \begin{vmatrix} 1\ 2 \end{vmatrix}\begin{vmatrix} -1\ 0 \end{vmatrix} = 0$$

or, shifting the notation by one unit:

$$\begin{vmatrix} 0\ 3 \end{vmatrix}\begin{vmatrix} 1\ 2 \end{vmatrix} - \begin{vmatrix} 1\ 3 \end{vmatrix}\begin{vmatrix} 0\ 2 \end{vmatrix} + \begin{vmatrix} 2\ 3 \end{vmatrix}\begin{vmatrix} 0\ 1 \end{vmatrix} = 0.$$

This is nothing but the Plücker identity which we used to represent the above Laplace expansion.

For $N \geq 3$ one has essentially the same Plücker identity. For example at $N = 3$ one starts with

$$\begin{vmatrix} f & a_0 & a_1 & 0 & a_2 & a_3 \\ g & b_0 & b_1 & 0 & b_2 & b_3 \\ h & c_0 & c_1 & 0 & c_2 & c_3 \\ 0 & 0 & a_1 & f & a_2 & a_3 \\ 0 & 0 & b_1 & g & b_2 & b_3 \\ 0 & 0 & c_1 & h & c_2 & c_3 \end{vmatrix} = 0,$$

which has a Laplace expansion which can still be represented by:

$$\begin{vmatrix} 0\ 1 \end{vmatrix}\begin{vmatrix} 2\ 3 \end{vmatrix} - \begin{vmatrix} 0\ 2 \end{vmatrix}\begin{vmatrix} 1\ 3 \end{vmatrix} + \begin{vmatrix} 0\ 3 \end{vmatrix}\begin{vmatrix} 1\ 2 \end{vmatrix} = 0,$$

suggesting that the same Plücker relation will hold for all sizes of determinants. This actually turns out to be the case and hence the Casorati determinants (23) solve the 2D Toda equation for all sizes. The importance of the Plücker relation will become apparent when we discuss Sato theory for the KP equation.

Remark.

The following periodic reduction of the 2D Toda equation:

$$V_n = V_{n+2} \quad \text{in} \qquad \frac{\partial^2}{\partial x \partial y} \log(1 + V_n) = V_{n-1} - 2V_n + V_{n+1},$$

is seen to be equivalent to the sinh-Gordon equation

$$u_{xy} = 2(e^v - e^u), v_{xy} = 2(e^u - e^v),$$

or

$$u_{xy} = 2(e^{-2} - e^u) = -4\sinh u,$$

if we introduce the variables $1 + V_{2n} = e^u, \quad 1 + V_{2n+1} = e^v$.

3.3 Two-dimensional Toda molecule equation

The two-dimensional Toda molecule equation is the system

$$\frac{\partial}{\partial x} V_n = V_n(I_n - I_{n+1}), \tag{24}$$

$$\frac{\partial}{\partial y} I_n = V_{n-1} - V_n, \tag{25}$$

with the boundary conditions $V_0 = V_M = 0$. The transformation (the same as for the Toda lattice) $V_n = \dfrac{\partial^2}{\partial x \partial y} \log \tau_n$ takes it to bilinear equation [6]

$$D_x D_y \tau_n \cdot \tau_n = 2\tau_{n+1}\tau_{n-1}, \quad \text{with } \tau_{-1}\tau_{M+1} = 0. \tag{26}$$

A solution is given by

$$\tau_n = \det\left(\frac{\partial^{i+j} f}{\partial_x^i \partial_y^j}\right), \qquad i = 0, \cdots, n-1, \; j = 0, \cdots, n-1, \quad \tau_0 = 1, \tag{27}$$

with $f(x,y) = \sum_{k=1}^{M} f_k(x) g_k(y)$, in which f_k and g_k are arbitrary functions. This requirement is necessary to satisfy the boundary conditions. The 2D Toda molecule equation has no continuum limit.

If one sets out to prove the existence of the above solutions, at $n = 1$ it is easy to see that:

$$D_x D_y \tau_1 \cdot \tau_1 = f_{xy} f - f_x f_y = \begin{vmatrix} f & f_x \\ f_y & f_{xy} \end{vmatrix} = \tau_2 \cdot \tau_0.$$

For $n = 2$, we find

$$\begin{vmatrix} f & f_{xx} \\ f_{yy} & f_{xxyy} \end{vmatrix} \begin{vmatrix} f & f_x \\ f_y & f_{xy} \end{vmatrix} - \begin{vmatrix} f & f_{xx} \\ f_y & f_{xxy} \end{vmatrix} \begin{vmatrix} f & f_x \\ f_{yy} & f_{xyy} \end{vmatrix} = \begin{vmatrix} f & f_x & f_{xx} \\ f_y & f_{xy} & f_{xxy} \\ f_{yy} & f_{xyy} & f_{xxyy} \end{vmatrix} f.$$

which is nothing but a Jacobi identity. For $n \geq 3$, one again applies the Jacobi identity, schematically denoted as:

$$D\begin{vmatrix} n \\ \\ n \end{vmatrix} D \begin{vmatrix} n+1 \\ \\ n+1 \end{vmatrix} - D \begin{vmatrix} n \\ \\ n+1 \end{vmatrix} D \begin{vmatrix} n+1 \\ \\ n \end{vmatrix} = D \begin{vmatrix} n\ n+1 \\ \\ n\ n+1 \end{vmatrix} D,$$

showing that the determinants (27) solve (26).

3.4 The Hirota-Miwa equation

Discretizing the 2D Toda equation one obtains [9],[10]:

$$\begin{aligned} &\tau_n(l+1,m+1)\tau_n(l,m) - \tau_n(l+1,m)\tau_n(l,m+1) \\ &= ab\{\tau_{n+1}(l,m+1)\tau_{n-1}(l+1,m) - \tau_n(l+1,m+1)\tau_n(l,m)\}. \end{aligned} \tag{28}$$

The variable x is discretized as l, y as m, the mesh sizes being denoted as a and b.

A typical solution is given by the Casorati determinant

$$\tau_n = \det(f^{(i)}_{n+j-1}), \quad i = 0, \cdots, N-1, \quad j = 1, \cdots, N, \tag{29}$$

in which the N functions $f^{(i)}$ obey the linear discrete system

$$\begin{aligned} \Delta_l f^{(i)}_k(l,m) &= \frac{1}{a}\left(f^{(i)}_k(l,m+1) - f^{(i)}_k(l,m)\right) = f^{(i)}_{k+1}(l,m), \\ \Delta_m f^{(i)}_k(l,m) &= \frac{1}{b}\left(f^{(i)}_k(l,m+1) - f^{(i)}_k(l,m)\right) = f^{(i)}_{k-1}(l,m). \end{aligned}$$

The one-soliton is given by

$$f^{(1)}_n = p^n(1+ap)^l(1+\frac{b}{p})^{-m} + q^n(1+aq)^l(1+\frac{b}{q})^{-m}.$$

It is worth remarking that Hirota actually started from this solution and obtained the equation from it.

An important remark is that the Hirota-Miwa equation (30) can be written in a symmetric way as

$$\left(z_1 e^{D_1} + z_2 e^{D_2} + z_3 e^{D_3}\right)\tau \cdot \tau = 0, \tag{30}$$

where z_1, z_2, z_3 are arbitrary parameters and

$$e^{D_1}\tau\cdot\tau = \tau(n_1+1,n_2,n_3)\tau(n_1-1,n_2,n_3), \ldots .$$

This difference equation is a "mother" equation for numerous soliton systems, as can be seen from the following identifications and reductions [9]:
1. discrete time Toda equation:
Let us rewrite $D_1 = \delta D_t,\ D_2 = D_n, z_1 = \delta^{-2},\ z_2 = -1,\ z_3 = 1 - \delta^{-2}$, ignore D_3 and introduce

$$V_n = \frac{\tau_{n+1}\tau_{n-1}}{\tau_n^2}.$$

Then we have from (30)

$$\Delta_t^2 \log(1 + V_n(t)) = \Delta_n^2 \delta^{-2} \log(1 + \delta^2 V_n),$$

where

$$\Delta_t a(t) = \frac{1}{2\delta}(a(t+\delta) - a(t-\delta)),$$
$$\Delta_n b_n = \frac{1}{2}(b(n+1) - b(n-1)).$$

2. KdV equation:
Let us rewrite $D_1 = \frac{1}{4}(3\delta D_\tau + D_n),\ D_2 = \frac{1}{4}(\delta D_\tau + 3D_n),\ z_1 = 1,\ z_2 = \delta,\ z_3 = -(1+\delta)$. Furthermore introduce x, t by $D_\tau = \frac{1}{3}\varepsilon^3 D_t - 2\varepsilon D_x,\ D_n = 2\varepsilon D_x$. Then by taking a limit of $\delta \to 0$ and $\varepsilon \to 0$, we have the KdV equation,

$$D_x(D_t + D_x^3)\tau \cdot \tau = 0.$$

3. Intermediate long wave (ILW) equation:
Let us rewrite

$$D_1 = \delta(i\varepsilon^2 D_t + \frac{1}{2}\varepsilon D_x) + ihD_x,$$
$$D_2 = \delta(i\varepsilon^2 D_t + \frac{1}{2}\varepsilon D_x) + ihD_x + 2\varepsilon D_x,$$
$$D_3 = -\delta(i\varepsilon^2 D_t + \frac{1}{2}\varepsilon D_x) + ihD_x$$
$$z_1 = 1, z_2 = -\delta, z_3 = -1 + \delta.$$

Then by taking a limit of $\delta \to 0$ and $\varepsilon \to 0$, we have

$$(iD_t - D_x^2)\tau(t, x+ih) \cdot \tau(t, x-ih) = 0.$$

This differential-difference equation is the bilinear form of the ILW equation [11],

$$u_t + uu_x + \left[P\frac{1}{2h}\int_{\mathcal{R}} \coth\frac{\pi}{2h}(y-x)u(y)\mathrm{d}y\right]_{xx} = 0, \tag{31}$$

which admits as limits the KdV (for $h \to 0$) and the Benjamin-Ono equation (for $h \to +\infty$).

4 Sato theory

In the previous sections we saw that the existence of solutions in the form of determinants (Wronskian or Casorati) for bilinear equations was always guaranteed by a similar type of algebraic identity: a so-called Plücker relation. Sato theory is a grand scheme in which this fact – as well as many other deep results concerning the algebraic structure of soliton equations – is revealed [12], [13]. A central element in this theory is the so-called *tau function*, which will provide solutions to equations in bilinear form.

4.1 Micro-differential operators

The starting point in this section is a "micro-differential" operator

$$W = 1 + w_1(x)\partial^{-1} + w_2(x)\partial^{-2} + \dots, \tag{32}$$

in which the operators ∂^{-n} are defined by

$$\partial^{-n} \equiv \left(\frac{\mathrm{d}}{\mathrm{d}x}\right)^{-n}, \quad n \in \mathcal{N},$$

in the sense of the following extension of the Leibniz rule ($n \in \mathcal{Z}$):

$$\partial^n f(x) = \sum_{r=0}^{\infty} \frac{n(n-1)\dots(n-r+1)}{r!} \frac{\mathrm{d}^r f}{\mathrm{d}x^r} \partial^{n-r}.$$

Examples.

$$\begin{aligned}
\partial f &= f_x + f\partial, \\
\partial^2 f &= f_{xx} + 2f_x\partial + f\partial^2, \\
\partial^{-1} f &= f\partial^{-1} - f_x\partial^{-2} + f_{xx}\partial^{-3} - \dots, \\
\partial^{-2} f &= f\partial^{-2} - 2f_x\partial^{-3} + 3f_{xx}\partial^{-4} - \dots.
\end{aligned}$$

For simplicity however we shall (for the time being) assume that

$$W = W_m = 1 + w_1(x)\partial^{-1} + \dots + w_m(x)\partial^{-m}, \tag{33}$$

for some finite m. Afterwards one can take a suitable limit $m \to \infty$, in order to cover the general case.

We now define the linear ODE

$$W_m\partial^m f(x) = (\partial^m + w_1\partial^{m-1} + \dots + w_m)f(x) = 0. \tag{34}$$

It has m independent solutions $f^{(1)}(x), f^{(2)}(x), \dots, f^{(m)}(x)$, which we assume to be regular, such that the following Taylor expansion exists:

$$f^{(j)}(x) = \xi_0^{(j)} + \frac{1}{1!}\xi_1^{(j)} x + \frac{1}{2!}\xi_2^{(j)} x^2 + \dots .$$

The $\infty \times m$ matrix

$$\Xi = \begin{pmatrix} \xi_0^{(1)} & \dots & \xi_0^{(m)} \\ \xi_1^{(1)} & \dots & \xi_1^{(m)} \\ \vdots & \ddots & \vdots \end{pmatrix},$$

is of rank m and satisfies

$$W_m \partial^m \left(1, \frac{x}{1!}, \frac{x^2}{2!}, \dots \right) \Xi = 0. \tag{35}$$

Remark that due to the linearity of (34) we have that for any regular $m \times m$ matrix R, ΞR also satisfies (35). Ξ is therefore unique up to a multiplicative factor in $\mathrm{GL}(m, \mathcal{C})$. Hence, Ξ can be regarded as an element of the quotient set

$$\{\text{matrices of order } (\infty \times m) \text{ and rank } m\} / \,\mathrm{GL}(m, \mathcal{C}), \tag{36}$$

which will serve here as the definition of the *Grassmannian manifold* $\mathrm{GM}(m, \infty)$.

Now, because the m solutions $f^{(j)}$ are supposed to be linearly independent, we can determine W_m from these functions. The set of equations

$$w_1 \partial^{m-1} f^{(j)} + \dots + w_m f^{(j)} = -\partial^m f^{(j)}, \quad j = 1, \cdots, m,$$

is solved as

$$w_j = \frac{\begin{vmatrix} \partial^{m-1} f^{(1)} & \dots & -\partial^m f^{(1)} & \dots & f^{(1)} \\ \vdots & & \vdots & & \\ \partial^{m-1} f^{(m)} & \dots & -\partial^m f^{(m)} & \dots & f^{(m)} \end{vmatrix}}{\begin{vmatrix} \partial^{m-1} f^{(1)} & \dots & -\partial^{m-j} f^{(1)} & \dots & f^{(1)} \\ \vdots & & \vdots & & \\ \partial^{m-1} f^{(m)} & \dots & -\partial^{m-j} f^{(m)} & \dots & f^{(m)} \end{vmatrix}}. \tag{37}$$

This yields

$$W = 1 + w_1\partial^{-1} + w_2\partial^{-2} + \cdots + w_m\partial^{-m}$$
$$= \frac{\begin{vmatrix} f^{(1)} & \dots & f^{(m)} & \partial^{-m} \\ \vdots & \ddots & \vdots & \vdots \\ \partial^{m-1}f^{(1)} & \dots & \partial^{m-1}f^{(m)} & \partial^{-1} \\ \partial^m f^{(1)} & \dots & \partial^m f^{(m)} & 1 \end{vmatrix}}{\begin{vmatrix} f^{(1)} & \dots & f^{(m)} \\ \vdots & \ddots & \vdots \\ \partial^{m-1}f^{(1)} & \dots & \partial^{m-1}f^{(m)} \end{vmatrix}}.$$

4.2 Introduction of an infinite number of time variables

Let us define the shift operator

$$\Lambda = \begin{pmatrix} 0\ 1 & 0 & \dots & 0 \dots \\ 0\ 0 & 1 & \dots & 0 \dots \\ \vdots\ \vdots & \ddots & \ddots & \vdots \dots \\ 0\ 0 & \dots & \dots & 1 \dots \end{pmatrix}.$$

Then we have

$$
\begin{aligned}
e^{x\Lambda} &= I + x\Lambda + \frac{x^2}{2!}\Lambda^2 + \dots \\
&= \begin{pmatrix} 1 & x & \frac{x^2}{2!} & \frac{x^3}{3!} & \dots \\ & 1 & x & \frac{x^2}{2!} & \dots \\ & & \dots & x & \dots \\ & O & & 1 & \dots \end{pmatrix},
\end{aligned}
$$

and

$$
\begin{aligned}
H(x) &\equiv e^{x\Lambda}\Xi \\
&= \begin{pmatrix} f^{(1)} & f^{(2)} & \dots & f^{(m)} \\ \partial f^{(1)} & \partial f^{(2)} & \dots & \partial f^{(m)} \\ \partial^2 f^{(1)} & \partial^2 f^{(2)} & \dots & \partial^2 f^{(m)} \\ \vdots & \vdots & \ddots & \vdots \end{pmatrix},
\end{aligned}
$$

that is, the shift operator Λ allows us to introduce a parameter x into an element of the Grassmanian (36).

We now define

$$
\begin{aligned}
&H(t) = (\exp \sum_{n=1}^{\infty} t_n \Lambda^n)\Xi, \\
&t \equiv (t_1, t_2, t_3, \dots), \quad t_1 = x, \\
&w_j = w_j(t), \quad f^{(j)} = f^{(j)}(t),
\end{aligned}
$$

and consider the formal expansion of $\exp \sum_{n=1}^{\infty} t_n \Lambda^n$,

$$
e^{t_1\Lambda + t_2\Lambda^2 + t_3\Lambda^3 + \dots} = \sum_{n=0}^{\infty} p_n(t)\Lambda^n,
$$

where the variables p_n defined by

$$
p_n = \sum_{\nu_1 + 2\nu_2 + 3\nu_3 + \dots = n, \nu_1, \nu_2, \dots \geq 0} \frac{t_1^{\nu_1} t_2^{\nu_2} t_3^{\nu_3} \dots}{\nu_1! \nu_2! \nu_3! \dots}
$$

are called the *Schur polynomials.*

Examples.

$$\begin{aligned}
p_0 &= 1,\\
p_1 &= t_1,\\
p_2 &= \frac{1}{2}t_1^2 + t_2,\\
p_3 &= \frac{1}{6}t_1^3 + t_1t_2 + t_3,\\
p_4 &= \frac{1}{24}t_1^4 + \frac{1}{2}t_1^2t_2 + \frac{1}{2}t_2^2 + t_1t_3 + t_4.
\end{aligned}$$

An important property of these polynomials is

$$\frac{\partial p_m}{\partial t_n} = p_{n-m}, \quad \text{with } p_n \equiv 0 \quad \text{for } n < 0,$$

and especially:

$$\frac{\partial p_n}{\partial x} = p_{n-1}.$$

Using these Schur polynomials we can define an element of the Grassmanian (36) which depends on infinitely many parameters (time variables):

$$\begin{aligned}
H(t) &= \sum_{n=0}^{\infty} p_n \Lambda^n \Xi\\
&= \begin{pmatrix} 1 & p_1 & p_2 & \dots \\ & 1 & p_1 & \dots \\ 0 & & 1 & \dots \end{pmatrix} \begin{pmatrix} \xi_0^{(1)} & \dots & \xi_0^{(m)} \\ \xi_1^{(1)} & \dots & \xi_1^{(m)} \\ \vdots & \ddots & \vdots \end{pmatrix}\\
&= \begin{pmatrix} h_0^{(1)}(t) & \dots & h_0^{(m)}(t) \\ h_1^{(1)}(t) & \dots & h_1^{(m)}(t) \\ \vdots & \ddots & \vdots \end{pmatrix},
\end{aligned}$$

where

$$\begin{aligned}
&h_0^{(j)}(x = t_1, 0, 0, 0, \dots) f^{(j)}(x),\\
&h_n^{(j)}(t) = \frac{\partial h_0^{(j)}(t)}{\partial t_n} = \frac{\partial^n h_0^{(j)}(t)}{\partial x^n}.
\end{aligned}$$

This means that all $h_0^{(j)}(t)$ satisfy

$$\left(\frac{\partial}{\partial t_n} - \frac{\partial^n}{\partial x^n}\right) h_0^{(j)} = 0, \; n = 1, 2, 3, \dots$$

with the initial condition $h_0^{(j)}(x, 0) = f^{(j)}(x)$.

4.3 The Sato equation

If we now assume that W_m depends on t as

$$W_m(t)\partial^m h_0^{(j)}(t) = \{\partial^m + w_1(t)\partial^{m-1} + \cdots + w_m(t)\} h_0^{(j)}(t) = 0, \quad (38)$$

we know from the above that:

$$w_j = \frac{\begin{vmatrix} h_{m-1}^{(1)} & \cdots & -h_m^{(1)} & \cdots & h_0^{(1)} \\ \vdots & \ddots & \vdots & \ddots & \vdots \\ h_{m-1}^{(m)} & \cdots & -h_m^{(m)} & \cdots & h_0^{(m)} \end{vmatrix}}{\begin{vmatrix} h_{m-1}^{(1)} & \cdots & h_{m-j}^{(1)} & \cdots & h_0^{(1)} \\ \vdots & \ddots & \vdots & \ddots & \vdots \\ h_{m-1}^{(m)} & \cdots & h_{m-j}^{(m)} & \cdots & h_0^{(m)} \end{vmatrix}}, \quad (39)$$

and

$$W_m = \frac{\begin{vmatrix} h_0^{(1)} & \cdots & h_0^{(m)} & \partial^{-m} \\ \vdots & \ddots & \vdots & \vdots \\ h_{m-1}^{(1)} & \cdots & h_{m-1}^{(m)} & \partial^{-1} \\ h_m^{(1)} & \cdots & h_m^{(m)} & 1 \end{vmatrix}}{\begin{vmatrix} h_0^{(1)} & \cdots & h_0^{(m)} \\ \vdots & \ddots & \vdots \\ h_{m-1}^{(1)} & \cdots & h_{m-1}^{(m)} \end{vmatrix}}.$$

Remark that the determinant in the numerator of w_1 is actually the x-derivative of the denominator, such that w_1 is nothing but the logarithmic derivative of the determinant:

$$\tau := \begin{vmatrix} h_{m-1}^{(1)} & \cdots & h_{m-j}^{(1)} & \cdots & h_0^{(1)} \\ \vdots & \ddots & \vdots & \ddots & \vdots \\ h_{m-1}^{(m)} & \cdots & h_{m-j}^{(m)} & \cdots & h_0^{(m)} \end{vmatrix}, \quad (40)$$

which we shall refer to as a "tau" function. Later on we will prove that τ satisfies the KP equation in bilinear form (19).

First of all however, let us try to see what kind of equation the micro-differential operator W_m satisfies. An obvious equation is

$$W\partial^m h_0^{(j)} = 0,$$

which is an m-th order ODE. If one applies ∂_{t_n} to it, bearing in mind that $\dfrac{\partial}{\partial t_n} = \partial^n$ when applied to $h_0^{(j)}$, one obtains

$$\left(\frac{\partial W}{\partial t_n}\partial^m + W\partial^{m+n}\right) h_0^{(j)} = 0, \tag{41}$$

i.e. an $(m+n)$-th order ODE which, by construction, should have the same solution set as $W\partial^m h_0^{(j)} = 0$. The differential operator in (41) should therefore factorize as

$$\left(\frac{\partial W}{\partial t_n}\partial^m + W\partial^{m+n}\right) = B_n W\partial^m,$$

for some differential operator B_n. Applying the inverse operator $\partial^{-m}W^{-1}$ to this expression one obtains

$$B_n = \frac{\partial W}{\partial t_n}W^{-1} + W\partial^n W^{-1}.$$

As the operator $\frac{\partial W}{\partial t_n}W^{-1}$ does not have a differential part (i.e. it only contains terms $\partial^{-n}, n \geq 1$) it is clear that the differential operator B_n equals the differential part of $W\partial^n W^{-1}$ (i.e. the part that only contains terms $\partial^n, n \geq 1$), which we denote as:

$$B_n = \left(W\partial^n W^{-1}\right)^+. \tag{42}$$

Examples.

$$B_1 = \partial,$$
$$B_2 = \partial^2 - 2w_{1,x},$$
$$B_3 = \partial^3 - 3w_{1,x}\partial - 3w_{2,x} + 3w_1 w_{1,x} - 3w_{1,xx}.$$

The *Sato equation* is defined as the system

$$\frac{\partial W}{\partial t_n} = B_n W - W\partial^n, \tag{43}$$

$$B_n = (W\partial^n W^{-1})^+, \tag{44}$$

or equivalently as

$$\frac{\partial W}{\partial t_n} = -\left(W\partial^n W^{-1}\right)^-, \tag{45}$$

where $\left(W\partial^n W^{-1}\right)^-$ denotes the purely micro-differential part of $W\partial^n W^{-1}$.

4.4 Generalized Lax equation

We introduce the fields u_j, linked to W through the operator:

$$L \equiv W\partial W^{-1} = \partial + u_2\partial^{-1} + u_3\partial^{-2} + \dots . \tag{46}$$

Examples.

$$u_2 = w_{1,x},$$
$$u_3 = -w_{2,x} + w_1 w_{1,x}.$$

Now, let us rewrite the Sato equation (45) in terms of the operator L. If one applies $\dfrac{\partial}{\partial t_n}$ to L, noting that

$$\frac{\partial(WW^{-1})}{\partial t_n} = \frac{\partial W}{\partial t_n}W^{-1} + W\frac{\partial W^{-1}}{\partial t_n},$$

one obtains

$$\begin{aligned}\frac{\partial L}{\partial t_n} &= \frac{\partial W}{\partial t_n}\partial W^{-1} + W\partial\frac{\partial W^{-1}}{\partial t_n}\\ &= (B_nW - W\partial^n)\partial W^{-1} - W\partial W^{-1}(B_nW - W\partial^n)W\\ &= B_nL - LB_n,\end{aligned}$$

or

$$\frac{\partial L}{\partial t_n} = [B_n, L]. \tag{47}$$

Equation (47) is called the *generalized Lax equation.*

The operator B_n can be seen to equal:

$$\begin{aligned}B_n &= \left(W\partial^n W^{-1}\right)^+\\ &= (W\partial W^{-1}W\partial W^{-1}\dots W\partial W^{-1})^+\\ &= (L^n)^+\end{aligned}$$

such that it can also be easily expresssed in terms of the variables u_j.

Examples.

$$\begin{aligned}B_1 &= \partial,\\ B_2 &= \partial^2 + 2u_2,\\ B_3 &= \partial^3 + 3u_2\partial + 3u_3 + 3u_{2,x}.\end{aligned}$$

Remark that this is essentially the same computation as in Magri [14].

The generalized Lax equation actually describes infinitely many (nonlinear) evolution equations which, taken together, make up the so called KP hierarchy. Let us for example calculate the KP equation from the generalized Lax equation.

We first compute the relation $\dfrac{\partial L}{\partial t_2} = [B_2, L]$.

$$\begin{aligned}
\frac{\partial L}{\partial t_2} &= \frac{\partial u_2}{\partial t_2}\partial^{-1} + \frac{\partial u_3}{\partial t_2}\partial^{-2} + \dots, \\
[B_2, L] &= (\partial^2 + 2u_2)(\partial + u_2\partial^{-1} + u_3\partial^{-2} + \dots) \\
&\qquad -(\partial + u_2\partial^{-1} + u_3\partial^{-2} + \dots)(\partial^2 + 2u_2) \\
&= (u_{2,xx} + 2u_{3,x})\partial^{-1} + (u_{3,xx} + 2u_{4,x} + 2u_2u_{2,x})\partial^{-2} + \dots.
\end{aligned}$$

Equating the lhs and the rhs yields the infinite set of equations

$$\frac{\partial u_2}{\partial t_2} = u_{2,xx} + 2u_{3,x}, \tag{48}$$

$$\frac{\partial u_3}{\partial t_2} = u_{3,xx} + 2u_{4,x} + 2u_2u_{2,x}, \tag{49}$$

$$\vdots$$

A similar calculation with the time t_3 instead of t_2, yields

$$\frac{\partial u_2}{\partial t_3} = u_{2,xxx} + 3u_{3,xx} + 3u_{4,x} + 6u_2u_{2,x}. \tag{50}$$

The elimination of u_3, u_4 among (48), (49), (50) yields the KP equation

$$\frac{\partial}{\partial x}\left(4\frac{\partial u_2}{\partial t_3} - 12u_2\frac{\partial u_2}{\partial x} - \frac{\partial^3 u_2}{\partial x^3}\right) - 3\frac{\partial^2 u_2}{\partial t^2} = 0.$$

4.5 Structure of tau functions

We saw before that, due to (39), the field $u_2 = -(w_1)_x$ which solves the KP equation is nothing but the second logaritmic derivative of the determinant (40), which we called a tau function. We will show now that such a τ-function satisfies specific bilinear equations. But first let us look a bit more careful at the tau function itself and especially at the way it is linked to elements of the Grassmanian. We have that:

$$\tau(t) = \begin{vmatrix} h_0^{(1)} & \dots & h_0^{(m)} \\ h_1^{(1)} & \dots & h_1^{(m)} \\ \vdots & \ddots & \vdots \\ h_{m-1}^{(1)} & \dots & h_{m-1}^{(m)} \end{vmatrix}$$

$$= \begin{vmatrix} 1 & p_1 & p_2 & \dots \\ & 1 & p_1 & \dots \\ 0 & & 1 & \dots \end{vmatrix} \begin{vmatrix} \xi_0^{(1)} & \dots & \xi_0^{(m)} \\ \xi_1^{(1)} & \dots & \xi_1^{(m)} \\ \vdots & & \vdots \end{vmatrix}$$

$$= \det\left(\Xi_0^t e^{\sum_{n=1}^{\infty} t_n \Lambda^n} \Xi\right),$$

in which the $m \times \infty$ matrix Ξ_0^t is defined by:

$$\Xi_0^t \equiv \begin{pmatrix} 1 & 0 & 0 & \dots & \dots \\ 0 & 1 & 0 & \dots & \dots \\ \vdots & \vdots & \ddots & \ddots & \dots \\ 0 & 0 & \dots & 1 & \dots \end{pmatrix}.$$

Furthermore, from an expansion theorem for the determinant of the product of two matrices, one has that

$$\tau(t) = \sum_{0 \le l_1 < l_2 \cdots < l_m} \begin{vmatrix} p_{l_1} & \dots & p_{l_m} \\ p_{l_1-1} & \dots & p_{l_m-1} \\ \vdots & & \vdots \\ p_{l_1-m+1} & \dots & p_{l_m-m+1} \end{vmatrix} \begin{vmatrix} \xi_{l_1}^{(1)} & \dots & \xi_{l_1}^{(m)} \\ \xi_{l_2}^{(1)} & \dots & \xi_{l_2}^{(m)} \\ \vdots & & \vdots \\ \xi_{l_m}^{(1)} & \dots & \xi_{l_m}^{(m)} \end{vmatrix}.$$

This expansion of $\tau(t)$ in terms of elements of the Grassmanian $GM(m, \infty)$, characterized by m numbers $(l_1, l_2, \dots, l_m)$, suggests a way of connecting such elements with differential expressions in the tau function. To see this, let us introduce the so called Maya-diagram corresponding to the set of numbers $(l_1, l_2, \dots, l_m)$.

We first prepare a chain of (empty) cells, ordered in numerical order. We then start to put particles (sometimes thought of as fermions) in some of the cells according to the following rules:

1. put a particle in each cell numbered less than $1 - m$ (i.e. from $-\infty$ all the way up to $-m$). (Think of these particles as part of a "Dirac sea")

2. put m supplementary particles in the cells numbered $l_1 - m + 1, l_2 - m + 1, \ldots, l_m - m + 1$.

The resulting diagram is called the Maya-diagram corresponding to the numbers $(l_1, l_2, \ldots, l_m)$. This correspondence is one to one for fixed m. Note however that the diagram for the vacuum case $(m = 0)$

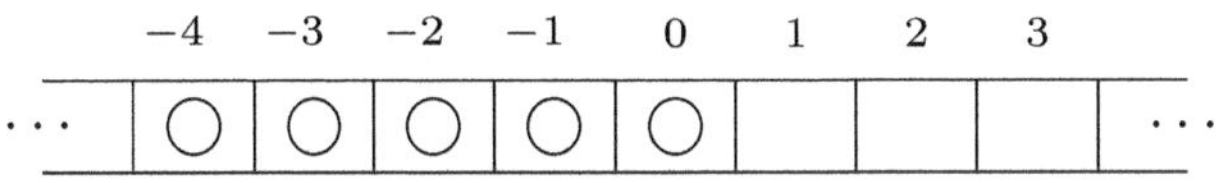

is identical to that for $(0, 1, 2, \ldots, m-1)$ for any $m \geq 1$.
Example.
The Maya-diagram corresponding to the sequence $(2, 3, 5, 7)$ is

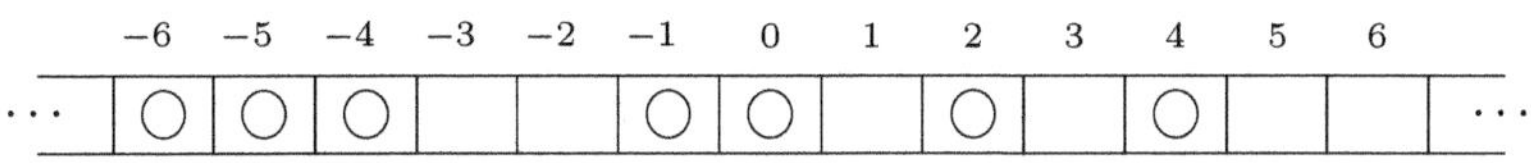

Next, we construct a Young-diagram corresponding to a given Maya-diagram. Going through the Maya-diagram from $-\infty$ to $+\infty$, we draw a (directed) continuous line by succesively:

- drawing a vertical arrow $\uparrow$ for each occupied cell we encounter in the Maya-diagram, or,
- drawing a horizontal arrow $\rightarrow$ for each unoccupied cell.

Clearly, the line obtained from the Maya-diagram corresponding to the vacuum state $m = 0$ has the shape of a single hook, first moving up and then turning right. The diagram enscribed by this "vacuum hook" and the line constructed from a particular Maya diagram can be interpreted as the Young diagram corresponding to that Maya-diagram. This correspondence is one to one for given m.
Example.
The Young-diagram corresponding to the Maya diagram for $(2, 3, 5, 7)$ has the shape:

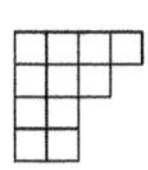

We use the symbol ϕ to denote the Young-diagram for the vacuum state.

In the section on the 2D Toda lattice it was shown that for typical determinant solutions, the bilinear equation took the form of a particular Plücker relation, irrespective of the size of these determinants. At the time we introduced a special notation for a Wronskian determinant, in which only the number of derivatives in each column was used (such a notation was first introduced by Freeman and Nimmo [15]). In this notation, the τ function (40) is represented as:

$$\tau = \begin{vmatrix} h_0^{(1)} & h_1^{(1)} & \ldots & h_{m-1}^{(1)} \\ h_0^{(2)} & h_1^{(2)} & \ldots & h_{m-1}^{(2)} \\ \vdots & \vdots & \ddots & \vdots \\ h_0^{(m)} & h_1^{(m)} & \ldots & h_{m-1}^{(m)} \end{vmatrix} := \begin{vmatrix} 0 \; 1 \ldots \; m-1 \end{vmatrix},$$

and for example, its x-derivative as:

$$\tau_x = \begin{vmatrix} h_0^{(1)} & \ldots & h_{m-2}^{(1)} & h_m^{(1)} \\ h_0^{(2)} & \ldots & h_{m-2}^{(2)} & h_m^{(2)} \\ \vdots & \ddots & \vdots & \vdots \\ h_0^{(m)} & \ldots & h_{m-2}^{(m)} & h_m^{(m)} \end{vmatrix} := \begin{vmatrix} 0 \; 1 \ldots \; m-2 \; m \end{vmatrix},$$

(remember that $h_n^{(j)} = \dfrac{\partial h_0^{(j)}}{\partial t_n} = \dfrac{\partial^n h_0^{(j)}}{\partial x_n}$).

We can now interpret the numbers appearing in this notation as defining a Maya-diagram and subsequently a Young-diagram. Hence the τ function itself corresponds to

$$\tau = \begin{vmatrix} 0 \; 1 \cdots \; m-1 \end{vmatrix}$$

$\sim$

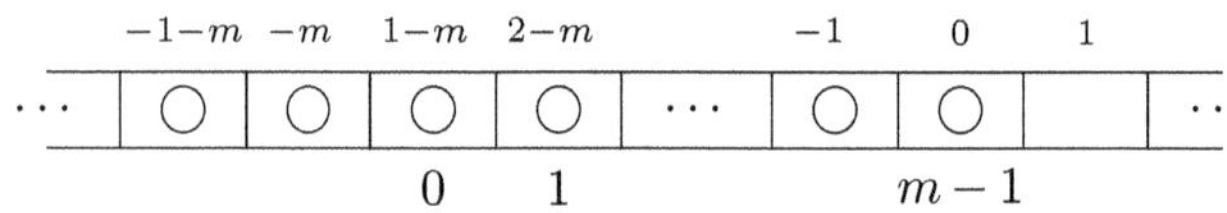

$\sim \phi,$

i.e., it corresponds to the vacuum state Maya-and/or Young-diagram (the larger numbers underneath the Maya-diagram indicate the corresponding entries in the Freeman-Nimmo notation).

Its x-derivative however can be seen to correspond to:

$$\tau_x = \begin{vmatrix} 0 \; 1 \cdots m-2 \; m \end{vmatrix}$$

$\sim$

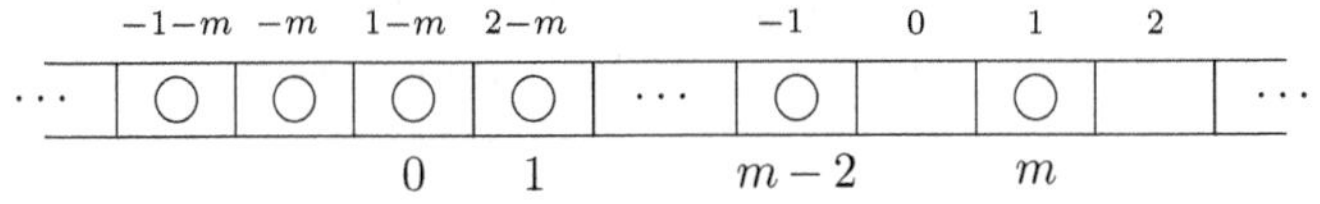

$$\sim \square \tag{51}$$

which allows us to attach the Young-diagram $\square$ to the x-derivative of τ:

$$\tau_{\square} := \tau_x. \tag{52}$$

In general, one can associate a particular differential operator with a Young-diagram Y, such that when this operator acts upon a tau function the result is a single determinant, denoted as τ_Y. If we wish to represent this determinant using Freeman-Nimmo notation we just need to deduce a sequence $(l_1, l_2, \ldots, l_m)$ – for some m – which will give rise to the correct Young-diagram. This procedure is implemented in the following way:

1. the height of the Young-diagram indicates the minimum value of m.
2. using e.g. this value of m (or any number larger than it) one can then (uniquely for each m) reconstruct a Maya-diagram from that Young-diagram.
3. from this Maya-diagram one can (again uniquely for a given m) reconstruct $(l_1, l_2, \ldots, l_m)$ for that value of m.
4. One then defines the Schur function $S_Y(x)$:

$$S(x) := \begin{vmatrix} p_{l_1}(x) & p_{l_2}(x) & \cdots & p_{l_m}(x) \\ p_{l_1-1}(x) & p_{l_2-1}(x) & \cdots & p_{l_m-1}(x) \\ \vdots & \vdots & \ddots & \vdots \\ p_{l_1-m+1}(x) & p_{l_2-m+1}(x) & \cdots & p_{l_m-m+1}(x) \end{vmatrix},$$

   which, it turns out, is independent of the chosen value of m (i.e. it is the same for all sets $(l_1, l_2, \ldots, l_m)$, which for different m give the same Maya-diagram). Please note that these Schur functions are the same as those which appear in the representation theory of symmetric groups [16].
5. We then define τ_Y as:

$$\tau_Y := S_Y(\tilde{\partial}_t)\tau, \tag{53}$$

   where $\tilde{\partial}_t := (\partial_x, \frac{1}{2}\partial_{t_2}, \frac{1}{3}\partial_{t_3}, \ldots)$.
6. If one wishes to derive the Freeman-Nimmo notation for this tau function, one first of all has to choose the size of the determinant large enough so as to fully capture the action of the differential operator $S_Y(\tilde{\partial}_t)$ on it (typically the size of the determinant must be chosen larger than the

height of the Young-diagram Y). If we think of the Young-diagram Y as corresponding to a particular partition $[\lambda] = [\lambda_1, \lambda_2, \lambda_3, \ldots, \lambda_m]$, where those λ_ℓ with an index higher than the height of the Young-diagram are taken to be zero, then the Freeman-Nimmo notation for τ_Y is:

$$\tau_Y = \begin{vmatrix} \lambda_m \; \lambda_{m-1}+1 \; \cdots \; \lambda_1+m-1 \end{vmatrix}.$$

4.6 Algebraic identities for tau functions

Let us now put these diagrams and special notations to use. For example, a first application is found in connection to the coefficients in the micro-differential operator $W = 1 + w_1(t)\partial^{-1} + w_2(t)\partial^{-2} + \ldots$, which in Freeman-Nimmo notation are given as:

$$w_j(t) = (-1)^j \frac{\begin{vmatrix} 0 \; 1 \; \cdots \; m-j-1 \; m-j+1 \; \cdots \; m \end{vmatrix}}{\begin{vmatrix} 0 \; 1 \; \cdots \; m-1 \end{vmatrix}}.$$

According to the above, we can represent the determinants appearing in the numerators of these coefficients as:

$$\begin{vmatrix} 0 \; 1 \; \cdots \; m-j-1 \; m-j+1 \; \cdots \; m \end{vmatrix}$$

$\sim$ 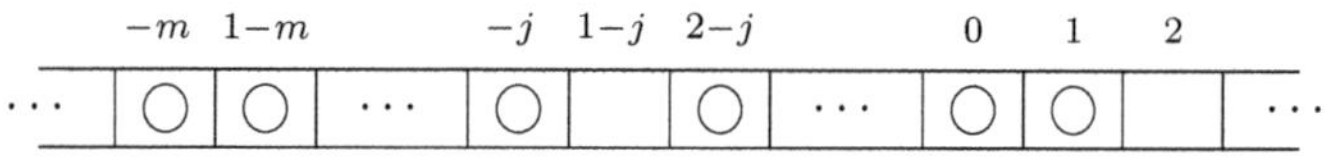

$\sim$  $j := Y_j$.

The Schur function S_{Y_j} which corresponds to this Young-diagram Y_j can be shown to be:

$$S_{Y_j} = p_j(-\tilde{\partial}_t),$$

and hence we find that the coefficients $w_j(t)$ can be expressed as

$$w_j(t) = \frac{1}{\tau} p_j(-\tilde{\partial}_t)\tau. \tag{54}$$

In particular we have:

$$w_1(t) = \frac{1}{\tau} p_1(-\tilde{\partial}_t)\tau \equiv \partial_x \log \tau,$$

(as was remarked before), and

$$w_2(t) = \frac{1}{\tau} p_2(-\tilde{\partial}_t)\tau = \frac{1}{2\tau}\left(\frac{\partial^2 \tau}{\partial x^2} - \frac{\partial \tau}{\partial t_2}\right).$$

As a second application, let us calculate the Maya and Young representations of the derivatives τ_{2x} and τ_{t_2}. The second x-derivative of τ can be easily seen to be:

$$\tau_{2x} = \left| 0 \cdots m-3\ m-1\ m \right| + \left| 0 \cdots m-2\ m+1 \right|$$

where:

$$\left| 0 \cdots m-3\ m-1\ m \right|$$

$\sim$

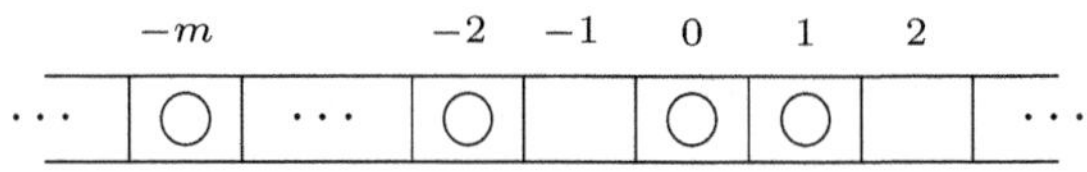

$\sim$  (55)

and

$$\left| 0 \cdots m-2\ m+1 \right|$$

$\sim$

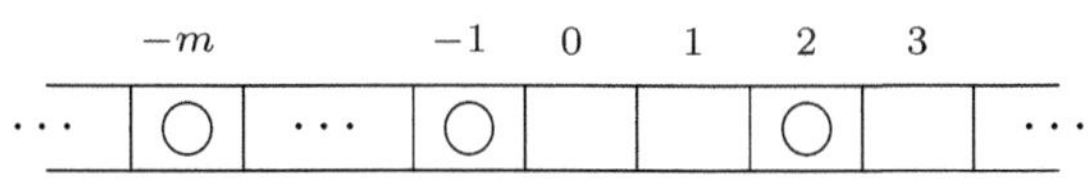

$\sim$ 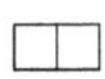 (56)

whereas the t_2-derivative is:

$$\begin{aligned} \tau_{t_2} &= \left| 0 \cdots m-2\ m+1 \right| + \left| 0 \cdots m-3\ m\ m-1 \right| \\ &= \left| 0 \cdots m-2\ m+1 \right| - \left| 0 \cdots m-3\ m-1\ m \right| \end{aligned}$$

such that one has the following expressions:

$$\tau_{2x} = \tau_{\square\square} + \tau_{\substack{\square\\\square}}, \qquad \tau_{t_2} = \tau_{\square\square} - \tau_{\substack{\square\\\square}}. \tag{57}$$

Conversely, if we want to calculate the differential operators $S_Y(\tilde{\partial}_t)$ which correspond to these Young-diagrams, it is best if we first look for a minimal Freeman-Nimmo representation for them (thus reducing the size of the determinant expressions for the corresponding Schur functions). For example, as the height of the Young- diagram $\substack{\square\\\square}$ is 2, it suffices to consider a 2×2 determinant $|l_1\ l_2|$ (in Freeman-Nimmo notation) to fully capture the action of the corresponding differential operator. From the Maya-diagram in (55) it is clear that $l_1 = 1$ and $l_2 = 2$ and hence we learn that $\tau_{\substack{\square\\\square}}$ can be represented as

$$\tau_{\substack{\square\\\square}} = |1\ 2|$$

(i.e. setting "$m = 2$" in (55), which is the minimal choice) and that correspondingly:

$$S_{\substack{\square\\\square}}(\tilde{\partial}_t) = \begin{vmatrix} p_1(\tilde{\partial}_t) & p_2(\tilde{\partial}_t) \\ 1 & p_1(\tilde{\partial}_t \end{vmatrix} \equiv \frac{1}{2}(\partial_x^2 - \partial_{t_2}).$$

For the Young-diagram $\square\square$ we consider, as a further example, two different representations: once as a 1×1 and once as a 2×2 determinant (both of which are apparent from (56)):

$$\tau_{\square\square} = |2| \text{ or } |0\ 3|.$$

It is easily verified that

$$S_{\square\square}(\tilde{\partial}_t) = p_2(\tilde{\partial}_t) \equiv \begin{vmatrix} p_0(\tilde{\partial}_t) & p_3(\tilde{\partial}_t) \\ 0 & p_2(\tilde{\partial}_t) \end{vmatrix} \equiv \frac{1}{2}(\partial_x^2 + \partial_{t_2}),$$

irrespective of the size of the representation. From the above expressions for these Schur functions we find that:

$$\tau_{\substack{\square\\\square}} = \frac{1}{2}(\tau_{2x} - \tau_{t_2}), \qquad \tau_{\square\square} = \frac{1}{2}(\tau_{2x} + \tau_{t_2}), \tag{58}$$

in accordance with (57).

Let us now finally prove that the tau functions as defined above, solve the KP equation. For this it is sufficient to use a representation as 2×2 determinants, for which we have the identity

$$0 \equiv \begin{vmatrix} h_0^{(1)} & h_1^{(1)} & h_2^{(1)} & h_3^{(1)} \\ h_0^{(2)} & h_1^{(2)} & h_2^{(2)} & h_3^{(2)} \\ 0 & h_1^{(1)} & h_2^{(1)} & h_3^{(1)} \\ 0 & h_1^{(2)} & h_2^{(2)} & h_3^{(2)} \end{vmatrix}$$

$$= \begin{vmatrix} 0 & 1 \end{vmatrix} \begin{vmatrix} 2 & 3 \end{vmatrix} - \begin{vmatrix} 0 & 2 \end{vmatrix} \begin{vmatrix} 1 & 3 \end{vmatrix} + \begin{vmatrix} 0 & 3 \end{vmatrix} \begin{vmatrix} 1 & 2 \end{vmatrix},$$

again by Laplace expansion. For some of the determinants appearing in this expansion we already know how to represent them in terms of Maya-and Young-diagrams, notably: $|0\ 1| = \tau_\phi$, $|0\ 2| = \tau_{\Box}$, $|1\ 2| = \tau_{\substack{\Box\\\Box}}$ and $|0\ 3| = \tau_{\Box\Box}$. It is an easy exercise to show that:

$$|1\ 3| = \tau_{\substack{\Box\Box\\\Box}} \qquad \text{and} \qquad |2\ 3| = \tau_{\substack{\Box\Box\\\Box\Box}}.$$

Hence, the above Laplace expansion is equivalent to the equation

$$\tau_\phi \tau_{\substack{\Box\Box\\\Box\Box}} - \tau_{\Box}\tau_{\substack{\Box\Box\\\Box}} + \tau_{\Box\Box}\tau_{\substack{\Box\\\Box}} = 0, \tag{59}$$

which is now completely general (irrespective of the size of the determinants considered): it is in fact the first non-trivial Plücker relation for the Grassmanian $GM(2,\infty)$.

It is also a simple exercise to show that:

$$\tau_{\substack{\Box\Box\\\Box}} = \frac{1}{3}(\partial_x^3 - \partial_{t_3})\tau \qquad \text{and} \qquad \tau_{\substack{\Box\Box\\\Box\Box}} = \frac{1}{12}(\partial_x^4 + 3\partial_{t_2}^2 - 4\partial_x\partial_{t_3})\tau,$$

such that we – bearing in mind (52) and (58) – can rewrite the Plücker relation (59) as:

$$(4D_xD_{t_3} - D_x^4 - 3D_{t_2}^2)\tau\cdot\tau = 0,$$

which is nothing but the bilinear form of the KP equation. As was mentioned before, the connection between the tau function and the solutions of the (usual nonlinear) KP equation is given by:

$$u_2 = -w_{1,x} = \partial_x^2 \log\tau,$$

which is the bilinearizing transformation of the KP equation

$$\frac{\partial}{\partial x}\left(4\frac{\partial u_2}{\partial t_3} - 12\frac{\partial u_2}{\partial x} - \frac{\partial^3 u_3}{\partial x^3}\right) - 3\frac{\partial^2 u_2}{\partial t_2^2} = 0.$$

4.7 Vertex operators and the KP bilinear identity

Let us give an explicit form for the τ function which corresponds to the N-soliton solution of the KdV equation. One first introduces

$$\xi(x,k) = \sum_{j=1}^{\infty} x_j k^j, \tag{60}$$

in which x_1 is like t_1, x_2 like t_2, etc. Then set

$$\xi_i = \xi(x,k_i) - \xi(x,-k_i) = 2\sum_{j=0}^{\infty} k_i^{2j+1} x_{2j+1}, \tag{61}$$

(which is in fact a reduction of the original $\xi(x,k)$) and finally, introduce

$$a_{ii'} = \left[\frac{k_i - k_{i'}}{k_i + k_{i'}}\right]^2,$$

which shall describe the interaction between two individual solitons in a multi-soliton solution.

Then we define

$$\tau_N(x_1, x_3, \dots) = \sum_{J\subset I}\prod_{i\in J} c_i \prod_{i,i'\in J, i<i'} a_{ii'} \exp\sum_{i\in J}\xi_i$$

in which J is an arbitrary subset of $I = \{1, 2, \dots, N\}$ and the c_i are arbitrary (phase) constants. These τ_N correspond to the N-soliton solutions for the KdV equation.

Examples.

$$\begin{aligned}
\tau_1 &= 1 + c_1 e^{\xi_1},\\
\tau_2 &= 1 + c_1 e^{\xi_1} + c_2 e^{\xi_2} + c_1 c_2 a_{12} e^{\xi_1+\xi_2},\\
\tau_3 &= 1 + c_1 e^{\xi_1} + c_2 e^{\xi_2} + c_3 e^{\xi_3} + c_1 c_2 a_{12} e^{\xi_1+\xi_2}\\
&\qquad + c_2 c_3 a_{23} e^{\xi_2+\xi_3} + c_3 c_1 a_{31} e^{\xi_3+\xi_1}\\
&\qquad\qquad + c_1 c_2 c_3 a_{12} a_{23} a_{31} e^{\xi_1+\xi_2+\xi_3}.
\end{aligned}$$

Let us now introduce so-called vertex operators $X(k)$, which can be shown to be intimately linked to the notion of a Darboux transformation for the KdV equation [17]. In particular, a vertex operator X will map an N-soliton solution to an $(N+1)$-soliton solution:

$$\tau_{N+1} = e^X \tau_N.$$

These vertex operators are expressed as

$$X(k) = \exp\left(2\sum_{j=0}^{\infty} k^{2j+1} x_{2j+1}\right) \exp\left(-2\sum_{j=0}^{\infty} \frac{1}{(2j+1)k^{2j+1}} \frac{\partial}{\partial x_{2j+1}}\right). \tag{62}$$

and enjoy the following properties:

$$X(k)f(x_1, x_3, \dots) = \exp\left(2\sum_{j=0}^{\infty} k^{2j+1} x_{2j+1}\right) f(x_1 - \frac{2}{k}, x_3 - \frac{2}{3}k^3, \dots),$$

$$X(k_1)X(k_2) = \left[\frac{k_1 - k_2}{k_1 + k_2}\right]^2 \exp\left(2\sum_{i=1}^{2}\sum_{j=0}^{\infty} k_i^{2j+1} x_{2j+1}\right)$$
$$\times \exp\left(-2\sum_{i=1}^{2}\sum_{j=0}^{\infty} \frac{1}{(2j+1)k_i^{2j+1}} \frac{\partial}{\partial x_{2j+1}}\right),$$

and thus

$$X(k)^2 = 0.$$

Then

$$\begin{aligned}
\tau_1 &= e^{c_1 X(k_1)} \cdot 1 \\
&= \left(1 + c_1 X + \frac{1}{2} c_1^2 X^2 + \cdots\right) \cdot 1 \\
&= (1 + c_1 X) \cdot 1 \\
&= 1 + c_1 e^{\xi_1},
\end{aligned}$$

and all higher N-soliton solutions can be obtained from τ_1 by successive applications of vertex operators.

Let us now consider the KP case, which is very similar to the KdV case, but for “un-doing” the reduction (61) as in:

$$\xi_i = \xi(x, p_i) - \xi(x, q_i) = \sum_{j=1}^{\infty} (p_i^j - q_i^j) x_j,$$

$$a_{ii'} = \frac{(p_i - p_{i'})(q_i - q_{i'})}{(p_i - q_{i'})(q_i - p_{i'})}.$$

From this we can define the so-called “solitonic-vertex operator” for the KP equation:

$$X(p, q) = \exp\left(\sum_{j=1}^{\infty} (p^j - q^j) x_j\right) \exp\left(-\sum_{j=1}^{\infty} \frac{1}{j}\left(\frac{1}{p_j} - \frac{1}{q_j}\right) \frac{\partial}{\partial x_j}\right). \tag{63}$$

As the name suggests, it acts on N-soliton solutions of the KP equation by increasing the number of solitons by 1, such that a general expression for these N-soliton solution can be given as:

$$\tau_N = e^{c_1 X(p_1,q_1)} \dots e^{c_N X(p_N,q_N)} \cdot 1 = \sum_{J\subset I} \prod_{i\in J} c_i \prod_{i,i'\in J, i<i'} a_{ii'} \exp \sum_{i\in J} \xi_i .$$

From their definition it can be checked that these τ-functions satisfy the bilinear identiy

$$\oint \frac{dk}{2\pi i} e^{\xi(x,k)-\xi(x',k)} \tau(x_1 - \frac{1}{k}, x_2 - \frac{1}{2k^2}, \dots) \tau(x'_1 + \frac{1}{k}, x'_2 + \frac{1}{2k^2}, \dots) = 0. \quad (64)$$

This identity can be (formally) calculated by introducing dummy variables y_j as in $x_j = x_j + y_j$ and $x'_j = x_j - y_j$, after which an expansion in Schur polynomials (of Hirota operators) yields:

$$\begin{aligned} 0 &= \oint \frac{dk}{2\pi i} e^{(2\sum_{j=1}^{\infty} k^j y_j)} \tau(x_1 + y_1 - \frac{1}{k},\ x_2 + y_2 - \frac{1}{2k^2},\ \dots) \\ &\qquad\qquad \times \tau(x_1 - y_1 + \frac{1}{k},\ x_2 - y_2 + \frac{1}{2k^2},\ \dots) \\ &= \oint \frac{dk}{2\pi i} e^{(2\sum_{j=1}^{\infty} k^j y_j)} \exp\left(\sum_{l=1}^{\infty} \left(y_l - \frac{1}{lk^l} \right) D_l \right) \tau \cdot \tau \\ &= \oint \frac{dk}{2\pi i} k^{i-j} \sum_{i=0}^{\infty} \sum_{j=0}^{\infty} p_i(2y)\, p_j(-\tilde{D})\, e^{\sum_{\ell=1}^{\infty} y_\ell D_\ell}\ \tau \cdot \tau \end{aligned}$$

and hence

$$\sum_{i=0}^{\infty} p_i(2y)\, e^{\sum_{\ell=1}^{\infty} y_\ell D_\ell}\ p_{i+1}(-\tilde{D})\ \tau \cdot \tau = 0, \quad (65)$$

where D_ℓ denotes the Hirota operator D_{x_ℓ} and where $\tilde{D}$ stands for the weighted Hirota operators $\tilde{D} = (D_1, \frac{1}{2}D_2, \frac{1}{3}D_3, \dots)$. This last bilinear identity is in fact a generating function for all bilinear equations the KP tau functions satisfy. For example, expanding (65) in terms of the dummy variables y one can calculate the term in y_ℓ ($\ell \geq 1$), to find:

$$[2p_{\ell+1}(-\tilde{D}) - D_1 D_\ell]\ \tau \cdot \tau = 0,$$

which gives at lowest orders in ℓ:

$$\begin{aligned} \ell = 1: &\quad [2p_2(-\tilde{D}) - D_1^2]\ \tau \cdot \tau \equiv -D_2\ \tau \cdot \tau \equiv 0, \\ \ell = 2: &\quad [2p_3(-\tilde{D}) - D_1 D_2]\ \tau \cdot \tau \equiv [-D_3 + \frac{1}{3} D_1^3]\ \tau \cdot \tau \equiv 0, \\ \ell = 3: &\quad [2p_4(-\tilde{D}) - D_1 D_3]\ \tau \cdot \tau \equiv \frac{-1}{12}[4D_1 D_3 - D_1^4 - 3D_2^2]\ \tau \cdot \tau = 0, \\ &\ \vdots \end{aligned}$$

showing that the bilinear KP equation (19) is the first non-trivial equation for the tau functions, contained in (64). The set of all equations generated in this way is often referred to as the *KP hierarchy*.

4.8 Fermion analysis based on an infinite dimensional Lie algebra

It was shown by Date, Jimbo, Kashiwara and Miwa that there is a fundamental interpretation of Sato theory in connection to the representation theory of infinite dimensional Lie algebras [17]. An important ingredient in their interpretation is a famous theorem called the *boson-fermion correspondence*.

Another important feature is a representation of vertex operators such as those introduced in the previous section, in terms of elements of a so-called *fermion* algebra. This is an algebra of (charged) *free fermion* creation and annihilation operators, which satisfy the anti commutation relations:

$$[\psi_i, \psi_j^*]_+ \equiv \psi_i \psi_j^* + \psi_j^* \psi_i = \delta_{i+j,0}$$

and

$$[\psi_i, \psi_j]_+ = [\psi_i^*, \psi_j^*]_+ = 0, \qquad \text{for } i,j \in \mathcal{Z} + \frac{1}{2}.$$

For this algebra one has the standard *Fock representation* , defining the Fock space $\mathcal{F}$ as well as its dual $\mathcal{F}^*$. The highest weight vectors in these representations ($\forall \ell \in \mathcal{N}_0$), can be expressed as:

$$\left.\begin{aligned} |\ell\rangle &= \psi_{1/2-\ell} \cdots \psi_{-1/2}\, |vac\rangle \\ |-\ell\rangle &= \psi^*_{1/2-\ell} \cdots \psi^*_{-1/2}\, |vac\rangle \end{aligned}\right. \in \mathcal{F}$$

and their dual states $\langle \ell|, \langle -\ell| \in \mathcal{F}^*$ are obtained from those in $\mathcal{F}$ by the duality relation: $\psi_j \leftrightarrow \psi^*_{-j}$. The cyclic vectors $|vac\rangle$ and $\langle vac|$ are defined in terms of states $|0\rangle$ and $\langle 0|$ which are respectively annihilated by all operators ψ_j and ψ_j^*. Expectation values are defined in the usual way ($\langle vac|1|vac\rangle = 1$) and can be calculated using Wick's theorem (w_j is any fermion operator):

$$\langle vac|w_1 \cdots w_r|vac\rangle = \begin{cases} 0 \\ \sum_\sigma \operatorname{sgn}(\sigma) \langle vac|w_{\sigma(1)} w_{\sigma(2)}|vac\rangle \cdots \\ \qquad\qquad\qquad \langle vac|w_{\sigma(r-1)} w_{\sigma(r)}|vac\rangle \end{cases}$$

depending on whether r is odd or even and where the sum $\sum_\sigma$ runs over all possible permutations satisfying $\sigma(1) < \sigma(2), \ldots, \sigma(r-1) < \sigma(r)$ and $\sigma(1) < \sigma(3) < \cdots < \sigma(r-1)$.

Two important observations are the following. It is first of all possible to construct bosonic operators H_n in terms of the fermion operators:

$$H_n := \sum_{j \in \mathcal{Z}+1/2} : \psi_{-j} \psi^*_{j+n} : \qquad \forall n \in \mathcal{Z}$$

(where : : is the usual normal ordering : $\psi_i \psi_j^*$: $= \psi_i \psi_j^* - \langle vac | \psi_i \psi_j^* | vac \rangle$) which satisfy the Heisenberg commutation relations:

$$[H_n, H_m] = \delta_{n+m,0}.$$

Conversely, it is also possible to give a representation of fermionic operators in terms of bosonic ones. This situation is summarized as
boson-fermion correspondence: *The map $\boldsymbol{\Phi}$ between the linear spaces $\mathcal{F}$ and $\mathcal{C}\left[z, z^{-1}; x_1, x_2, \ldots\right]$ (i.e. the space of formal power series in $(x_1, x_2, x_3, \ldots)$ and z, z^{-1}, polynomial in z and z^{-1}):*

$$\begin{aligned} \boldsymbol{\Phi} : \quad \mathcal{F} &\longrightarrow \mathcal{C}\left[z, z^{-1}; x_1, x_2, x_3, \cdots\right] \\ |u\rangle &\longrightarrow \sum_{\ell \in \mathcal{Z}} z^\ell \langle \ell | e^{H(x)} | u \rangle \end{aligned}$$

is an isomorphism of Fock spaces. Here $H(x)$ stands for $H(x) = \sum_{n=1}^{\infty} x_n H_n$.
In particular, introducing formal extensions of fermion operators:

$$\psi(k) \equiv \sum_{j \in \mathcal{Z}+1/2} \psi_j k^{-j-1/2}, \quad \psi^*(k) \equiv \sum_{j \in \mathcal{Z}+1/2} \psi_j^* k^{-j-1/2}, \tag{66}$$

the vertex operators

$$\Gamma(x,k) := e^{\xi(x,k)} e^{-\xi(\partial_x, 1/k)}, \qquad \Gamma^*(x,k) := e^{-\xi(x,k)} e^{\xi(\partial_x, 1/k)},$$

can be realized in terms of fermion operators as:

$$\Gamma(x,k) \simeq \boldsymbol{\Phi} \cdot \psi(k) \cdot \boldsymbol{\Phi}^{-1}, \qquad \Gamma^*(x,k) \simeq \boldsymbol{\Phi} \cdot \psi^*(k) \cdot \boldsymbol{\Phi}^{-1},$$

(i.e. up to a constant multiple). As the KP solitonic vertex operator (63) can be thought of (up to a constant multiple) as the product $\Gamma(x,p) \cdot \Gamma^*(x,q)$, one might start wondering what the exact connection with the KP hierarchy (and Sato theory) is.

To see this connection we need to introduce the infinite dimensional Lie algebra $gl(\infty)$

$$\begin{aligned} gl(\infty) := \{ \sum_{i,j \in \mathcal{Z}+1/2} a_{ij} : \psi_i \psi_j^* : + a_0 \mid \exists\, R :\ a_{ij} = 0, \\ \forall\ |i-j| > R, \text{ with } a_{ij}, a \in \mathcal{C} \}, \end{aligned}$$

and its corresponding Lie group $GL(\infty)$:

$$GL(\infty) := \{ g | g = e^X,\ X \in gl(\infty) \}.$$

It is then proved by straightforward calculation that the following algebraic identity holds :

$$\forall g \in GL(\infty): \sum_{j\in\mathcal{Z}+1/2} \psi_{-j} g \otimes \psi_j^* g = \sum_{j\in\mathcal{Z}+1/2} g\psi_{-j} \otimes g\psi_j^* .$$

If we now define a tau function as an element in the $GL(\infty)$ orbit of the cyclic vector $|vac\rangle$:

$$\tau(x) := \langle vac|e^{H(x)} g|vac\rangle, \tag{67}$$

it can be derived from the above identity that this $\tau(x)$ necessarily satisfies the KP bilinear identity (64). Conversely it can be shown that all functions which solve the KP bilinear identity correspond, by formula (67), to elements in (a completion) of the $GL(\infty)$ orbit of $|vac\rangle$.

When calculating tau functions from a given element in $GL(\infty)$, the following formulae are very useful:

$$\begin{aligned} e^{H(x)}\ \psi(k)\ e^{-H(x)} &= \psi(k)\ e^{\xi(x,k)}, \\ e^{H(x)}\ \psi^*(k)\ e^{-H(x)} &= \psi^*(k)\ e^{-\xi(x,k)}, \end{aligned} \tag{68}$$

$$\begin{aligned} &\langle vac|\psi(p_1)\psi(p_2)\cdots\psi(p_N)\psi^*(q_N)\psi^*(q_{N-1})\cdots\psi^*(q_1)|vac\rangle \\ &\qquad = \det\left[\frac{1}{p_i-q_j}\right]_{1\le i,j\le N}, \end{aligned}$$

with $\xi(x,k)$ as in (60). Note also that $H_n|vac\rangle = 0$ ($\forall n \geq 1$).

Another important feature of this method is that it provides immediate access to discrete soliton equations as well. For, if in the bilinear identity (64) (and accordingly in the definition of the tau function (67) and the above formulae) we use a so-called (generalised) Miwa transformation [10], [18]

$$x = \sum_i^{\ell} \varepsilon[a(i)] + \sum_j^{m} \varepsilon[b(j)] + \sum_k^{n} \varepsilon[c(k)], \tag{69}$$

where

$$\varepsilon[\lambda] := (\frac{1}{\lambda}, \frac{1}{2\lambda^2}, \frac{1}{3\lambda^3}, \dots)$$

and

$$\sum_i^{l} \equiv \begin{cases} \sum_{i=1}^{l} & \text{for } l \geq 1 \\ 0 & \text{for } l = 0 \\ -\sum_{i=l+1}^{0} & \text{for } l \leq -1 \end{cases},$$

which relates the (infinite) set of continuous variables $x = (x_1, x_2, x_3, \dots)$, to a set of 3 discrete variables ℓ, m and n, one obtains a nonautonomous version of the discrete KP equation:

$$(b_m - c_n)\tau_\ell \tau_{mn} + (c_n - a_l)\tau_m \tau_{\ell n} + (a_\ell - b_m)\tau_n \tau_{\ell m} = 0 \tag{70}$$

(a subscript indicates an increment in that variable, with respect to a "reference" – the explicit dependence on these variables being omitted).

Soliton solutions for the equations in the KP hierarchy are easily expressed in terms of elements of (the completion) of $GL(\infty)$:

$$g^{(N)} \equiv \prod_{j=1}^{N} (1 + c_j \psi(p_j) \psi^*(q_j)),$$

gives rise to the N-soliton tau function

$$\begin{aligned} \tau_G^{(N)}(x) &\equiv \langle vac | e^{H(x)} g^{(N)} | vac \rangle \\ &= \det \left[\delta_{i,j} + \frac{c_i}{p_i - q_j} e^{\xi(x,p_i) - \xi(x,q_j)} \right]_{1 \le i,j \le N} . \end{aligned}$$

Of course, the variable transformation (69) immediately produces corresponding soliton solutions for the nonautonomous discrete KP equation (70). It suffices to note that the "discrete" version of the exponentials $\exp \xi(x,p)$ which appear in the soliton solutions (or in the evolutions (68)), take on the form:

$$e^{\xi(\ell,m,n;p)} = \left[\prod_i^{\ell} (1 - \frac{p}{a(i)}) \prod_j^{m} (1 - \frac{p}{b(j)}) \prod_k^{n} (1 - \frac{p}{c(k)}) \right]^{-1},$$

using the following multiplication convention:

$$\prod_i^k F(i,j,\cdots) \equiv \begin{cases} \prod_{i=1}^k F(i,j,\cdots) & \text{for } k \ge 1 \\ 1 & \text{for } k = 0 \\ \prod_{i=k+1}^0 F(i,j,\cdots)^{-1} & \text{for } k \le -1, \end{cases}$$

for each product $\prod_i^k$.

5 Extensions of the bilinear method

5.1 q-discrete equations

An important lesson we learned from the fermionic treatment of the KP hierarchy is that as long as the specific algebraic structure of a τ function is kept, that particular solution can be adapted to (almost) any dispersion relation. For example, in the continuous case we typically had $\partial_{x_k} f = \partial_x^k f$, in the semi-discrete case $\partial_x f_n = f_{n+1}$ and in the fully discrete case $\Delta_l f_n = f_{n+1}$ as dispersion relations for the entries in the determinants representing the tau function. We shall see now that is also possible to introduce q-type dispersion relations, allowing for an extension of the bilinear method to the case of q-difference equations.

If we take the following dispersion relations, defined in terms of q-difference operators $\delta_{q^\alpha,x}$, $\delta_{q^\beta,y}$,

$$\delta_{q^\alpha,x} f_n(x,y) := \frac{f_n(x,y) - f_n(q^\alpha x, y)}{(1-q)x} = -f_{n+1}(x,y), \tag{71}$$

$$\delta_{q^\beta,y} f_n(x,y) := \frac{f_n(x,y) - f_n(x, q^\beta y)}{(1-q)y} = f_{n-1}(x,y), \tag{72}$$

it can be shown that the τ-function

$$\tau_n = \begin{vmatrix} f_n^{(1)} & \cdots & f_{n+N-1}^{(1)} \\ \vdots & \ddots & \vdots \\ f_n^{(N)} & \cdots & f_{n+N-1}^{(N)} \end{vmatrix} \tag{73}$$

satisfies the q-discrete two-dimensional Toda lattice equation [19]

$$\begin{aligned}(\delta_{q^\alpha,x}\delta_{q^\beta,y}\tau_n)\tau_n &- (\delta_{q^\alpha,x}\tau_n)(\delta_{q^\beta,y}\tau_n) \\ &= \tau_{n+1}(x,q^\beta y)\tau_{n-1}(q^\alpha x,y) - \tau_n(q^\alpha x, q^\beta y)\tau_n(x,y).\end{aligned} \tag{74}$$

In the reduced case

$$xy = r^2, \qquad \alpha = \beta = 2, \tag{75}$$

one has

$$\begin{aligned}\{(\frac{1}{r}\delta_{q,r} + q\delta_{q,r}^2)\tau_n(r)\}\tau_n(r) &- \{\delta_{q,r}\tau_n(r)\}^2 \\ &= \tau_{n+1}(qr)\tau_{n-1}(qr) - \tau_n(q^2 r)\tau_n(r),\end{aligned} \tag{76}$$

which is the q-discrete cylindrical Toda lattice equation, so named after its continuum limit $q \to 1$:

$$\{(\frac{1}{r}\partial_r + \partial_r^2)\tau_n(r)\}\tau_n(r) - \{\partial_r\tau_n(r)\}^2 = \tau_{n+1}(r)\tau_{n-1}(r) - \{\tau_n(r)\}^2.$$

Under (75) the dispersion relations (71) and (72) reduce to the contiguity relations of the q-Bessel function

$$\{q^{-n}\delta_{q,r} + \frac{[-n]}{r}\}J_{q,n}(r) = -J_{q,n+1}(r),$$
$$\{q^{n}\delta_{q,r} + \frac{[n]}{r}\}J_{q,n}(r) = J_{q,n-1}(r),$$

in which the notation $[n]$ is defined as:

$$\frac{1-q^n}{1-q}.$$

We thus find that the determinant

$$\tau_n(r) = \begin{vmatrix} J_{q,n}(r) & \dots & J_{q,n+N-1}(r) \\ \vdots & \ddots & \vdots \\ J_{q,n+p_N}(r) & \dots & J_{q,n+N+p_{N-1}}(r) \end{vmatrix} \tag{77}$$

constitutes a solution to the q-discrete cylindrical Toda lattice equation (76).

In a similar way we can obtain a q-discrete version of the Toda molecule equation (as well as of its Bäcklund transformation and Lax pair, see [20]), which has a particular solution expressed in terms of Airy functions.

5.2 Special function solution for soliton equations

Many soliton equations possess (determinant) solutions expressed in terms of special functions. In fact, one can tackle this problem the other way around. Instead of trying to find particular solutions to a given soliton equation one can start from a particular type of special function and try to construct an integrable (solitonic) system that will allow for such solutions. Let us explain this procedure on a couple of simple examples, starting from the so-called discrete KP (dKP) equation [21]:

$$\begin{aligned}(b-a)\tau(\ell{+}1,m,n)\tau(\ell,m{+}1,n{+}1)+(c-a)\tau(\ell,m{+}1,n)\tau(\ell{+}1,m,n+1)\\ +(a-b)\tau(\ell,m,n+1)\tau(\ell+1,m+1,n) = 0.\end{aligned} \tag{78}$$

This equation is obtained as an autonomous reduction (a, b, c : constant) of the general dKP equation (70). It possesses an associated linear system, consisting of

$$\varphi(\ell+1,m+1,n) = \frac{1}{b-a}\frac{\tau(\ell+1,m,n)\tau(\ell,m+1,n)}{\tau(\ell,m,n)\tau(\ell+1,m+1,n)} \times\{b\varphi(\ell+1,m,n) - a\varphi(\ell,m+1,n)\},$$

and two similar equations obtained from the above one by the (cyclic) changes $(a,b;\ell,m) \mapsto (b,c;m,n)$ and $\mapsto (c,a;n,\ell)$.

For $\tau =$ constant, these linear equations reduce to the dispersion relations:

$$a\{\varphi(\ell,m,n)-\varphi(\ell-1,m,n)\}=b\{\varphi(\ell,m,n)-\varphi(\ell,m-1,n)\}$$
$$=c\{\varphi(\ell,m,n)-\varphi(\ell,m,n-1)\},$$

or

$$\Delta_\ell^-\varphi=\Delta_m^-\varphi=\Delta_n^-\varphi, \tag{79}$$

in terms of the backward-shift operators Δ^-.

It can be shown that the dKP equation admits the solution

$$\tau=\begin{vmatrix}\varphi^{(1)} & \Delta^-\varphi^{(1)} & \ldots & (\Delta^-)^{N-1}\varphi^{(1)}\\ \vdots & & \ddots & \vdots\\ \varphi^{(N)} & \Delta^-\varphi^{(N)} & \ldots & (\Delta^-)^{N-1}\varphi^{(N)}\end{vmatrix}, \tag{80}$$

where the functions $\varphi^{(j)}$ obey (79) (any reference to specific variables in the backward shift operators in the determinant can therefore be omitted).

It is easily seen that the functions

$$\Psi_\lambda=\left(1-\frac{\lambda}{a}\right)^{-\ell}\left(1-\frac{\lambda}{b}\right)^{-m}\left(1-\frac{\lambda}{c}\right)^{-n} \tag{81}$$

satisfy the dispersion relations (79) and that they can therefore be used in the determinant (80). For example, the N-soliton solution for the dKP equation is obtained by taking

$$\varphi^{(j)}=\psi_{p_j}+\psi_{q_j},$$

where p_j and q_j $(j=1,\ldots,N)$ are arbitrary constants.

It is important to notice that in the above we can in fact consider general superpositions

$$\varphi^{(\cdot)}=\int_C d\mu_\lambda\psi_\lambda$$

for any contour C and any measure $d\mu_\lambda$. This fact can then be exploited to obtain special function solutions, not only for the dKP equation (or reductions thereof) but also for many other soliton equations related to the KP hierarchy.

For example, if for the dKP equation (78) we transform the solutions to its linear problem (79) as $\Phi_\lambda=(-a)^\ell\Psi_\lambda$, after which we take the limit $a\to 0, b\to 1$ and $c\to 1/z$:

$$\Phi_\lambda==\lambda^{-\ell}(1-\lambda)^{-m}(1-z\lambda)^{-n}.$$

The dKP equation itself is reduced to:

$$\tau_\ell \tau_{mn} + \frac{1}{z-1}\tau_m \tau_{\ell n} - \frac{z}{z-1}\tau_n \tau_{\ell m} = 0. \tag{82}$$

Considering now general superpositions of these elementary solutions, we choose the measure and contour as $h(\lambda)d\lambda/\lambda = \lambda^{\beta-1}(1-\lambda)^{\gamma-\beta-1}(1-z\lambda)^{-\alpha}\, d\lambda$ and $\int_C = \int_0^1$, such as to obtain the following solution to the dispersion relations *as well as* to the reduced dKP equation (82):

$$\begin{aligned}\Phi &= \int_0^1 d\lambda\, \lambda^{\beta-\ell-1}(1-\lambda)^{\gamma-\beta-m-1}(1-z\lambda)^{-(\alpha+n)},\\ &= \frac{\Gamma(\beta-\ell)\Gamma(\gamma-\beta-m)}{\Gamma(\gamma-\ell-m)} F(\alpha+n, \beta-\ell, \gamma-\ell-m; z)\end{aligned}$$

(Γ denotes the gamma function and $F(\alpha,\beta,\gamma;z)$ is the hypergeometric function). Furthermore, arbitrary sized determinants of type (80) containing such eigenfunctions will yield tau functions for (82) entirely expressed as nonlinear combinations of hypergeometric functions.

As a second example we shall construct an equation possessing solutions expressed in terms of the q-hypergeometric function

$${}_2\varphi_1(\alpha,\beta;\gamma;q,z) \equiv \sum_{n=0}^{\infty} \frac{(\alpha;q)_n(\beta;q)_n}{(\gamma;q)_n(q;q)_n} z^n,$$

where $(\alpha;q)_n \equiv \prod_{j=1}^{n}(1-\alpha q^{j-1})$. It has the the q-integral representation:

$${}_2\varphi_1(q^a,q^b;q^c;q,z) = \frac{\Gamma_q(c)}{\Gamma_q(b)\Gamma_q(c-b)} \int_0^1 d_q\lambda\, \lambda^{b-1} \frac{(\lambda q;q)_\infty(\lambda z q^a;q)_\infty}{(\lambda q^{c-b};q)_\infty(\lambda z;q)_\infty},$$

for $\Gamma_q(\nu) \equiv (1-q)^{1-\nu} \prod_{r=0}^{\infty} \left(\frac{1-q^{r+1}}{1-q^{r+\nu}}\right)$, and $\int_0^1 d_q\lambda\, f(\lambda) \equiv (1-q)\sum_{j=0}^{\infty} q^j f(q^j)$. It is then a case of straightforward arithmetic to show that for a choice

$$\Psi_\lambda^* = \lambda^\ell(\lambda q^{c-b};q)_m(\lambda z q^{a-1};q^{-1})_n,$$

we obtain a solution

$$\Phi^* = \int_0^1 d_q\lambda\, \lambda^{b-1} \frac{(\lambda q;q)_\infty(\lambda z q^a;q)_\infty}{(\lambda q^{c-b};q)_\infty(\lambda z;q)_\infty}\Psi_\lambda^*, \tag{83}$$

$$= \frac{\Gamma_q(b+\ell)\Gamma_q(c-b+m)}{\Gamma_q(c+\ell+m)} {}_2\varphi_1(q^{a-n},q^{b+\ell};q^{c+\ell+m};q,z), \tag{84}$$

for the special reduction of the nonautonomous dKP equation obtained by choosing $a(\ell) \to 1$, $b(m) \to q^{b-c-m}$, $c(n) \to q^{-a+n+1}/z$:

$$(q^{-m+b-c-1} - \frac{1}{z}q^{n+2-a})\tau(\ell+1,m,n)\tau(\ell,m+1,n+1) \tag{85}$$

$$+\frac{1}{z}q^{n+2-a}\tau(\ell,m+1,n)\tau(\ell+1,m,n+1) \tag{86}$$

$$-q^{-m+b-c-1}\tau(\ell,m,n+1)\tau(\ell+1,m+1,n) = 0. \tag{87}$$

It can also be shown that general sized determinants (80), consisting of such q-hypergeometric functions solve this equation as well. It is worth pointing out that as the q-Bessel function is obtained from the q-hypergeometric function by specialization of the free parameters which appear in it, the above determinant solutions encompass the q-Bessel type solutions (77) and that, accordingly, the q-discrete cylindrical Toda equation (76) is a special case of the q-difference equation (87) we constructed here.

Finally, we would like to stress that the same procedure can be used in the continuous case as well. For example, for the 2D Toda equation (22)

$$\frac{1}{2}D_xD_y(\tau(l;x,y)\cdot\tau(l;x,y)) = \tau(l+1;x,y)\tau(l-1;x,y) - \{\tau(l;x,y)\}^2,$$

we choose

$$\Psi_\lambda = \lambda^{-1}e^{-\lambda x+y/\lambda},$$

from which we can obtain the solution

$$\varphi^{(j)}(l;x,y) = \frac{(-1)^{\nu_j}}{2\pi i}\int \lambda^{-\nu_j-l-1}e^{-\lambda x-y/\lambda}d\lambda = (-1)^l\left(\frac{x}{y}\right)^{\frac{l+\nu_j}{2}} J_{l+\nu_j}(2\sqrt{xy}),$$

where $J_\nu(z)$ is the Bessel function of index ν.

5.3 Ultra discrete soliton system

Another possible extension of the bilinear method is to fully discrete or *ultra discrete* soliton systems, i.e.: systems where all variables, including the dependent ones, are discrete. In practice such systems take the form of so-called *soliton cellular automaton* (SCA). The original example of a SCA was proposed in [22] as a fully discrete, 1+1 dimensional, evolutionary system in which every state is a soliton. Each cell in this system takes a value 0 or 1 and the value of the jth cell at time t (denoted by u_j^t) is given by the rule:

$$u_j^{t+1} = \begin{cases} 1 & \text{if } u_j^t = 0 \text{ and } \sum_{i=-\infty}^{j-1} u_i^t > \sum_{i=-\infty}^{j-1} u_i^{t+1} \\ 0 & \text{otherwise.} \end{cases} \tag{88}$$

This rule can be put into words in the following way: at time t one starts from a state consisting of an infinite sequence of 0's and 1's, the number of 1's being finite. The state at time $t + 1$ is then decided by the rules:

1. Move every 1 only once.
2. Exchange the leftmost 1 with its nearest 0 neighbour on the right.
3. Repeat this procedure for the 1's that have not moved yet, each time using the leftmost one, until all 1's have moved.

An example of a typical evolution would be:

$$t=0: \ \cdots 0\,1\,1\,1\,1\,0\,0\,0\,0\,0\,1\,1\,0\,0\,1\,0\,0\,0\,0\,0\,0\,0\,0\,0\,0\,0\,0\,0\,0\,0\,0\,0\,0 \cdots$$

$$t=1: \ \cdots 0\,0\,0\,0\,0\,1\,1\,1\,1\,0\,0\,0\,1\,1\,0\,1\,0\,0\,0\,0\,0\,0\,0\,0\,0\,0\,0\,0\,0\,0\,0\,0\,0 \cdots$$

$$t=2: \ \cdots 0\,0\,0\,0\,0\,0\,0\,0\,0\,1\,1\,1\,0\,0\,1\,0\,1\,1\,1\,0\,0\,0\,0\,0\,0\,0\,0\,0\,0\,0\,0\,0\,0 \cdots$$

$$t=3: \ \cdots 0\,0\,0\,0\,0\,0\,0\,0\,0\,0\,0\,0\,1\,1\,0\,1\,0\,0\,0\,1\,1\,1\,1\,0\,0\,0\,0\,0\,0\,0\,0\,0\,0 \cdots$$

$$t=4: \ \cdots 0\,0\,0\,0\,0\,0\,0\,0\,0\,0\,0\,0\,0\,0\,1\,0\,1\,1\,0\,0\,0\,0\,0\,1\,1\,1\,1\,0\,0\,0\,0\,0\,0 \cdots$$

$$t=5: \ \cdots 0\,0\,0\,0\,0\,0\,0\,0\,0\,0\,0\,0\,0\,0\,0\,1\,0\,0\,1\,1\,0\,0\,0\,0\,0\,0\,0\,1\,1\,1\,1\,0\,0 \cdots$$

The SCA has the interesting features that groups of (separated) 1's interact as genuine solitons and that there exist infinitely many time invariants for it. These facts can be proven by using combinatorial techniques [23]. In fact, this SCA has a deep connection with usual (continuous or discrete) soliton systems [24]. This can be seen by introducing the new variable S_j^t in (88)

$$S_j^t := \sum_{i=-\infty}^{j} u_i^t,$$

such as to find:

$$S_{j+1}^{t+1} - S_j^t = -\max[S_{j+1}^t - S_j^{t+1} - 1, 0].$$

If we then further transform the variables (dependent as well as independent ones) as:

$$h_\xi^\tau := f_{j-t}^t, \qquad f_j^t = S_{j+1}^t - S_j^{t+1}, \tag{89}$$

we obtain the equation:

$$h_\xi^{\tau+1} - h_\xi^\tau = -\max[h_{\xi+1}^{\tau+1} - 1, 0] + \max[h_{\xi-1}^\tau - 1, 0], \tag{90}$$

which is closely related to the KdV equation.

To explain this fact, we notice that the KdV equation (2) (written in the field a in order to avoid confusion)

$$a_t = a_{xxx} + 6aa_x,$$

can be seen to be a continuum limit of the so-called Lotka-Volterra system:

$$\dot{b}_j = b_j(b_{j+1} - b_{j-1}) \tag{91}$$

(use the transformation: $b_j(t) = 1 + \varepsilon^2 a((j+2t)\varepsilon, \frac{1}{3}\varepsilon^3 t)$ and set $\varepsilon \to 0$). From the continuum limit (after a change of variables similar to the one in (89))

$$C_j^t = b_j(-\delta t), \quad \delta \to 0,$$

it is also clear that the equation

$$C_\xi^{\tau+1}(1 + \delta C_{\xi+1}^{\tau+1}) = C_\xi^\tau(1 + \delta C_{\xi-1}^\tau), \tag{92}$$

is a valid time-discretization of this Lotka-Volterra system (and hence, it can also be considered to be a space-and time-discrete KdV equation). If we now introduce variables d_ξ^τ by:

$$C_\xi^\tau = \exp d_\xi^\tau,$$

we obtain a final equation

$$d_\xi^{\tau+1} - d_\xi^\tau = \log \frac{1 + \delta \exp d_{\xi-1}^\tau}{1 + \delta \exp d_{\xi+1}^{\tau+1}}, \tag{93}$$

which allows for a very special limiting procedure

$$\delta = e^{-1/\varepsilon}, \quad d_\xi^\tau = \varepsilon^{-1} h_\xi^\tau, \quad \varepsilon \to 0,$$

called the "ultra-discretization" of the equation (93).
By using the (key) formula:

$$\lim_{\varepsilon \downarrow 0} \varepsilon \log \left(e^{A/\varepsilon} + e^{B/\varepsilon} \right) = \max[A, B], \tag{94}$$

the result of the "ultra discrete" limit of (93) is easily seen to coincide with equation (90).

Although this is not immediately clear from the above reasoning, it can be shown that the soliton solutions appearing in the SCA (88) are, intrinsically, those of the KdV equation. Whereas the link between both systems might seem a bit circuitous when explained on their usual forms (as was done above) if one goes to their respective bilinear forms however, the correspondence becomes much clearer. First of all, remark that as the discrete-time Lotka-Volterra equation (92) is a discrete version of the KdV equation, it can be said that their soliton solutions are actually the same (the only difference between the corresponding tau functions is a Miwa-transformation-like transition between the continuous – KdV – case and the discrete case). Now, the bilinear form of the discrete-time Lotka-Volterra system (92) is

$$g_{\xi+1}^{\tau+1} g_{\xi}^{\tau} + \delta g_{\xi+2}^{\tau+1} g_{\xi-1}^{\tau} - (1+\delta) g_{\xi}^{\tau+1} g_{\xi+1}^{\tau} = 0, \tag{95}$$

with

$$C_{\xi}^{\tau} = \frac{g_{\xi+2}^{\tau+1} g_{\xi-1}^{\tau}}{g_{\xi+1}^{\tau+1} g_{\xi}^{\tau}}$$

as the bilinearizing transformation.

Introducing the parameter ε

$$\delta = e^{-1/\varepsilon},$$

in the bilinear equation (95), we can take its $\varepsilon \to 0$ limit by calculating

$$\varepsilon \log[(1+e^{-1/\varepsilon}) g_j^{t-1} g_{j+1}^{t+1}] = \varepsilon \log[g_{j+1}^{t} g_j^{t} + e^{-1/\varepsilon} g_j^{t+1} g_{j+1}^{t-1}],$$

(where we undid the change of independent variables (89)) and by introducing the "ultra-discrete" limits of the tau functions g_j^t:

$$\rho_j^t := \lim_{\varepsilon \downarrow 0} \varepsilon \log g_j^t.$$

The resulting equation for this variable

$$\rho_{j+1}^{t+1} + \rho_j^{t-1} = \max[\rho_{j+1}^{t} + \rho_j^{t}, \rho_{j+1}^{t-1} + \rho_j^{t+1} - 1],$$

is nothing but the "bilinear form" of the SCA (88) (here to be understood as its "tau function expression", as it is not self-evidient that this is actually a quadratic expression) whose solutions are related to the "ultra-discrete tau functions" ρ_j^t by the dependent variable transformation:

$$u_j^t = \rho_j^t - \rho_j^{t+1} - \rho_{j-1}^{t} + \rho_{j-1}^{t+1}.$$

Hence the claim that the soliton solutions it exhibits are fundamentally connected to those of the KdV equation.

Several extensions of this elementary SCA exist, some of which exhibit deep connections with quantum integrable models. In all these cases, genuine understanding of the properties of these equations (be it the nature of their soliton solutions, their conserved quantities, etc.) can only be gained through the study of their bilinear forms, especially when these are interpreted as ultra-discretizations of soliton equations related to the KP hierarchy. The reader is referred to [25] for a review of soliton cellular automata.

5.4 Trilinear equations

A final extension we would like to discuss is that of so-called "trilinear equations" [26], [27]. The observation which lies at the origin of the "trilinear" idea is that the determinant identities which guarantee the existence of certain types of determinant solutions (Wronski, Casorati, etc...) are not restricted to quadratic ones. For example, in terms of the Schur polynomials $p_n(x)$, one can consider the trilinear form:

$$\begin{vmatrix} p_i(\tilde{\partial}_x)p_\ell(-\tilde{\partial}_y)\tau & p_i(\tilde{\partial}_x)p_m(-\tilde{\partial}_y)\tau & p_i(\tilde{\partial}_x)p_n(-\tilde{\partial}_y)\tau \\ p_j(\tilde{\partial}_x)p_\ell(-\tilde{\partial}_y)\tau & p_j(\tilde{\partial}_x)p_m(-\tilde{\partial}_y)\tau & p_j(\tilde{\partial}_x)p_n(-\tilde{\partial}_y)\tau \\ p_k(\tilde{\partial}_x)p_\ell(-\tilde{\partial}_y)\tau & p_k(\tilde{\partial}_x)p_m(-\tilde{\partial}_y)\tau & p_k(\tilde{\partial}_x)p_n(-\tilde{\partial}_y)\tau \end{vmatrix} = 0, \tag{96}$$

($\tilde{\partial}_x = (\frac{\partial}{\partial_{x_1}}, \frac{1}{2}\frac{\partial}{\partial_{x_2}}, \frac{1}{3}\frac{\partial}{\partial_{x_3}}, \ldots)$ and $\tilde{\partial}_y = (\frac{\partial}{\partial_{y_1}}, \frac{1}{2}\frac{\partial}{\partial_{y_2}}, \frac{1}{3}\frac{\partial}{\partial_{y_3}}, \ldots)$ as before) which can be shown to possess the determinant solutions:

$$\tau = \begin{vmatrix} f & f_{x_1} & \cdots & f_{(n-1)x_1} \\ f_{y_1} & f_{x_1y_1} & \cdots & f_{y_1(n-1)x_1} \\ \vdots & \vdots & \ddots & \vdots \\ f_{(n-1)y_1} & f_{x_1(n-1)y_1} & \cdots & f_{(n-1)x_1(n-1)y_1} \end{vmatrix}. \tag{97}$$

The dispersion relations the function f appearing in this determinant has to satisfy are:

$$\frac{\partial f}{\partial x_k} = \frac{\partial^k f}{\partial x_1^k}, \qquad \frac{\partial f}{\partial y_k} = \frac{\partial^k f}{\partial y_1^k}.$$

Example.

At $(i,j,k) = (\ell,m,n) = (0,1,2)$ we obtain the 4-dimensional (exactly solvable) system:

$$w_{y_2} = w_{2y_1} + 2ww_{y_1} + r_{y_1}, \tag{98}$$

$$(r_{x_2} + r_{2x_1})w_{x_1} = r_{x_1}(w_{x_2} + w_{2x_1}), \tag{99}$$

in the variables

$$w = \frac{\partial}{\partial y_1}\log\tau, \qquad r = -2p_2(-\tilde{\partial}_y)\log\tau - (\frac{\partial}{\partial y_1}\log\tau)^2,$$

which for example contains the Broer-Kaup or Classical-Boussinesq system as a 1+1 dimensional reduction [29].

As before, if we now change the dispersion relation so as to include differential-difference equations into this scheme, we can also consider determinants

$$\tau_{m,n} = \begin{vmatrix} f_{m,n} & f_{m,n+1} & \cdots & f_{m,n+N-1} \\ \vdots & \vdots & \ddots & \vdots \\ f_{m+N-1,n} & f_{m+N-1,n+1} & \cdots & f_{m+N-1,m+N-1} \end{vmatrix}, \tag{100}$$

where f obeys

$$\frac{\partial}{\partial x} f_{m,n} = f_{m,n+1}, \quad \frac{\partial}{\partial y} f_{m,n} = f_{m+1,n}.$$

For example, in this case it can be shown that $\tau_{m,n}$ satisfies the trilinear equation:

$$\begin{vmatrix} \partial_y \tau_{m,n-1} & \tau_{m,n-1} & \tau_{m+1,n-1} \\ \partial_y \tau_{m,n} & \tau_{m,n} & \tau_{m+1,n} \\ \partial_y \partial_x \tau_{m,n} & \partial_x \tau_{m,n} & \partial_x \tau_{m+1,n} \end{vmatrix} = 0, \tag{101}$$

which in a reduced case turns into the relativistic Toda equation proposed by Ruijsenaars [30]:

$$\partial_x^2 q_n = -\partial_x q_n \left(\frac{\partial_x q_{n-1}}{e^{q_n - q_{n-1}} - 1} - \frac{\partial_x q_{n+1}}{e^{q_{n+1} - q_n} - 1} \right). \tag{102}$$

It is remarked that (98), (99) and (101) are four dimensional systems. It is also noted that we can construct four dimendional fully discrete equation by a similar procedure [28].

Acknowledgments

The author would like to thank Prof. A. Greco for his kind invitation to present this series of lectures on the Hirota bilinear method at the occasion of the CIME school. He would also like to express his gratitude for the patience he was shown while completing these lecture notes. He would specifically like to thank Dr. G. Gambino for providing the first draft of this contribution, Dr. R. Willox for helping the author in completing and fine-tuning the manuscript and finally, Mr. S. Isojima for valuable assistance in producing the illustrations contained in it.

References

1. Hirota, R.: Exact solution of the Korteweg-deVries equation for multiple collisions of solitons. Phys. Rev. Lett. **27** (1971), 1192–1194
2. Hirota, R.: Exact envelope soliton solutions of a nonlinear wave equation. J. Math. Phys. **14** (1973), 805–809
3. Hirota, R.: Exact N-soliton solution of a nonlinear lumped network equation. J. Phys. Soc. Jpn. **35** (1973), 289–294
4. Kajiwara, K., Ohta, Y., Satsuma, J., Grammaticos, B., Ramani, A.: Casorati determinant solutions for the discrete Painlevè-II equation. J. Phys. A **27** (1994), 915–922
5. Satsuma, J.: N-Soliton Solution of the Two-Dimensional Korteweg-deVries Equation. J. Phys. Soc. Jpn. **40** (1976), 286–290
6. Hirota, R., Ohta, Y., Satsuma, J.: Wronskian Structures of Solutions for Soliton Equations. Prog. Theor. Phys. Suppl. No.94 (1988), 59–72
7. Darboux, G.: *Leçons sur la théorie générale des surfaces et les applications géométriques du calcul infinitésimal* (Gauthier-Villars, 4 volumes, Paris, 1894). Reprinted, *Théorie générale des surfaces* (Chelsea, New York, 1972). Reprinted (Gabay, Paris 1993).
8. Hirota, R., Ito, M., Kako, F.: Two-Dimensional Toda Lattice Equation Prog. Theor. Phys. Suppl. No.94 (1988), 42–58
9. Hirota, R.: Discrete analogue of the generalized Toda equation. J. Phys. Soc. Jpn. **50** (1981), 3785–3791
10. Miwa, T.: On Hirota's difference equations. Proc. Jpn. Acad. **58A** (1982), 8–11
11. Satsuma, J., Ablowitz, M. J.: Solutions of an Internal Wave Equation Describing a Stratified Fluid with Finite Depth. *Nonlinear Partial Differential Equations in Engineering and Applied Sciencc*, Ed. R.L.Sternberg, A.J.Kalinowski and J.S.Papadakis (Dekker, New York 1980), 393–414
12. Sato, M., Sato, Y.: Soliton equations as dynamical systems on infinite dimensional Grassmann manifolds. Lect. Notes in Num. Appl. Anal. **5** (1982), 259–271
13. Ohta, Y., Satsuma, J., Takahashi, D., Tokihiro, T.: An Elementary Introduction to Sato Theory. Prog. Theor. Phys. Suppl. No.94 (1988), 210–241; Prog. Theor. Phys. **80** (1988), 742
14. Magri, F.: this volume.
15. Freeman, N. C., Nimmo, J. J. C.: Soliton solutions of the Korteweg-de Vries and the Kadomtsev-Petviashvili equations: the Wronskian technique. Proc. Roy. Soc. London Ser. A **389** (1983), 319–329
16. Littlewood, D. E.: *The Theory of Group Characters and Matrix Representations of Groups, 2nd Edition.* (Oxford, London 1958)
17. Date, E., Jimbo, M., Kashiwara, M., Miwa, T.: Transformation Groups for soliton equations. Proc. RIMS Symp. Nonlinear integrable systems, Classical and quantum theory (World Scientific, Singapore 1983), 39–119
18. Willlox, R., Tokihiro, T., Satsuma, J.: Darboux and binary Darboux transformations for the nonautonomous discrete KP equation. J. Math. Phys. **38** (1997), 6455–6469
19. Kajiwara, K., Satsuma, J.: q-Difference Version of the Two-dimensional Toda Lattice Equation. J. Phys. Soc. Jpn. **60** (1991), 3986–3989
20. Kajiwara, K., Ohta, Y., Satsuma, J.: q-Discrete Toda Molecule Equation. Phys. Letter A **180** (1993), 249–256

21. Tokihiro, T., Satsuma, J., Willox, R.: On special function solutions to nonlinear integrable equations. Phys. Lett. A **236** (1997), 23–29
22. Takahashi, D., Satsuma, J.: A Soliton Cellular Automaton. J. Phys. Soc. Jpn. **59** (1990), 3514–3519
23. Torii, M., Takahashi, D., Satsuma, J.: Combinatrial Representation of Invariants of a Soliton Cellular Automaton. Physica D **92** (1996), 209–220
24. Tokihiro, T., Takahashi, D., Matsukidaira, J., Satsuma, J.: From Soliton Equations to Integrable Cellular Automata through a Limiting Procedure. Phys. Rev. Lett. **76** (1996), 3247–3250
25. Satsuma, J., Tokihiro, T., Willlox, R.: Scattering in Cellular Automata. in *Scattering* ed. R.Pike and P.Sabatier (Academic Press, London 2001), 1754–1772
26. Matsukidaira, J., Satsuma, J., Strampp, W.: Soliton Equations Expressed by Trilinear Forms and Their Solutions. Phys. Letter A **147** (1990), 467-471
27. Matsukidaira, J., Satsuma, J.: Integrable Four-Dimensional Nonlinear Lattice Expressed by Trilinear Form. J. Phys. Soc. Jpn. **59** (1990), 3413–3416
28. Matsukidaira, J., Satsuma, J.: Exactly Solvable Four-Dimensional Discrete Equation Expressed by Trilinear Form. Phys. Letter A **154**366-372 (1991), 366–372
29. Satsuma, J., Kajiwara, K., Matsukidaira, J., Hietarinta, J.: Solutions of the Brouer-Kaup System through Its Trilinear Form. J. Phys. Soc. Jpn. **61** (1992), 3096–3102
30. Ruijsenaars. S. N. M.: Relativistic Toda Systems. Commun. Math. Phys. **133** (1990), 217–247

Lie groups, singularities and solutions of nonlinear partial differential equations

Pavel Winternitz

Centre de recherches mathématiques, Université de Montréal, Case postale 6128, Succ. Centre ville, Montréal, Canada; *Wintern@crm.umontreal.ca*

Summary. It is shown how Lie group and Lie algebra theory can be used to solve partial differential equations. A method for calculating the symmetry group of a set of differential equations is presented. Special attention is devoted to algorithms for classifying subalgebras of Lie algebras. The concept of conditional symmetries is introduced and applied to perform dimensional reduction.

1 Introduction

Virtually all fundamental equation of physics are nonlinear, e.g. the Einstein equations, Yang-Mills equations, Navier-Stokes equations, etc, and are difficult to solve.

A systematic method is to use Lie group theory (that's how Lie group theory started in the 19th century).

Recently there has been a new boom of activities, with many new books and appearing articles. So what is new in this field?

1. Developments in physics and other sciences, nonlinear phenomena have really caught up with us.
2. Developments in mathematics, like the thory of infinite dimensional Lie algebras and Lie groups, new results concerning the structure of Lie algebras and their subalgebras and a resurgence of singularity analysis.
3. Developments in computer science: symbol manipulating languages like MACSYMA, MATHEMATICA, SCRATCHPAD and REDUCE. Lots of calculations involved are algorithmic, conceptually simple, often quite "horrendous". The computer can now be used at all stages to perform the calculations involved in applications of Lie group theory.

What does Lie group theory do for us in this context? Among many things, let us mention a few.

1. It allows us to transform known solutions into new ones and thus to obtain finite, or even infinite families of solutions.
2. It allows us to perform symmetry reduction. For ordinary differential equations (ODEs) this means a reduction of the order of the equation. This is done with no loss of generality. If the symmetry group is large

P. Winternitz, Lie groups, singularities and solutions of nonlinear partial differential equations, Lect. Notes Phys. **632**, 223–273 (2003)
http://www.springerlink.com/

enough it provides the general solution. For partial differential equations (PDEs) symmetry reduction reduces the number of independent variables in the equation.This only provides particular solutions, but very often these are physically important ones.

3. It allows us to classify equations into equivalence classes and to recognize special types of equations. Thus, certain types of nonlinear equations may be equivalent to linear ones and this provides a method for solving them.

The method of symmetry reduction for PDEs consists of several steps that can be summed up as follows:

1. Find the symmetry group G of the system of PDEs and its Lie algebra L.
2. Identify the Lie algebra L as an abstract Lie algebra, i.e. transform it to a canonical basis in which its basis independent properties are manifest []. Thus, if L is decomposable, it should be presented as a direct sum of indecomposable Lie algebras: $L = L_1 \oplus L_2 \oplus \cdots \oplus L_k$. Each indecomposable component should be further identified as one of the complex or real classical Lie algebras, or one of the exceptional ones. If L_i is solvable, its nilradical [] should be identified. If it is neither simple,nor solvable, its Levi decomposition should be presented explicitly.
3. The subalgebras of L should be classified into conjugacy classes under the action of the group G leaving the considered PDE invariant. The corresponding subgroups of G should be identified and a representative list of subgroups created.
4. For each subgroup $G_0 \subset G$ in the representative list find a basis for the invariants of G_0 in the space $X \otimes U$ of independent and dependent variables. Let us denote the invariants $I_\mu(x,u),\ \mu = 1,\ldots,N$. The simplest and most favorable situation arises if the invariants I_μ can be divided into two sets:
$$\{\xi_1(x),\ldots,\xi_k(x)\} \text{ and } \{J_1(x,u),\ldots,J_q(x,u)\}, \tag{1}$$
$$k+q=N,\ 1 \le k < p$$
satisfying
$$\det J = \det\left(\frac{\partial J_\alpha}{\partial u_\beta}\right) \neq 0, \tag{2}$$
where $\dfrac{\partial J_\alpha}{\partial u_\beta}$ is the Jacobian of the transformation $\{u_\alpha\} \to \{J_\mu\}$.
5. On the solution set of the considered equations set
$$J_\mu(x,u) = F_\mu(\xi_1,\ldots,\xi_k) \tag{3}$$

and use condition (2) to solve equation (3) for u_α. We obtain the expressions

$$u_\alpha(x) = U_\alpha(F_1, \ldots, F_q, x_1, \ldots, x_p). \tag{4}$$

In equation (4) U_α are known functions, as are the variables ξ_i. The functions $\{F_1, \ldots, F_q\}$ are, so far, arbitrary.

6. Substitute $u_\alpha(x)$ of (4) into the PDEs and obtain the reduced equations. The equations are invariant under G (and hence also $G_0 \subset G$). Hence the variables x_i will drop out and the reduced equations will involve only ξ_i, F_μ and derivatives of F_μ with respect to ξ_i. Since we have $k < p$ we obtain a dimensional reduction. In particular, for $k = 0$ we obtain an algebraic equation. For $k = 1$ an ODE.
7. Solve the reduced system to obtain $F_\mu(\xi_1, \ldots, \xi_k)$, $\mu = 1, \ldots, q$. This can be done by a further use of group theory, or by using the tools of integrability theory, in particular Painlevé analysis (singularity analysis).
8. Apply the group G to the obtained solutions. They are invariant under g_0, not however under the entire group G. This will enlarge the class of obtained solutions.
9. Do "physics" with nthe solutions. In particular investigate their stability, asymptotic behavior, etc.

The conditions (1) and (2) can be relaxed. If some of the variables ξ_i depend on u_α, we proceed in the same manner and obtain implicit solutions. If condition (2) is not satisfied, we may be able to obtain "partially invariant solutions" [], or even invariant solutions by a different method [].

2 The symmetry group of a system of differential equations

2.1 Formulation of the problem

We are given the system of differential equations:

$$\Delta^i(x, u, u_{(1)}, u_{(2)}, \ldots, u_{(n)}) = 0, \tag{5}$$

in which $x \in \mathcal{R}^p$, $u \in \mathcal{R}^q$, $i = 1, \ldots, m$, $u_{(k)}$ are all partial derivatives of order k and p, q, m, n are arbitrary positive integers.

G is a local Lie group of local point transformations acting on $M \subset X \times U$, with $X \sim \mathcal{R}^p$ and $U \sim \mathcal{R}^q$.

"Local group": defined in the neighbourhood of $e \in G$.

"Local transformation": defined in the neighbourhood of the origin (x, u)

$$(x,u) \to (\tilde{x}_g(x,u), \tilde{u}_g(x,u)) \qquad (6)$$
$$f(x) \to \tilde{f}(\tilde{x}) = g \circ f(x).$$

"Point": $(\tilde{x}, \tilde{u})$ depends on (x,u) only , $u = f(x)$ solution $\Rightarrow$ $\tilde{u}(\tilde{x}) = g \circ f(x)$ solution.

Prolongation

Prolongations of a function $f(x)$: the function together with its derivatives

$$f : X \to U$$

$$\mathrm{pr}^{(n)} f(x) : \{f(x), f_{x_i}(x), \dots, f_{x_{j1},\dots,x_{jn}}(x)\}.$$

Prolongation of a transformation: the transformation of variables, functions and their derivatives

$$G : (x,u) \to (\tilde{x}(x,u), \tilde{u}(x,u)) \in M$$

$$f(x) \to \tilde{f}(\tilde{x}) = g \circ f(x)$$

$$\mathrm{pr}^{(n)} G : (x, u = f(x), f_{x_i}(x), \dots, f_{x_{j1},\dots,x_{jn}}(x)) \mapsto (\tilde{x}, \tilde{f}(\tilde{x}), \tilde{f}_{\tilde{x}_i}(\tilde{x}), \dots, \tilde{f}_{\tilde{x}_{j1},\dots,\tilde{x}_{jn}}(\tilde{x})).$$

If a function f is given then its prolongation $\mathrm{pr}^{(n)} f$ is known.

Transformation G given $\Rightarrow$ $\mathrm{pr}^{(n)} G$ known.

More general transformations can also depend on first derivatives:

$$\tilde{x} = \tilde{x}(x, u, u_{x_i}),$$
$$\tilde{u} = \tilde{u}(x, u, u_{x_i}),$$

on higher derivatives

$$\tilde{x} = \tilde{x}(x, u, u_{x_i}, u_{x_i x_k}, \dots),$$
$$\tilde{u} = \tilde{u}(x, u, u_{x_i}, u_{x_i x_k}, \dots),$$

or also in integrals of the dependent variables.

and the same generalization for $\tilde{u}$.

Symmetry group: Global approach, use the chain rule

$$
\begin{aligned}
&\tilde{u} = \Omega_g(x,u), \quad \tilde{x} = \Lambda_g(x,u) \\
&\tilde{u}_{\tilde{x}} = (\Omega_u u_x + \Omega_x)(x_{\tilde{x}} + x_{\tilde{u}}\tilde{u}_{\tilde{x}}) \\
&\tilde{u}_{\tilde{x}} = \frac{(\Omega_u u_x + \Omega_x)x_{\tilde{x}}}{(1 - \Omega_u u_x - \Omega_x)x_{\tilde{u}}} \\
&\tilde{u}_{\tilde{x}\tilde{x}} = \cdots \Rightarrow
\end{aligned} \tag{7}
$$

substitute into the original system of equations and require that the transformed quantities satisfy the same equations. This leads to a system of nonlinear equations for $\Omega(x,u)$, $\Lambda(x,u)$ (or $\tilde{\Lambda}(\tilde{x},\tilde{u})$). Solving them is usually more difficult than solving the original system.

Symmetry group: Infinitesimal approach

The essence of Lie group theory: the Lie algebra catches all continuous phenomena.

$$
\begin{aligned}
&\tilde{u}(\tilde{x}) = u(x) + \varepsilon\phi(x,u) \\
&\tilde{x} = x + \varepsilon\xi(x,u) \\
&\text{or } x = \tilde{x} - \varepsilon\xi(\tilde{x},\tilde{u}) \\
&\tilde{u}_{\tilde{x}} = [u_x + \varepsilon(\phi_x + \phi_u u_x)][1 - \varepsilon(\xi_x + \xi_u \tilde{u}_{\tilde{x}})]
\end{aligned} \tag{8}
$$

Keep only order $\varepsilon \Rightarrow$ linear equations for $\xi(x,u)$ and $\phi(x,u)$.

Reformulation

Lie group G , its Lie algebra L, $G \simeq \exp L$.

Local Lie group of local point transformations $\simeq$ Lie algebra of vector fields:

$$
\hat{v} = \sum_{i=1}^{p} \xi^i(x,u)\partial_{x_i} + \sum_{\alpha=1}^{q} \phi_\alpha(x,u)\partial_{u_\alpha}. \tag{9}
$$

The transformations are obtained by integrating:

$$
\begin{aligned}
&\frac{\partial \tilde{x}^i}{\partial \lambda} = \xi^i(\tilde{x},\tilde{u}) \text{ with } \tilde{x}^i|_{\lambda=0} = x^i \text{ and} \\
&\frac{\partial \tilde{u}_\alpha}{\partial \lambda} = \phi(\tilde{x},\tilde{u}), \text{ with } \tilde{u}_\alpha|_{\lambda=0} = u_\alpha.
\end{aligned} \tag{10}
$$

Conversely,

$$\frac{\partial \tilde{x}^i}{\partial \lambda} = \frac{\partial \Lambda(x,u)}{\partial \lambda}|_{\lambda=0} = \xi^i(x,u) \qquad (11)$$
$$\frac{\partial \tilde{u}_\alpha}{\partial \lambda} = \frac{\partial \Omega(x,u)}{\partial \lambda}|_{\lambda=0} = \phi_\alpha(x,u).$$

The above gives the relation between a one parameter group of local transformations and the one-dimensional Lie algebra that corresponds to it.

Entire group: compose the one-parameter groups

$$g = g_1 g_2 \cdots g_n.$$

2.2 Prolongation of vector fields and the symmetry algorithm

The vector fields

$$\hat{v} = \xi^i \partial_{x_i} + \phi^\alpha \partial_{u_\alpha}$$

act on functions $F(x,u)$ of the independent and dependent variables. The n-th prolongation of a vector field $\mathrm{pr}^{(n)}\,\hat{v}$ acts on $F(x,u,u_{x_i},\dots,u_{x_n})$ and has the form

$$\mathrm{pr}^{(n)}\,\hat{v} = \hat{v} + \sum_{\alpha=1}^{q}\sum_{k=1}^{n}\sum_{J} {\phi_\alpha}^J \frac{\partial}{\partial {u_J}^\alpha}$$
$$J \equiv J(k) = (j_1, j_2, \dots, j_k) \text{ and } 1 \le j_k \le p \qquad (12)$$
$$k = j_1 + j_2 + \dots + j_k.$$

An explicit formula for the coefficients is given by Olver []. Here we just give a recursive formula:

$$\mathrm{pr}^{(1)}\,\hat{v} = \hat{v} + \sum_{\alpha=1}^{q}\sum_{i=1}^{p} {\phi_\alpha}^i(x,u,u_{x_i}) \frac{\partial}{\partial {u_{x_i}}^\alpha}$$
$${\phi_\alpha}^i = D_{x_i}\phi_\alpha - \sum_{k=1}^{p} (D_{x_i}\xi^k) {u_{x_k}}^\alpha \qquad (13)$$
$${\phi_\alpha}^{J,k}(x,u,u^{(1)},\dots,u^{(k)}) = D_{x_k}{\phi_\alpha}^J - \sum_{a=1}^{p} (D_k \xi^a) {u_{J,x_a}}^\alpha$$

Here $D_i \equiv D_{x_i}$ is the total derivative

$$D_x = \frac{\partial}{\partial x_i} + \sum_\alpha \frac{\partial u_\alpha}{\partial x_i}\frac{\partial}{\partial u_\alpha} + \sum_{\alpha,J} \frac{\partial u_{\alpha,x_J}}{\partial x_i}\frac{\partial}{\partial u_{\alpha,x_J}} + \dots$$

The integration of $\hat{v}$ provides us with the transformation g.
The integration of $pr^{(n)}\hat{v} \Rightarrow \mathrm{pr}^{(n)}\,g$.

The basic properties of the prolongation of vector fields are

$$\begin{aligned} \mathrm{pr}^{(n)}[\hat{v},\hat{w}] &= [\mathrm{pr}^{(n)}\,\hat{v},\mathrm{pr}^{(n)}\,\hat{w}] \\ \mathrm{pr}^{(n)}(\alpha\hat{v}+\beta\hat{w}) &= \alpha\,\mathrm{pr}^{(n)}\,\hat{v}+\beta\,\mathrm{pr}^{(n)}\,\hat{w}. \end{aligned} \tag{14}$$

This ensures that the prolongations provide a Lie algebra isomorphic to that of the vector fields themselves.

Calculation of $\mathrm{pr}^{(n)}\,\hat{v}$: trivial with computer algebra.

Algorithm for determining symmetry algebra L: $\hat{v}$ is in L if:

$$\mathrm{pr}^{(n)}\,\hat{v}\Delta^i|_{\Delta^l=0} = 0 \text{ with } i,l = 1,\dots,m. \tag{15}$$

In practice the system of equations (5) should be viewed as a system of algebraic equations for m of the highest derivatives of u_α. Let us call them $v_1,\dots,v_m$.

Replace v_i in equations (15). The coefficients ξ_i and ϕ^α depend on x and u, but not on the derivatives u_{x_i}, so $\mathrm{pr}^{(n)}\,\hat{v}\Delta^i|_{\Delta^l=0} = 0$ yields a set of "determining equations".

Determining equations: a system of linear PDEs for $\xi_i(x,u),\phi_\mu(x,u)$.

They are linear because of the infinitesimal approach in which ε^2 and higher powers of ε are ignored.

Usually (except for $1-st$ order ODEs) the system of determining equations is overdetermined.

The order d of equations in the determining system satisfies

$$1 \le d \le N = \text{ order of studied equation.}$$

For instance if the equation $E(x,t,u,u_x,u_t,u_{xx},u_{xt},u_{tt}) = 0$ is polynomial in the derivatives, then

$$\mathrm{pr}^{(2)}\,\chi E|_{E=0} = \sum_{abcd}(\psi_{abcd})({u_x}^a{u_t}^b{u_{xx}}^c{u_{xt}}^d) = 0 \Rightarrow \psi_{abcd} = 0\ \forall a,b,c,d \tag{16}$$

The expressions ψ_{abcd} used are linear in $\xi(x,u)$ and $\phi(x,u)$ and in their first and second derivatives with respect to x,t and u.

The following possibilities occur:

1. The determining equations only admit the trivial solution $\Rightarrow \xi_i = 0,\ \phi_\alpha = 0$ and symmetry approach is of no use.
2. Their general solution depends on r integration constants, $1 \le r < \infty$.

$$\dim L = \dim G = r$$

3. Their general solution depends on arbitrary functions of x and u: the symmetry algebra is infinite-dimensional.

Computer programs exist in MACSYMA, REDUCE,MATHEMATICA, MAPLE and other languages. They produce the determining equations and partly, or completely, solve them.

The use of differential Gröbner bases provides a systematic method of solving the determining equations (and any overdetermined system of differential equations).

2.3 First example: Variable coefficient KdV equation

$$E \equiv u_t + f(x,t)uu_x + g(x,t)u_{xxx} = 0 \tag{17}$$

in which $f \neq 0, g \neq 0$ are smooth functions and we have, $p = 1, q = 2, N = 3$

$$\begin{aligned}
X &= \tau(x,t,u)\partial_t + \xi(x,t,u)\partial_x + \phi(x,t,u)\partial_u \\
\mathrm{pr}^{(3)}\, X &= \tau\partial_t + \xi\partial_x + \phi\partial_u + \phi^t\partial_{u_t} + \phi^x\partial_{u_x} + \phi^{xx}\partial_{u_{xx}} + \phi^{tt}\partial_{u_{tt}} + \\
&\quad \phi^{tx}\partial_{u_{xx}} + \phi^{xxx}\partial_{u_{xxx}} + \phi^{xxt}\partial_{u_{xxt}} + \phi^{xtt}\partial_{u_{xtt}} + \phi^{ttt}\partial_{u_{ttt}} \\
\mathrm{pr}^{(3)}\, XE|_{E=0} &= g\phi^{xxx} + \phi^t + fu\phi^x + fu_x\phi + (\xi f_x + \tau f_t)uu_x + \\
&\quad (\xi g_x + \tau g_t)u_{xxx}|_{E=0} = 0
\end{aligned}$$

Choose

$$v_1 \equiv u_{xxx} = \frac{1}{g}(-u_t + fuu_x) \tag{18}$$

$$\begin{aligned}
\phi^t &= D_t\phi - (D_t\tau)u_t - (D_t\xi)u_x \\
&= \phi_t - \xi_t u_x + (\phi_u - \tau_t)u_t - \xi_u u_x u_t - \tau_u u_t^2, \\
\phi^x &= \phi_x + (\phi_u - \xi_x)u_x - \tau_x u_t - \xi_u u_x^2 - \tau_u u_x u_t, \\
\phi^{xx} &= D_x\phi^x - (D_x\xi)u_{xx} - (D_x\tau)u_{xt}, \\
\phi^{xxx} &= D_x\phi^{xx} - (D_x\xi)u_{xxx} - (D_x\tau)u_{xxt} = \phi_{xxx} + \\
&\quad 3(\phi_{xxu} - \xi_{xxx})u_x - \tau_{xxx}u_t + 3(\phi_{xuu} - \xi_{xxu})u_x^2 \\
&\quad -3\tau_{xxu}u_x u_t + (\phi_{uuu} - 3\xi_{xuu})u_x^3 - 3\tau_{xuu}u_x^2 u_t + \\
&\quad -\xi_{uuu}u_x^4 - \tau_{uuu}u_x^3 u_t + 3(\phi_{xu} - \xi_{xx})u_{xx} - 3\tau_{xx}u_{xt} + \\
&\quad 3(\phi_{uu} - 3\xi_{xu})u_x u_{xx} - 6\tau_{xu}u_x u_{xt} - 3\tau_{xu}u_t u_{xx} + \\
&\quad -6\xi_{uu}u_x^2 u_{xx} - 3\tau_{uu}u_x u_t u_{xx} - 3\tau_{uu}u_x^2 u_{xt} + \\
&\quad -3\xi_u u_{xx}^2 - 3\tau_u u_{xt}u_{xx} + (\phi_u - 3\xi_x)u_{xxx} + \\
&\quad -3\tau_x u_{xxu} - \tau_u u_t u_{xxx} - 4\xi_u u_x u_{xxx} - 3\tau_u u_x u_{xxt}
\end{aligned}$$

Throughout u_{xxx} must be replaced using equation (18).
Coefficients of $u_{xxt}u_x, u_{xxt}, (u_{xx})^2, u_x u_{xx}$ and u_{xx} must vanish:

$$\begin{aligned}
\tau_u &= 0 \\
\tau_x &= 0 \\
\xi_u &= 0 \\
\phi_{uu} &= 0 \\
\phi_{xu} - \xi_{xx} &= 0.
\end{aligned}$$

The two conditions $\tau_u = 0$ and $\xi_u = 0$ imply that the transformation is fiber preserving.

Thus, $\forall f(x,t), g(x,t)$:

$$\tau = \tau(t), \ \xi = \xi(x,t), \ \phi = A(x,t)u + B(x,t). \tag{19}$$

$\phi_{uu} = 0 \Leftrightarrow$ linear transformations.

Substitute into remaining equations $\Rightarrow$ great simplification.

The dependence on u is now explicit. Hence the coefficient of u^2 must vanish

$$\phi_{xu} = 0 \Rightarrow \xi_{xx} = 0.$$

So for all functions $f(x,t), g(x,t)$ we have:

$$\tau = \tau(t), \ \xi = b(t)x + c(t), \ \phi = a(t)u + R(x,t). \tag{20}$$

Thus the three functions $\tau(x,t,u), \xi(x,t,u)$ and $\phi(x,t,u)$ have been expressed in terms of four functions of one variable and one function of two variables.

The remaining determining equations are

$$\begin{aligned}
u_x &: -\dot{b}x - \dot{c} + Rf = 0 \\
u &: \dot{a} + fR_x = 0 \\
uu_x &: (bx+c)f_x + \tau f_t + (\dot{\tau} - b + a)f = 0 \\
u_t &: (bx+c)g_x + \tau g_t + (\dot{\tau} - 3b)g = 0 \\
1 &: gR_{xxx} + R_t = 0.
\end{aligned} \tag{21}$$

To proceed further, we can either make specific assumptions on f and g or perform a complete analysis, a symmetry classification of variable coefficient KdV equations.

Assume: $f \equiv f_0 =$ const, $g \equiv g_0 =$ const.

For definiteness we put $f_0 = g_0 = 1$. We then obtain from (21)

$$\begin{aligned} \xi &= c_1 + c_2 x + c_3 t \\ \tau &= c_4 + 3c_2 t \\ \phi &= -2c_2 u + c_3. \end{aligned} \tag{22}$$

Lie algebra L is:

$$\begin{aligned} &P_0 = \partial_t \Rightarrow \text{ time translations} \\ &P_1 = \partial_x \Rightarrow \text{ space translations} \\ &D = x\partial_x + 3t\partial_t - 2u\partial_u \text{ dilatations} \\ &B = t\partial_x + \partial_u \text{ Galilei transformations.} \end{aligned} \tag{23}$$

The nonzero commutation relations are

$$\begin{aligned} &[P_0, B] = P_1, \qquad [P_1, D] = P_1 \\ &[P_0, D] = 3P_0, \qquad [B, D] = -2B. \end{aligned} \tag{24}$$

The next step is to identify L as an abstract Lie algebra.
We have:
$L \sim \{D, B, P_0, P_1\}$ is solvable and its nilradical (maximal nilpotent ideal) is $NR(L) \sim \{B, P_0, P_1\}$.

The corresponding finite transformations are:

$$\begin{aligned} \tilde{t} &= e^{3\lambda}(t - t_0) \\ \tilde{x} &= e^{\lambda}[x - x_0 + v(t - t_0)] \\ \tilde{u}(\tilde{x}, \tilde{t}) &= u(x, t) + v. \end{aligned} \tag{25}$$

Thus from any solution $u(x,t) =$ we obtain a 4-parameter family of solutions:

$$\begin{aligned} \tilde{u}(\tilde{x}, \tilde{t}) &= u(x, t) + v \\ x &= e^{-\lambda}\tilde{x} + x_0 - ve^{-3\lambda}\tilde{t} \\ t &= e^{-3t}\tilde{t} + t_0. \end{aligned} \tag{26}$$

We have found the symmetries group G and its Lie algebra L. So what to do with them?

2.4 Symmetry reduction for the KdV

The aim of symmetry reduction for partial differential equations is to obtain particular solutions by first imposing that the solution be invariant under some subgroup of the symmetry group. For the KdV equation (two independent variables) it suffices to use a one dimensional subgroup. A subgroup classification yields the following representatives of subgroup classes:

$$\begin{aligned} &D,\ B + aP_0 \quad (a = 0, \pm 1), \\ &P_0 + aP_1, \quad (a = 0, \pm 1), \\ &P_1. \end{aligned} \tag{27}$$

Reductions

Let X be the generator of a one dimensional subgroup G_0.To perform symmetry reduction, we must solve the equation

$$\xi u_x + \tau u_t - \phi = 0, \tag{28}$$

together with the KdV equation

$$u_t + uu_x + u_{xxx} = 0. \tag{29}$$

Equation (28) is first solved by the method of characteristics. This will provide the general form of the solution, namely

$$u(x,t) = \alpha(x,t)F(z) + \beta(x,t), \quad z = z(x,t) \tag{30}$$

where α, β and z are known functions.

Equation (30) is then substituted into the KdV equation (29) and we obtain an ordinary differential equation for $F(z)$. This must then be solved.

Let us run through the individual cases in equation (27).

1. $X = P_1$: The reduction formula (30) is $u(x,t) = u(t)$ and the result is only the constant solution

$$u = u_0 \tag{31}$$

2. $X = P_0 + aP_1 = \partial_t + a\partial_x$: The reduction formula (30) is

$$u = u(z), \quad z = x - at \tag{32}$$

and the KdV reduces to

$$-au_\xi + uu_\xi + u_{\xi\xi\xi} = 0. \tag{33}$$

We integrate twice and obtain

$$-au + \frac{1}{2}u^2 + u_{\xi\xi} = b$$
$$-\frac{1}{2}au^2 + \frac{1}{6}u^3 + \frac{1}{2}u_z^2 = bu + c$$
$$u_z^2 = -\frac{1}{3}u^3 + au^2 + 2bu + 2c$$
$$u_z^2 = -\frac{1}{3}(u-u_1)(u-u_2)(u-u_3). \tag{34}$$

Equation (34) is the well known equation for elliptic functions. Its solutions are well known. If the constant roots u_1, u_2, u_3 are all different we obtain finite or singular periodic solutions in terms of elliptic functions. If two, or all the roots coincide, we obtain solutions in terms of elementary functions. In particular, for $b = c = 0$ we obtain the famous one soliton solution:

$$u(x,t) = \frac{3a}{\cosh^2 \frac{1}{2}\sqrt{a}[(x-at)-\xi_0]} \tag{35}$$
$$u \to 0;\ \xi \to \pm\infty$$
$$u \to 3a;\ \xi \to \xi_0$$

Slightly more generally, for $u_1 = u_2 < u_3,\ u_i \in \mathbb{R}$, we have

$$\begin{aligned} u(x,t) &= u_1 + \frac{u_3 - u_1}{\cosh^2 \omega(\xi - \xi_0)} \\ \xi &= x - at;\ \omega = \frac{1}{2\sqrt{3}\sqrt{u_3 - u_1}} \\ \xi &\to \xi_0 \Rightarrow u(x,t) = u_3 \\ \xi &\to \pm\infty \Rightarrow u(x,t) = u_1 \end{aligned} \tag{36}$$

3. $X = B + aP_0 = t\partial_x + \partial_u + a\partial_t$, with $a \neq 0$,
$(B + aP_0)F(x,t,u) = 0, \quad \frac{\mathrm{d}x}{t} = \mathrm{d}u = \frac{\mathrm{d}t}{a}$.
The reduction formula is

$$z = x - \frac{t^2}{2a};\ u = \frac{t}{a} + F(z)$$
$$F_{zzz} + FF_z + \frac{1}{a} = 0$$

Integrating once we obtain the second order equation

$$F_{zz} + \frac{1}{2}F^2 + \frac{1}{a}\xi + b = 0 \tag{37}$$

This is equivalent to the first Painlevé equation, i.e. the equation for the first Painlevé transcendent P_I.

4. $X = B = t\partial_x + \partial_u$: the reduction formula is

$$u = \frac{x}{t} + F(t)$$
$$\frac{\mathrm{d}F}{\mathrm{d}t} + \frac{F}{t} = 0 \Rightarrow F = \frac{c}{t}$$

and the Galilei invariant solution is

$$u(x,t) = \frac{x+c}{t}. \tag{38}$$

5. $X = D = 3t\partial_t + x\partial_x - 2u\partial_u$.
We have

$$z = xt^{-1/3}; \; u = t^{-2/3}F(z)$$
$$F_{zzz} + FF_z - \frac{1}{3}zF_z - \frac{2}{3}F = 0$$

To solve this equation we put

$$F = \frac{\mathrm{d}\omega}{\mathrm{d}z} - \frac{1}{6}\omega^2 \tag{39}$$

and obtain

$$\left(\frac{\mathrm{d}}{\mathrm{d}z} - \frac{1}{3}\omega\right)\left(\omega_{zzz} - \frac{1}{6}\omega^2\omega_z - \frac{1}{3}z\omega_z - \frac{1}{3}\omega\right) = 0.$$

A family of solutions of this equation is obtained by solving

$$\omega_{zz} = \frac{1}{18}\omega^3 + \frac{1}{3}\xi\omega + a. \tag{40}$$

Equation (40) is the equation for the second Painlevé transcendent P_{II}.

2.5 Second example: Modified Kadomtsev-Petviashvili equation

The MKP equation is

$$(u_{xxx} - 2u_x^3 - 4u_t)_x - 6u_{xx}u_y + 3u_{yy} = 0. \tag{41}$$

We search for the symmetry algebra in the form

$$X = \xi\partial_x + \eta\partial_y + \tau\partial_t + \phi\partial_u, \tag{42}$$

in which ξ, τ, η, ϕ are functions of x, y, t, u.

$$\begin{aligned}\mathrm{pr}^{(4)} X = & \xi\partial_x + \tau\partial_t + \eta\partial_u + \phi^x\partial_{u_x} + \phi^y\partial_{u_y} + \phi^t\partial_{u_t} + \phi^{xx}\partial_{u_{xx}} + \phi^{yy}\partial_{u_{yy}} + \\ & \phi^{xt}\partial_{u_{xt}} + \cdots + \phi^{xxx}\partial_{u_{xxx}} + \cdots + \phi^{xxxx}\partial_{u_{xxxx}}\end{aligned}$$

$$\mathrm{pr}^{(4)} XE|_{u_{tx} = \frac{1}{4}[(u_{xxx} - 2u_x^3)_x - 6u_{xx}u_y + 3u_{yy}]} = 0.$$

We used a MACSYMA Package and after simplification we obtain 20 determining equations.

The first ten of them are one term equations, namely

$$\begin{aligned}&\tau_u = 0,\ \tau_x = 0,\ \tau_y = 0,\ \eta_u = 0,\ \eta_x = 0,\ \xi_u = 0\\ &\phi_{uu} = 0,\ \phi_{ux} = 0,\ \phi_{xx} = 0,\ \xi_{xx} = 0.\end{aligned}$$

They imply the following form of the coefficients of the vector field (42)

$$\begin{aligned}&\tau \equiv \tau(t)\\ &\eta \equiv \eta(y,t)\\ &\xi \equiv \xi(x,y,t) = A(y,t)x + B(y,t)\\ &\phi \equiv \phi(x,y,t,u) = R(y,t)u + S(y,t)x + T(y,t).\end{aligned} \tag{43}$$

Substituting (43) into the remaining determining equations, we are left with

$$\begin{aligned}&\tau_t - 2\eta_y + \xi_x = 0\\ &\tau_t + 2\phi_u - 3\xi_x = 0\\ &4\phi_{ut} + 3\xi_{yy} - 4\xi_{xt} = 0\\ &\tau_t - \eta_y + \phi_u - \xi_x = 0 \Rightarrow E_{17}, \ldots, E_{20}.\end{aligned} \tag{44}$$

Solving (44) we obtain the final result

$$X = T(f) + X(h) + Y(g) + U(k), \tag{45}$$

in which f, g, h, k are arbitrary functions of t and

$$\begin{aligned}
T(f) &= f\partial_t + \left(\frac{1}{3}x\dot{f} + \frac{2}{9}y^2\ddot{f}\right)\partial_x + \frac{2}{3}\dot{f}y\partial_y + \left(\frac{2}{9}\ddot{f}xy + \frac{4}{81}y^3\dddot{f}\right)\partial_u \\
X(h) &= h\partial_x + \frac{2}{3}\dot{h}y\partial_u \qquad (46)\\
Y(g) &= g\partial_y + \frac{2}{3}y\dot{g}\partial_x + \frac{1}{3}\left(x\dot{g} + \frac{2}{3}y^2\ddot{g}\right)\partial_u \\
U(k) &= k\partial_u
\end{aligned}$$

The commutation relations are those of a Kac-Moody-Virasoro algebra:

$$\begin{aligned}
[T(f_1), T(f_2)] &= T(f_1\dot{f}_2 - \dot{f}_1 f_2), \qquad (47)\\
[X(h), Y(g)] &= \frac{1}{3}U(h\dot{g} - \dot{h}g),\\
[Y(g_1), Y(g_2)] &= \frac{2}{3}X(g_1\dot{g}_2 - \dot{g}_1 g_2),\\
[T(f), X(h)] &= X\left(f\dot{h} - \frac{1}{3}\dot{f}h\right),\\
[T(f), Y(g)] &= Y\left(f\dot{g} - \frac{2}{3}\dot{f}g\right),\\
[T(f), U(k)] &= U(f\dot{k}).
\end{aligned}$$

Thus, $\{T(f)\}$ is a Virasoro algebra and $\{X(h), Y(g), U(k)\}$ forms a Kac-Moody ideal. If we take f, h and g constant, we get the subalgebra of translations

$$\begin{aligned}
P_0 &= T_1 = \partial_t \\
P_1 &= X(1) = \partial_x \qquad (48)\\
P_2 &= Y(1) = \partial_y.
\end{aligned}$$

If f, g and h are linear, we get dilations, Galilei boosts and "pseudorotations", respectively:

$$\begin{aligned}
D &= T(t) = t\partial_t + \frac{1}{3}x\partial_x + \frac{2}{3}y\partial_y,\\
B &= X(t) = t\partial_x + \frac{2}{3}y\partial_u, \qquad (49)\\
R &= Y(t) = t\partial_y + \frac{2}{3}y\partial_x + \frac{1}{3}x\partial_u.
\end{aligned}$$

For any $k(t)$, $U(k)$ generates gauge transformations:

$$U(k):\ \tilde{u}(x,y,t) = u(x,y,t) + \lambda k(t). \qquad (50)$$

We mention that all integrable equations in $2+1$ dimensions, like the KP equations and all the equations in its hierarchy, the potential KP equation, the Davey-Stewartson equation, the three-wave resonant interaction equation and the generalized Toda field theory equations

$$\begin{aligned} u_{n,xy} &= e^{u_{n-1}-u_n} - e^{u_n-u_{n+1}} \\ u_{nxy} &= \sum_l K_{nl} e^{\sum_m H_{lm} u_m}, \end{aligned}$$

all have an infinite-dimensional Lie point symmetry algebras with a loop algebra structure.

3 Classification of the subalgebras of a finite dimensional Lie algebra

3.1 Formulation of the problem

As we saw in the example of the KdV equation, the classification of the subalgebras of the symmetry algebra L is an important step in symmetry reduction.

More generally, let L be a Lie algebra and G the corresponding Lie group. We wish to classify all subalgebras $L_i \subset L$ into conjugacy classes under the action of the group G. Thus we have

$$\begin{aligned} &L_i \subset L \\ &[L_i, L_i] \subseteq L_i \end{aligned} \tag{51}$$

and two algebras L_i and L_i' are in the same class if they satisfy

$$GL_iG^{-1} \sim L_i'. \tag{52}$$

Two approaches to the subgroup classification have been developed:

1. A structural one: Restrictions on L and L_i are imposed. For instance $L \sim$ simple, L_i maximal , or L_i semisimple , or L_i abelian, [] The dimension $\dim L = n$ is arbitrary.
2. Enumerative: Given a Lie algebra L and an automorphism group G . Needed: a representative list of all subalgebra of L : $L_i \subseteq L$, such that every subalgebra of L is conjugate(under G) to precisely one in the list. Thus: $GLG^{-1} \sim L$. In this case the algebra L and its dimension $\dim L$ is a priori fixed (e. g. $L \sim O(4,2)$, or $L = sl(3,\mathbb{R})$). The classifying group G can be e.g. the group of inner automorphism:

$$G = G_0 \sim \exp L$$

or G can be a larger group, containing G_0, e.g.

$$G = G_D \supset G_0, \quad G_D \text{ discrete.}$$

For our purposes, i.e. symmetry reduction, G must leave the solution set of the equation invariant.

The representative list will consist of the algebras L_i^R such that every subalgebra $L_i \subset L$ is conjugate to precisely one algebra L_i^R in the list:

$$GL_iG^{-1} = L_i^R.$$

Three different algorithms can be used, depending on the structure of L

1. L simple,
2. L direct sum $L \sim L_1 \oplus L_2$,
3. L semidirect sum $L \sim F \oplus N$, $[F,F] \subseteq F$, $[N,N] \subseteq N$, $[F,N] \subseteq N$, $F \neq \emptyset$, $N \neq \emptyset$.

3.2 Subalgebras of a simple Lie algebra

Let L be simple ($\leftrightarrow$ no ideals, i.e. $I \subset L, [L,I] \subseteq I \to I \sim \emptyset$ or $I \sim L$).

The first step is to find all maximal subalgebras $L_M \subset L$. The subalgebra L_M is maximal in L if

$$\begin{aligned} &L_M \subseteq \tilde{L} \subseteq L, \quad [\tilde{L}, \tilde{L}] \subseteq \tilde{L} \\ &\Rightarrow \tilde{L} = L_M \text{ or } \tilde{L} = L. \end{aligned} \tag{53}$$

To proceed further we consider a finite dimensional faithful representation $E(L)$ of L: $E(L)$ are matrices acting on space V, $\dim V = N_0 < \infty$. The matrices $E(L_M) \subset E(L)$ can be embedded reducibly or irreducibly in $E(L)$.

1. Reducible case
In this case there exists a subspace V_0 of V left invariant by all matrices $E(L_M)$:
$$V_0 \subset V, \quad E(L_M)V_0 \subseteq V_0. \tag{54}$$
We classify the subspaces V_0 into equivalence classes under the action of $E(G)$ (matrices representing the classifying group G) and choose a representative of each space

$$E(G)V_0 \sim V_0^R. \tag{55}$$

We then find the representative maximal subalgebra L_M^R leaving each V_0^R invariant by imposing

$$E(L_M^R)V_0^R \subseteq V_0^R. \tag{56}$$

This has reduced the problem of finding all reducibly embedded maximal subalgebras of the simple Lie algebra L to a problem of linear algebra. The classification of subspaces of V is actually very easy. If the classifying group G is $SL(n,\mathbb{R})$, or $SL(n,\mathbb{C})$, then each subspace is completely characterized by its dimension. If the group G allows an invariant metric in the space V, then each subspace is completely characterized by its dimension and signature (e.g. for $G \sim O(p,q)$ or $G \sim U(p,q)$).

2. Irreducible case

If a subalgebra $L_M \subset L$ is embedded irreducibly in a given representation $E(L)$, then L_M must itself be simple, semisimple or reductive (a direct sum of simple and abelian Lie algebras). All semisimple subalgebras of the complex simple Lie algebras have been classified by Dynkin []. For the real case a classification is due to Cornwell [] and to Komrakov []. The reductive ones are obtained from the simple and semisimple ones L_0 by adding the centralizer $\mathrm{cent}(L_0, L)$ of L_0 in L:

$$\mathrm{cent}(L_0, L) = \{x \in L | [x, L_0] = 0\}. \tag{57}$$

Once we have found all the maximal subalgebras of a simple Lie algebra L, we proceed recursively. If the algebra L_M^R is simple, we apply the same algorithm. If not, we apply one of the other algorithms, mentioned above.

3.3 Example: Maximal subalgebras of $o(4,2)$

The algebra $o(4,2)$ is of physical interest for many reasons. One of them is that is isomorphic to the Lie algebra of the conformal group $C(3,1)$ of compactified Minkowski space.

As the representation $E(L)$, let us consider the defining representation of $o(4,2)$, namely real 6×6 matrices X, satisfying

$$\begin{aligned} &X \in \mathbb{R}^{6\times 6}, \quad XK + KX^T = 0 \\ &K = K^T, \quad \mathrm{sgn}\ K = (4,2). \end{aligned} \tag{58}$$

We shall need three different realizations of the metric matrix K, namely

$$K_0 \equiv I_{4,2} = \mathrm{diag}(1,1,1,1,-1,-1) \tag{59}$$

$$K_1 = \begin{pmatrix} & & 1 \\ & I_{3,1} & \\ 1 & & \end{pmatrix}, \quad K_2 = \begin{pmatrix} & & I_2 \\ & I_2 & \\ I_2 & & \end{pmatrix}.$$

The subspaces V_i of $V \sim \mathbb{R}^6$ to be considered can have the following dimensions and signatures:

$$\begin{aligned} &\dim V_i = 1 \quad \mathrm{sgn}\ V_i : (+), (-), (0) \\ &\dim V_i = 2 \quad \mathrm{sgn}\ V_i : (++), (--), (00), (+0), (+-) \\ &\dim V_i = 3 \quad \mathrm{sgn}\ V_i : (+++), (++-), (++0), (+--), (+-0), (+00). \end{aligned} \tag{60}$$

Since we are only interested in maximal subalgebras of $0(4,2)$ we need only consider nondegenerate spaces (no isotropic vectors in any orthogonal basis), or completely isotropic spaces. The isotropic spaces are of dimension 1 or 2, i.e. (0) and (00) in equation (60). The two dimensional space $(+-)$ can be omitted, since it leads to a nonmaximal subalgebra. We are left with the spaces $(+), (-), (++), (--), (+++)$ and $(++-)$. Higher dimensional nondegenerate spaces need not be considered, since any subalgebra that leaves a nondegenerate subspace invariant also leaves its orthogonal complement invariant.

For nondegerate spaces we use the metric K_0 of equation (59) and choose the representative spaces as

$$(+) \sim \begin{pmatrix} x \\ 0 \\ 0 \\ 0 \\ 0 \\ 0 \end{pmatrix} \quad (-) \sim \begin{pmatrix} 0 \\ 0 \\ 0 \\ 0 \\ 0 \\ x \end{pmatrix} \quad (++) \sim \begin{pmatrix} x \\ y \\ 0 \\ 0 \\ 0 \\ 0 \end{pmatrix} \quad (--) \sim \begin{pmatrix} 0 \\ 0 \\ 0 \\ 0 \\ x \\ y \end{pmatrix} \tag{61}$$

$$(+++) \sim \begin{pmatrix} x \\ y \\ z \\ 0 \\ 0 \\ 0 \end{pmatrix} \quad (++-) \sim \begin{pmatrix} x \\ y \\ 0 \\ 0 \\ z \\ 0 \end{pmatrix} .$$

It is easy to see that the spaces $(+)$ and $(-)$ are left invariant by the simple Lie algebras $o(3,2)$ and $o(4,1)$, respectively (the two de Sitter algebras). The other spaces in the list (61) are left invariant by the following semisimple Lie algebras:

$$\begin{aligned} (++) &\sim o(2) \oplus o(2,2) \\ (--) &\sim o(4) \oplus 0(2) \\ (+++) &\sim o(3) \oplus o(1,2) \\ (++-) &\sim o(2,1) \oplus o(2,1). \end{aligned} \tag{62}$$

The one-dimensional isotropic subspace (0) is best visualized using the metric K_1. In this metric the space (0) and the subalgebra leaving this space invariant can be written as

$$(0) \sim \begin{pmatrix} x \\ 0 \\ 0 \\ 0 \\ 0 \\ 0 \end{pmatrix} \quad E(L(0)) \sim \begin{pmatrix} a & \beta & 0 \\ 0 & \Lambda & -I_{3,1}\beta^T \\ 0 & 0 & -a \end{pmatrix} \quad \Lambda I_{3,1} + I_{3,1}\Lambda^T = 0 \qquad (63)$$

$$\Lambda \in \mathbb{R}^{4\times 4}, \quad a \in \mathbb{R}, \quad \beta \in \mathbb{R}^{1\times 4}.$$

The algebra (63) is one of the two different maximal parabolic subalgebras of $o(4,2)$ and it is isomorphic to the Lie algebra of the similitude group (Poincaré group extended by dilations). We denote this algebra sim$(3,1)$. The two-dimensional isotropic space (00) and the maximal parabolic subalgebra of $0(4,2)$ leaving it invariant are best seen using the metric K_2 of equation (59). We have

$$(00) \sim \begin{pmatrix} x \\ y \\ 0 \\ 0 \\ 0 \\ 0 \end{pmatrix} \quad E(L(0)) \sim \begin{pmatrix} A & B & C \\ 0 & D & -B^T \\ 0 & 0 & -A^t \end{pmatrix} \qquad (64)$$

$$A, B, C, D \in \mathbb{R}^{2\times 2}, \quad C + C^T = 0, \quad D + D^T = 0.$$

This Lie algebra has been called opt$(3,1)$, the "optical" algebra of $3+1$-dimensional Minkowski space.

Now let us turn to irreducibly imbedded maximal subalgebras of $o(4,2)$. The only semisimple algebra that has a six-dimensional real pseudo-orthogonal irreducible representation with the signature $(4,2)$ is the simple Lie algebra su$(2,1)$. Adding the centralizer u(1) we obtain u$(2,1)$ which in the realization of $o(4,2)$ using the metric K_0 can be written as

$$\begin{pmatrix} 0 & a & d & e & f & g \\ -a & 0 & -e & d & -g & f \\ -d & e & 0 & b & h & j \\ -e & -d & -b & 0 & -j & h \\ f & -g & h & -j & 0 & c \\ g & f & j & h & -c & 0 \end{pmatrix}, \quad a,b,c,d,e,f,g,h \in \mathbb{R}. \tag{65}$$

Thus, $o(4,2)$ has nine nonisomorphic maximal subalgebras, listed above.

To proceed further, we would apply the same method to find the maximal subalgebras of $o(4,1)$ and $o(3,2)$, a lower dimensional task. To find the subalgebras of $o(2,2) \oplus o(2)$, $o(4) \oplus o(2)$, $o(3) \oplus o(1,2)$ and $o(2,1) \oplus o(2,1)$, we would apply a method due to Goursat []. The subalgebras of the nonsemisimple algebras $\mathrm{sim}(3,1)$ and $\mathrm{opt}(3,1)$ are found using the algorithm for semidirect sums, which we now proceed to describe.

3.4 Subalgebras of semidirect sums

Let us consider a Lie algebra $L = F \oplus N$ with the structure

$$[F,F] \subseteq F, \ [N,N] \subseteq N, \ [F,N] \subseteq N, \quad F \neq \emptyset, \ N \neq \emptyset. \tag{66}$$

This could for instance be a Levi decomposition. Then F would be semisimple and N would be the maximal solvable ideal, the radical. Alternatively, F could be solvable and N its nilradical. The decomposition (66) for F general is not necessarily unique. Unless F is simple, such a decomposition exists (it may reduce to a direct sum, if we have $[F,N]=0$). Once a decomposition $L \sim F \oplus N$ is chosen, two types of subalgebras exist.

1. Splitting They are themselves semidirect sums: $L_i = F_i \oplus N_i$, $F_i \subseteq F$, $N_i \subseteq N$ (up to conjugacy)
2. Nonsplitting Subalgebras of L, not conjugate under G to a splitting one.

The classification algorithm consists of several steps:

1. The first step is to form a representative list $S(F)$ of subalgebras of F classified under $G_F =< \exp F >$

$$S(F) \equiv \{F_1 = \{0\}, F_2, F_3, \ldots, F_N \equiv F\}. \tag{67}$$

The list should be "normalized", for each F_i its normalizer $\mathrm{nor}(F_i, F)$ is included in $S(F)$

$$\text{nor}(F_i, F) = \{x \in F | [x, F_i] \subseteq F_i\}. \tag{68}$$

For each $F_i \in S(F)$ construct the normalizer of F_i in the group G_F

$$\text{Nor}(F_i, G_F) = \{g \in G_F | gF_i g^{-1} \subseteq F_i\}.$$

Comment: We always have

$$\text{Nor}(F_i, G_F) \supseteq < \exp(\text{nor}(F_i, F)) >$$

2. The second step is to classify all splitting subalgebras of L

$$L_i \sim F_i \oplus N_i, \;\; F_i \subseteq F, N_i \subseteq N. \tag{69}$$

Procedure:

a) For each $F_i \in S(F)$ find all invariant subalgebras $\tilde{N}_{i,a} \subseteq N$

$$[F_i, \tilde{N}_{i,a}] \subseteq \tilde{N}_{i,a}, \tag{70}$$

$\tilde{N}_{i,a} \subseteq N \; [\tilde{N}_{i,a}, \tilde{N}_{i,a}] \subseteq \tilde{N}_{i,a}$

Comments.

Include $\tilde{N}_{i,0} = \{0\}, \;\; N_{i,n} = \{N\}$ in all cases.

N abelian $\rightarrow$ invariant subspaces $\equiv$invariant subalgebras.

b) For each $F_i \in S(F)$ classify the invariant s.a. $\tilde{N}_{i,a}$ into conjugacy classes under $\text{Nor}(F_i, G_F)$. Choose a representative $N_{i,a}$ of each class.

c) Form a representative list of all splitting subalgebras of L:

$$S_1(L) = \{L_{i,a} \subseteq L | L_{i,a} = F_i \oplus N_{i,a}\}. \tag{71}$$

Comments.

$S(F) \subset S_1(L)$

The list $S_1(L)$ should be normalized.

d) For each $L_{i,a} \in S_1(L)$ find its normalizer in the entire classification group G

$$\text{Nor}(L_{i,a}, G). \tag{72}$$

3. The third step is to classify all nonsplitting subalgebras of L

Starting point: List $S_1(L)$ of splitting subalgebras.

Procedure

a) For each s.a. $L_{i,a} \subset S_1(L)$ choose a basis:

$$L_{i,\alpha} = \{B_a, X_j\},\ B_a \in F, X_j \in N \qquad (73)$$
$$1 \le a \le f_i = \dim F_i, 1 \le j \le r = \dim N_{i,\alpha}.$$

Complement basis to one of N:

$$N = \{X_1, \dots, X_r, Y_1, \dots, Y_s\},\ r + s = \dim N$$

b) For each $L_{i,\alpha} \subset S_1(F)$ construct
$V = \{B_a + \sum_{\mu=1}^{s} c_{a\mu} Y_\mu, X_j\}$, and $a = 1, \dots, f_i; j = 1, \dots, r$
V: a vector space
Require: it should be a Lie algebra

$$[V, V] \subseteq V. \qquad (74)$$

Equation (74) is a system of equations for $c_{a\mu}$
Given: commutation relations for L:

$$\begin{aligned}
&[B_a, B_b] = f_{ab}^c B_c,\\
&[B_a, X_k] = \alpha_{ak}^l X_l,\\
&[X_i, X_j] = \omega_{ij}^m X_m,\\
&[B_a, Y_\mu] = \rho_{a\mu}^\nu Y_\nu + \sigma_{a\mu}^m X_m,\\
&[Y_\mu, Y_\nu] = \beta_{\mu\nu}^\sigma Y_\sigma + \gamma_{\mu\nu}^m X_m,\\
&[X_i, Y_\mu] = \lambda_{i\mu}^\nu Y_\nu + \tau_{i\mu}^m X_m.
\end{aligned} \qquad (75)$$

The system (74) implies a system of equations for $c_{a\mu}$

$$c_{b\nu}\rho_{a\nu}^\alpha - c_{a\mu}\rho_{b\mu}^\alpha - c_{c\alpha} f_{ab}^c = -c_{a\mu}c_{b\nu}\beta_{\mu\nu}^\alpha \qquad (76)$$
$$c_{a\mu}\lambda_{j\mu}^\nu = 0.$$

In general solving the equations (76) is a problem in algebraic geometry. This reduces to a problem in linear algebra if we have $\beta_{\mu\nu}^\alpha = 0$. A specially simple case occurs if the ideal N is abelian: $[N, N] = 0$. Then we have

$$\beta_{\mu\nu}^\sigma = \gamma_{\mu\nu}^m = \lambda_{i\mu}^\nu = \tau_{i\mu}^m = \omega_{ij}^m = 0. \qquad (77)$$

Equations for $c_{a\mu}$ reduce to

$$c_{b\nu}\rho^{\alpha}_{a\nu} - c_{a\mu}\rho^{\alpha}_{b\mu} - c_{c\alpha}f^{c}_{ab} = 0. \tag{78}$$

This implies that the coefficients $c_{a\mu}$ are 1-cocycles. Thus, if $c_{a\mu}$ are solutions of equation (78) and N is abelian, then V is a Lie algebra.

c) We must now classify the Lie algebras V under the action of the group

$$\begin{aligned}\tilde{G} &= \mathrm{Nor}(L_{i\alpha}, G) \oplus \mathrm{Nor}(N_{i\alpha}, G_N) \\ G_N &= \exp N.\end{aligned} \tag{79}$$

If N is abelian, then $\mathrm{Nor}(N_{i\alpha}, G_N) = G_N$
Further we just consider the abelian case:
First eliminate trivial coboundaries from the cocycles:

$$\begin{aligned}e^{\lambda_\alpha Y_\alpha} B_a e^{-\lambda_\alpha Y_\alpha} &= B_a + \lambda_\alpha [Y_\alpha, B_a] \\ e^{\lambda_\alpha Y_\alpha} X_j e^{-\lambda_\alpha Y_\alpha} &= X_j.\end{aligned} \tag{80}$$

The algebra (80) is a splitting one.
Coboundaries:

$$S_{a\mu} = -\lambda_\alpha \rho^{\mu}_{a\alpha}. \tag{81}$$

Simplify cocycles by adding coboundaries:

$$c_{a\mu} \to c_{a\mu} + S_{a\mu} \quad . \tag{82}$$

By choice of λ_α arrange $c_{a\mu} \to 0$ whenever possible. If all $c_{a\mu} \to 0$ then the algebra is splitting.
Once coboundaries are eliminated: classify remaining cocycles under the group

$$\mathrm{nor}(L_{i\alpha}, G)$$

.

d) Form final list:

$$S(L) = S_1(L) \cup S_2(L). \tag{83}$$

$S_1(L)$: splitting
$S_2(L)$: nonsplitting

Comments:

It is usually convenient to reorder the final list by dimension and isomorphism classes:

(i) Decomposable algebras: decompose into direct sums.
(ii) Find Levi decomposition $L_{i\alpha} = S \oplus R$ for all indecomposable subalgebras.
(iii) Solvable: find nilradical.

Decomposition $L = F \oplus N$: usually not unique. If possible: N abelian
Procedure iterative: apply first to F.

3.5 Example: All subalgebras of $sl(3,\mathcal{R})$ classified under the group $SL(3,\mathcal{R})$

A. Preliminaries

The algebra $L \sim sl(3,\mathcal{R})$ is realized as the algebra of 3×3 traceless real matrices

$$X = \begin{vmatrix} a+b & c & d \\ e & -a+b & f \\ g & h & -2b \end{vmatrix} \tag{84}$$

Basis:

$$K = \begin{vmatrix} 1 & 0 & 0 \\ 0 & -1 & 0 \\ 0 & 0 & 0 \end{vmatrix}, \quad A_1 = \begin{vmatrix} 0 & 1 & 0 \\ 0 & 0 & 0 \\ 0 & 0 & 0 \end{vmatrix}, \quad A_2 = \begin{vmatrix} 0 & 0 & 0 \\ 1 & 0 & 0 \\ 0 & 0 & 0 \end{vmatrix},$$

$$D = \begin{vmatrix} 1 & 0 & 0 \\ 0 & 1 & 0 \\ 0 & 0 & -2 \end{vmatrix}, \quad X_1 = \begin{vmatrix} 0 & 0 & 1 \\ 0 & 0 & 0 \\ 0 & 0 & 0 \end{vmatrix}, \quad X_2 = \begin{vmatrix} 0 & 0 & 0 \\ 0 & 0 & 1 \\ 0 & 0 & 0 \end{vmatrix}, \tag{85}$$

$$Y_1 = \begin{vmatrix} 0 & 0 & 0 \\ 0 & 0 & 0 \\ 1 & 0 & 0 \end{vmatrix}, \quad Y_2 = \begin{vmatrix} 0 & 0 & 0 \\ 0 & 0 & 0 \\ 0 & 1 & 0 \end{vmatrix}.$$

It can also be realized as the Lie algebra of the group of projective transformations of $\mathbb{R}^2$. The basis corresponding to (85) is given by the following vector fields:

$$\begin{aligned} &X_1 = \partial_x,\ X_2 = \partial_y,\ K = -x\partial_x + y\partial_y, \\ &D = -3(x\partial_x + y\partial_y),\ A_1 = -y\partial_x,\ A_2 = -x\partial_y, \\ &Y_1 = -x(x\partial_x + y\partial_y),\ Y_2 = -y(x\partial_x + y\partial_y) \end{aligned} \tag{86}$$

B. Maximal subalgebras of $sl(3,\mathcal{R})$

B.1. Irreducibly embedded in defining representation

$$O(3): \begin{vmatrix} 0 & a & b \\ -a & 0 & c \\ -b & -c & 0 \end{vmatrix}, \; A_1 - A_2, \; X_1 - Y_1, \; X_2 - Y_2$$

$$O(2,1): \begin{vmatrix} 0 & a & b \\ -a & 0 & c \\ b & c & 0 \end{vmatrix}, \; A_1 - A_2, \; X_1 + Y_1, \; X_2 + Y_2$$

B.2. Reducibly embedded $\Leftrightarrow$ invariant subspace V

i) $\dim V = 2: \; V \sim \begin{vmatrix} x \\ y \\ 0 \end{vmatrix}$

$$PV \subseteq V \Rightarrow P_1 \sim \begin{vmatrix} a+b & c & d \\ e & -a+b & f \\ 0 & 0 & -2b \end{vmatrix} \tag{87}$$

P_i : maximal parabolic $\sim$ gaff$(2,\mathcal{R})$, $\{K, A_1, A_2\} \oplus \{D, X_1, X_2\}$

ii) $\dim V = 1, V \sim \begin{vmatrix} 0 \\ 0 \\ z \end{vmatrix}$

$$PV \subseteq V \Rightarrow P_2 = \begin{vmatrix} a+b & c & 0 \\ e & -a+b & 0 \\ g & h & -2b \end{vmatrix}. \tag{88}$$

Thus, the algebra $L \sim \mathrm{sl}(3,\mathbb{R})$ has precisely 4 maximal subalgebras. Two are simple: $0(3)$ and $o(2,1)$. Two are parabolic, mutually isomorphic and isomorphic to the general affine algebra aff$(2,\mathbb{R})$. P_1 and P_2 are conjugate under an outer automorphism of $\mathrm{sl}(3,\mathbb{R})$, namely transposition (in the matrix realization). Notice however that the vector fields $\{X_1, X_2\}$ in equation (86) cannot be transformed into $\{Y_1, Y_2\}$.

C. All subalgebras of $P_1 \sim$aff$(2,\mathbb{R})$

We shall use the algorithm of Section 3.4. The decomposition $L = F \oplus N$ with $L \sim$aff$(2,\mathbb{R})$ is

$$\begin{aligned} F &\sim \text{gl}(2,\mathbb{R}) \sim \{K, A_1, A_2\} \oplus \{D\} \\ N &\sim \{X_1, X_2\}. \end{aligned} \tag{89}$$

We have chosen the decomposition so that N is abelian.

Step 1. All subalgebras of $F \sim$gl$(2,\mathbb{R})$.

The algebra $F \sim$gl$(2,\mathbb{R})$ has 16 classes of subalgebras, when classified under gl$(2,\mathbb{R})$, three of them depend on a continuous parameter α. A representative list $S(F)$ is

$$\begin{aligned} &F_1 \approx \{K, A_1, A_2, D\}, \quad F_2 \approx \{K, A_1, A_2\}, \quad F_3 \approx \{K, A_1, D\} \\ &F_4 \approx \{K, A_1\}, \quad F_5 \approx \{K, D\}, \quad F_6 \approx \{A_1 - A_2, D\} \\ &F_7 \approx \{A_1, D\}, \quad F_8(\alpha) \approx \{A_1, K + \alpha D, \alpha \neq 0\} \\ &F_9 \approx \{K\}, \quad F_{10} \approx \{A_1 - A_2\} \\ &F_{11} \approx \{A_1\}, \quad F_{12} \approx \{K + \alpha D, \alpha > 0\}, \\ &F_{13}(\alpha) \approx \{A_1 - A_2 + \alpha D, \alpha > 0\}, \quad F_{14} \approx \{A_1 + D\} \\ &F_{15} \approx \{D\}, \quad F_{16} \approx \{\emptyset\}. \end{aligned} \tag{90}$$

Step 2. Splitting subalgebras of aff$(2,\mathbb{R})$ are obtained from those of (90), by adding to F_i all invariant subspaces of F_i in N. In all cases we can add all of N, i.e. $\{X_1, X_2\}$, or the empty space $\{\emptyset\}$. This provides a list of 32 "trivial" splitting subalgebras. To those we add the "nontrivial" splitting, consisting of F_i and one additional element of N. Our notation will be

$$S_{i,1} \sim F_i, \quad S_{i,2} \sim \{F_i; X_1, X_2\}, \quad i = 1, \ldots, 16 \tag{91}$$

for the "trivial" splitting subalgebras.

The "nontrivial" ones are

$$\begin{aligned} &S_{3,3} \sim \{K, A_1; X_1\} \quad S_{4,3} \sim \{K, A_1, D; X_1\} \\ &S_{5,3} \sim \{K, D; X_1\} \quad S_{7,3} \sim \{A_1, D; X_1\} \\ &S_{8,3}(\alpha) \sim \{A_1, K + \alpha D; X_1, \alpha \neq 0\} \quad S_{9,3} \sim \{K, X_1\} \\ &S_{11,3} \sim \{A_1; X_1\} \quad S_{12,3}(\alpha) \sim \{K + \alpha D; X_1, \alpha > 0\} \\ &S_{12,4}(\alpha) \sim \{K + \alpha D; X_2, \alpha > 0\} \quad S_{14,3} \sim \{A_1 + D; X_1\} \\ &S_{15,3} \sim \{D; X_1\} \quad S_{16,3}\} \sim \{X_1\}. \end{aligned} \tag{92}$$

Step 3. Nonsplitting subalgebras of $P_1 \sim \mathrm{aff}(2,\mathbb{R})$.

We start from the list of splitting subalgebras of P_1, more precisely from (90) and (92). First of all we notice that subalgebras containing D as an element, and reductive subalgebras, will not allow nonsplitting extensions.

Let us consider the subalgebras $F_8(\alpha)$ and $F_4(= F_8(0))$. We consider

$$\begin{aligned}\tilde{K} &= K + \alpha D + c_1 X_1 + c_2 X_2 \\ \tilde{A}_1 &= A_1 + c_3 X_1 + c_4 X_2.\end{aligned} \tag{93}$$

We have

$$[\tilde{K}, \tilde{A}_1] = 2\tilde{A}_1 + (c_3(-1+3\alpha) - c_2)X_1 + 3c_4(-1+\alpha)X_2. \tag{94}$$

If the algebra does not contain X_1, we must have

$$c_4(\alpha - 1) = 0, \quad c_2 = c_3(3\alpha - 1). \tag{95}$$

Moreover, for $\alpha \neq \pm\frac{1}{3}$ we can use the coboundaries to set $c_1 = c_2 = 0$. We are left with the following nonsplitting subalgebras

$$\begin{aligned}S_{8,4}(\varepsilon) &\sim \{K + D, A_1 + \varepsilon X_2, \varepsilon = \pm 1\} \\ S_{8,5}(\varepsilon) &\sim \{K + D, A_1 + \varepsilon X_2; X_1, \varepsilon = \pm 1\} \\ S_{8,6} &\sim \{K - \frac{1}{3}D + X_1, A_1\} \\ S_{8,7} &\sim \{K + \frac{1}{3}D + X_2, A_1; X_1\}.\end{aligned} \tag{96}$$

The only other subalgebras that allow nonsplitting extensions are F_{11} and F_{12}.From Them we obtain

$$\begin{aligned}S_{11,4} &\sim \{A_1 + X_2\} \\ S_{11,5} &\sim \{A_1 + X_2; X_1\} \\ S_{12,5} &\sim \{K + \frac{1}{3}D + X_2\} \\ S_{12,6} &\sim \{K + \frac{1}{3}D + X_2; X_1\}.\end{aligned} \tag{97}$$

Finally, the complete list of all subalgebras of $\mathrm{aff}(2,\mathbb{R})$, classified under $\mathrm{Aff}(2,\mathbb{R})$, consists of the algebras (90), the algebras (90) with $\{X_1, X_2\}$ added; (92), (96) and (97).

D. <u>All subalgebras of sl(3, $\mathbb{R}$)</u>

A complete list of representatives of all subalgebras of sl(3, $\mathbb{R}$), classified under SL(3, $\mathbb{R}$), will contain the four maximal subalgebras $o(3)$, $o(2,1)$, P_1 and P_2. Further it will contain subalgebras of P_1 that we have already classified under $< \exp P_1 >$. Some subalgebras of P_2 must be added, if they are not conjugate to those in P_1. Some subalgebras of P_1 may be conjugate under SL(3, $\mathbb{R}$), even if they are not conjugate under $< \exp P_1 >$. This does not occur for any of the one-dimensional subalgebras, but for instance we have

$$S_{12,4}(\alpha = \frac{1}{2} \sim F_7 \tag{98}$$

under $SL(3, \mathbb{R})$.

We shall not go into these matters here.

3.6 Generalizations

The subgroup classification method for semidirect sums has been generalized to the case of certain infinite dimensional Lie algebras. In particular this has been done for the classification of finite dimensional subalgebras of Kac-Moody-Virasoro algebras. We refer to the original articles for the exposition.

4 The Clarkson-Kruskal direct reduction method and conditional symmetries

4.1 Formulation of the problem

Let us consider a PDE of the form

$$E(x, t, u, u_x, u_t, u_{xx}, u_{xt}, u_{tt}, \dots) = 0 \tag{99}$$

i.e. involving one dependent variable u and two independent ones. Our aim is to express $u(x,t)$ in terms of one function $\omega(z)$ of one variable z in such a manner that $\omega(z)$ satisfies an ordinary differential equation (ODE)

$$\tilde{E}(z, \omega, \omega_z, \omega_{zz}, \dots) = 0. \tag{100}$$

Every solution $\omega(z)$ of equation (100) should provide a particular solution of the PDE (99).

One way of achieving this goal is to use symmetry reduction, as discussed above in Section 2. That is one finds a Lie algebra of vector fields

$$X = \xi(x,t,u)\partial_x + \tau(x,t,u)\partial_t + \phi(x,t,u)\partial_u, \tag{101}$$

the N-th prolongation of which annihilates equation (99) on its solution surface. In the case of two independent variables one then classifies all one-dimensional subalgebras.

Invariance under the corresponding one-dimensional groups will provide a reduction to an ODE. The question is: will this method provide all reductions? The answer is: not necessarily!

The direct reduction method [] was proposed precisely to answer the above question and was first tested on the example of the Boussinesq equation []

$$u_{tt} + uu_{xx} + u_x^2 + u_{xxxx} = 0. \tag{102}$$

The solution of equation (102) was postulated to have the form

$$u(x,t) = U(x,t,\omega(z)), \quad z = z(x,t). \tag{103}$$

The functions U and z must be determined from the requirement that $\omega(z)$ satisfy an ODE.

4.2 Symmetry reduction for Boussinesq equation

Let us first look at symmetry reduction for equation (102). The Lie point symmetry algebra is

$$\{D = x\partial_x + 2t\partial_u - 2u\partial_u, P_1 = \partial_x, P_0 = \partial_t\}. \tag{104}$$

Equation (102) also invariant under the reflections

$$II : x \to -x, \quad T : t \to -t.$$

The one-dimensional subalgebras of the symmetry algebra are

$$\{D\}, \quad \{P_0 + aP_1, \ a = 0, 1\}, \quad \{P_1\}. \tag{105}$$

The individual reductions are:

$$\underline{P_1} : \ u(x,t) = \omega(z), \quad z = t. \tag{106}$$

Equation (102) then implies $\omega_{zz} = 0$

$$\underline{P_0 + aP_1} : \ u(x,t) = \omega(z), \quad z = x - at. \tag{107}$$

Equation (102) reduces to

$$\frac{\partial^4 \omega}{\partial z^4} + \omega \frac{\partial^2 \omega}{\partial z^2} + \left(\frac{\partial \omega}{\partial z}\right)^2 + a^2 \frac{\partial^2 \omega}{\partial z^2} = 0. \tag{108}$$

Equation (108) can be integrated twice and we obtain

$$\omega_{zz} + \frac{1}{2}\omega^2 + a^2 \omega = Az + B. \tag{109}$$

For $A = 0$ we can integrate once more and obtain the equation for elliptic functions. For $A \neq 0$ equation (109) is solved in terms of the first Painlevé transcendent P_I.

$$\underline{D}: \; u(x,t) = \frac{1}{t}\omega(z), \quad z = \frac{x}{\sqrt{t}}. \tag{110}$$

The reduced ODE is

$$\frac{\mathrm{d}^4 \omega}{\mathrm{d}z^4} + \omega \frac{\mathrm{d}^2 \omega}{\mathrm{d}z^2} + \left(\frac{\mathrm{d}\omega}{\mathrm{d}z}\right)^2 + \frac{1}{4} z^2 \frac{\mathrm{d}^2 \omega}{\mathrm{d}z^2} + \frac{1}{4} z \frac{\mathrm{d}\omega}{\mathrm{d}z} + 2\omega = 0. \tag{111}$$

Equation (111) can be solved in terms of the fourth Painlevé transcendent P_{IV}.

4.3 The direct method

Using the direct method, Clarkson and Kruskal found these reductions, plus five more. In their article they expressed the "hope that a group theoretical explanation will be possible in due course". This was provided in [] and the group theoretical explanation turned out to be related to the nonclassical method of Bluman and Cole.

In the direct method one assumes that $u(x,t)$ has the form (103) and substitutes into equation (102). We require an ODE for $\omega(z)$. Assuming $z_x \neq 0$ we obtain

$$\Omega_{\omega\omega} = 0 \Leftrightarrow u(x,t) = \beta(x,t)\omega(z(x,t)) + \alpha(x,t). \tag{112}$$

Substituting (112) into (102) again we obtain

$$\begin{aligned} u(x,t) &= \theta^2(t)\omega(z) - \frac{1}{\theta^2(t)}\left(x\frac{d\theta}{dt} + \frac{d\phi}{dt}\right)^2 \\ z(x,t) &= x\theta(t) + \phi(t) \\ \frac{d^2\theta}{dt^2} &= A\theta^5 \\ \frac{d^2\phi}{dt^2} &= (A\phi + B)\theta^4 \end{aligned} \tag{113}$$

where A, B are constants. Finally $\omega(z)$ satisfies the ODE

$$\frac{\partial^4\omega}{\partial^4 z} + \omega\frac{\partial^2\omega}{\partial^2 z} + \left(\frac{\partial\omega}{\partial z}\right)^2 + (Az + B)\frac{\partial\omega}{\partial z} + 2A\omega = 2(Az + B)^2. \tag{114}$$

The result obtained by Clarkson and Kruskal is that there are six reductions to standard forms:

$$u(x,t) = \omega_1(z),\ z = x + \mu_1 t. \tag{115}$$

This is the Lie group reduction (107).

$$\begin{aligned} u(x,t) &= t^2\omega_2(z) - \frac{x^2}{t^2},\ z = xt \\ u(x,t) &= \omega_3(z) - 4\mu_3^2 t^2,\ z = x + \mu_3 t^2. \end{aligned}$$

This is the Galilei group reduction by

$$B = t\partial_x - 2\mu_3\partial_u + \frac{1}{2\mu_3}\partial_t,$$

though the equation is not Galilei invariant.

$$\begin{aligned} u(x,t) &= t^2\omega_4(z) - (x + 6\mu_4 t^5)\frac{1}{t^2},\ z = xt + \mu_4 t^6 \\ u(x,t) &= t^{-1}\omega_5(z) - (x - 3\mu_5 t^2)\frac{1}{4t^2},\ z = \frac{x}{\sqrt{t}} + \mu_5 t^{3/2} \end{aligned}$$

For $\mu_5 = 0$ this is the Lie reduction (110).

$$\begin{aligned} u(x,t) &= \frac{1}{\wp(t)}\left\{\omega_6(z) - \left[\frac{1}{2}z\frac{\mathrm{d}\wp(t)}{\mathrm{d}t} + \mu_6\wp(t)^{3/2}(t)\right]^2\right\}, \\ z &= \wp^{-1/2}(t)[x + \mu_6\zeta(t)] \end{aligned}$$

Here $\wp(t)$ denotes the Weierstrass elliptic function, and $\zeta(t)$ the Weierstrass zeta function.

The solutions $\omega_i(z)$ are expressed in terms of Pailevé transcendents as follows

$$\begin{aligned} &\omega_1(z),\ \omega_2(z) \to \mathrm{P1}. \\ &\omega_3(z),\ \omega_4(z) \to \mathrm{P2}. \\ &\omega_5(z),\ \omega_6(z) \to \mathrm{P4}. \end{aligned} \tag{116}$$

Thus, the direct method provided lots of new solutions of the Boussinesq equation.

4.4 Conditional symmetries

To each solution we can associate a vector field $\hat{X}$:

$$\hat{X} = \xi\partial_x + \tau\partial_t + \phi\partial_u$$

and a Lie group transformation G. However, G is not a symmetry group of the equation, though $\hat{X}$ does annihilate the solution. Thus, we have

$$\xi(x,t,u)u_x + \tau(x,t,u)u_t - \phi(x,t,u) = 0 \tag{117}$$

(on the particular solutions (113)).

This gives rise to the idea of "conditional symmetries". These are symmetries that you can apply only to a subset of solutions. Thus, we are given a PDE

$$E(x,t,u,u_x,u_t,\ldots) = 0 \tag{118}$$

and add a "condition"

$$C(x,t,u,u_x,u_t,\ldots) = 0. \tag{119}$$

We then look for transformations leaving both equations (118) and (119) invariant on their common solution set

$$\mathrm{pr}\,XE\Big|_{\substack{E=0\\C=0}} = 0,\ \ \mathrm{pr}\,XC\Big|_{\substack{E=0\\C=0}} = 0,\ \ X = \xi\partial_x + \tau\partial_t + \phi\partial_u \tag{120}$$

How to choose C?
As unconstraining as possible.
Give a Lie point symmetry generator

$$\hat{X} = \xi\partial_x + \tau\partial_t + \phi\partial_u$$

we perform symmetry reduction by imposing

$$\hat{X}\phi(x,t,u) = 0 \Leftrightarrow \frac{\mathrm{d}x}{\xi} = \frac{\mathrm{d}t}{\tau} = \frac{\mathrm{d}u}{\phi}.$$

The idea was to take $C = 0$ as the first order PDE:

$$C = \xi(x,t,u)u_x + \tau(x,t,u)u_t - \phi(x,t,u) = 0, \tag{121}$$

with ξ, τ, ϕ the same as in vector field $\hat{X}$. []
Then we have

$$\mathrm{pr}^1\,\hat{X}C = -(\xi_u u_x + \tau_u u_t - \phi_u)C \Rightarrow \mathrm{pr}^1\,\hat{X}C|_{C=0} = 0. \tag{122}$$

This is an identity and does not impose any constraint on ξ, τ, ϕ.
The algorithm for finding such conditional symmetries is

$$\mathrm{pr}^{(n)}\,XE|_{E=0,C=0} = 0 \quad .$$

Thus, we will have fewer determining equations than in the case of ordinary Lie symmetries. Hence, in principal we may get more solutions.

Since $C = 0 \Leftrightarrow f(x,t,u)\ C = 0 \Rightarrow$ for $\tau \neq 0$ *we can set* $\tau = 1$, i.e. the condition (119) reduces to

$$u_t + \xi u_x - \phi - 0 \tag{123}$$

and the vector field representing the conditional symmetries is

$$\hat{X} = \partial_t + \xi\partial_x + \phi\partial_u$$

Then we eliminate:

$$u_t = -\xi u_x + \phi$$

and all differential consequences $(u_{tt}, u_{tx}, \dots)$ from the determining equations.

Then use the same program that calculates Lie point symmetries. Differential consequences: built in in the Champagne-Winternitz program [].

Caveat!

1. The determining equations are now nonlinear.
2. Conditional symmetries do not form a vector space, nor a Lie algebra (each X: different condition)
3. Not useful to integrate $X \to G$ the group either leaves a solution invariant or takes it out of the solution space.

For symmetry reduction: "conditional" symmetries are just as good as ordinary Lie point symmetries.

History: G. Bluman, J. Cole, J. Math. Mech. 18, 1025 (1969) "Nonclassical method" for heat equation.

Linear equation → nonlinear determining equations.

P. Clarkson and M. D. Kruskal, J. Math. Phys. 30, 2201 (1989)

D. Levi, P. Winternitz, J. Phys. A, 22, 2915 (1989).

Now let us apply the method of conditional symmetries to the Boussinesq equation (102).

Ordinary symmetries: the determining equations are the coefficients of

$$u_x^{n_1}, u_{xx}^{n_2}, u_{xxx}^{n_3}, u_t^{n_4}, \ u_{tt}^{n_5}, u_{tx}^{n_6}, u_{txx}^{n_7}, u_{ttx}^{n_8}, u_{ttt}^{n_9}, u_{txxx}^{n_{10}}, u_{ttxx}^{n_{11}}, u_{tttx}^{n_{12}}, u_{4t}^{n_{13}}$$

Conditional symmetries:

$$\begin{aligned} v_1 &\equiv u_{4x} = -u_{tt} - uu_{xx} - u_x^2 \\ v_2 &\equiv u_t = -\xi u_x + \phi \end{aligned}$$

Determining equations are coefficient of:

$$u_x^{n_1}, \ u_{xx}^{n_2}, \ u_{xxx}^{n_3}$$

Our program generates 14 equations for ξ and ϕ $(\tau = 1)$. Their solution is

$$\begin{aligned} \hat{X} = \partial_t + [\alpha(t)x + \beta(t)]\partial_x - [2\alpha(t)u + 2\alpha(\dot{\alpha} + 2\alpha^2)x^2 + \\ 2(\alpha\dot{\beta} + \dot{\alpha}\beta + 4\alpha^2\beta)x + 2\beta(\dot{\beta} + 2\alpha\beta)]\partial_u \end{aligned} \tag{124}$$

where $\alpha(t)$ and $\beta(t)$ satisfy

$$\begin{aligned} \ddot{\alpha} + 2\alpha\dot{\alpha} - 4\alpha^3 = 0 \\ \ddot{\beta} + 2\alpha\dot{\beta} - 4\alpha^2\beta = 0 \end{aligned} \tag{125}$$

Reductions of the Boussinesq equation are obtained ny imposing

$$\hat{X}\phi(x,t,u) = 0 \Rightarrow \mathrm{d}t = \frac{\mathrm{d}x}{\alpha(t)x + \beta(t)} = -\frac{\mathrm{d}u}{[\ldots]}$$

The general reduction formula is

$$
\begin{aligned}
u(x,t) &= \omega(z)K^2(t) - (\alpha x + \beta)^2 \\
z(x,t) &= xK(t) - \int_0^t \beta(s)K(s)\,ds \\
K(t) &= \exp\left(-\int_0^t \alpha(s)\,ds\right)
\end{aligned}
\tag{126}
$$

Substitute into the Boussinesq equation

$$
\frac{\partial^4\omega}{\partial z^4} + \omega\frac{\partial^2\omega}{\partial z^2} + \left(\frac{\partial\omega}{\partial z}\right)^2 + (Az+B)\frac{\partial\omega}{\partial z} + 2A\omega = 2(Az+B)^2 \tag{127}
$$

$$
A = \frac{\alpha^2 - \dot{\alpha}}{k^4},\ B = \frac{\alpha\beta - \dot{\beta}}{k^3} + \frac{\alpha^2 - \dot{\alpha}}{k^4}\int_0^t \beta(s)K(s)\,ds.
$$

The equations for α, β imply that A and B are constants

$$
\frac{dA}{dt} = 0,\ \frac{dB}{dt} = 0.
$$

Only now: solve for α and β. The first of equations (125) is one of the reducible equations with the Painlevé property P10 in Ince's list []). It can be solved in terms of elliptic functions:

$$
\begin{aligned}
\alpha &= \frac{\dot{H}}{2H},\ \dot{H}^2 = h_0H^3 + h_1 \\
\beta &= \beta_1\frac{\dot{H}}{H} + \beta_2\frac{\dot{H}}{H}\int_0^t \frac{H(s)}{\dot{H}^2(s)}\,ds \text{ for } \dot{H} \neq 0,
\end{aligned}
\tag{128}
$$

where h_0, h_1, β_1 and β_2 are constants.

Analyze equations for α and β:

1. $h_0 = h_1 = 0$
   $\alpha = 0, \beta = \beta_0 + \beta_1 t, K(t) = 1, A = 0, B = -\beta_1$
   $\beta_1 = 0 \Rightarrow$ translations:

$$
\hat{X} = \partial_t + \beta_0\partial_x. \tag{129}
$$

   $\beta_1 \neq 0$ simplify equations using symmetry group

$$
\hat{X} = \partial_t - 2t\partial_u + t\partial_x. \tag{130}
$$

Galilei: not a symmetry

$$
\begin{aligned}
&z = x - \frac{1}{2}t^2, \ \omega = \omega(z) - t^2 \\
&\frac{\partial^3 \omega}{\partial z^3} + \omega \frac{\partial \omega}{\partial z} - \omega = 2z + C_1 .
\end{aligned} \tag{131}
$$

Equation (131) is solved in terms of P_{II}.

2. $h_0 \neq 0, h_1 = 0$
 $\alpha = -\frac{1}{t}, \ \beta = \beta_1 t^4 + \frac{\beta_2}{t}, \ K = t, \ A = 0, \ B = -5\beta_1$
 Use symmetry group: $\beta_2 \to 0, \ \beta_1 \to 1$ (unless we have $\beta_1 = 0$).

$$
\begin{aligned}
&\hat{X} = \partial_t + \left(-\frac{x}{t} + \beta_1 t^4\right)\partial_x + \left(\frac{2}{t}u + \frac{6}{t^3}x^2 - 2\beta_1 t^2 x - 4\beta^2 t^2\right)\partial_u \\
&z = xt - \frac{1}{6}\beta_1 t^6, \ \ u(x,t) = \omega(z)t^2 - \left(\frac{x}{t} - \beta_1 t^4\right)^2 \\
&\beta_1 = 0 \Rightarrow \frac{\partial^2 \omega}{\partial z^2} + \frac{1}{2}\omega^2 = c_1 z + c_0 \\
&\qquad c_1 = 0: \ \text{elliptic function } c_1 \neq 0: \mathrm{P1} \\
&\beta_1 \neq 0 \Rightarrow \beta_1 = 1
\end{aligned} \tag{132}
$$

$$
\frac{\partial^3 \omega}{\partial z^3} + \omega \frac{\partial \omega}{\partial z} - 5\omega = 50z + c_0 \to \mathrm{P2} \tag{133}
$$

3. $h_0 = 0, h_1 \neq 0$
 $\alpha = \frac{1}{2t}, \ \beta = \beta_1 t + \frac{\beta_2}{t}, \ K = \frac{1}{\sqrt{t}}, \ A = \frac{3}{4}, \ B = 0$

$$
\begin{aligned}
&\beta_1 = 0 \Rightarrow \ \text{dilatations} \\
&\beta_1 \neq 0, \ \text{transform } \beta_2 \to 0, \ \beta_1 = 1 \\
&\qquad \hat{X} = \partial_t + \left(\frac{x}{2t} + t\right)\partial_x - \frac{1}{t}(u + 2x + 4t^2)\partial_u \\
&\qquad z = \frac{x}{\sqrt{t}} - \frac{2}{3}t^{3/2}, \ u = \frac{1}{t}\omega(z) - \left(\frac{x}{2t} + t\right)^2 \\
&\qquad \frac{\partial^4 \omega}{\partial z^4} + \omega \frac{\partial^2 \omega}{\partial z^2} + \left(\frac{\partial \omega}{\partial z}\right)^2 + \frac{3}{2}z\frac{\partial \omega}{\partial z} + \frac{3}{2}\omega = \frac{9}{8}z^2 \to \mathrm{P4}
\end{aligned} \tag{134}
$$

4. $h_0 \neq 0, h_1 \neq 0 \Rightarrow$ Weierstrass elliptic function.
 $\alpha = \frac{\dot{\wp}}{2\wp}, \ \beta = \beta_1 \frac{\dot{\wp}}{2\wp} + \beta_2 \frac{\dot{\wp}}{2\wp}\int_0^t \frac{\wp(s)}{\dot{\wp}^2}\,\mathrm{d}s$

$$\dot{\wp}^2 = 4\wp^3 - g_3, \ \wp = \wp(t, 0, g_3)$$

Translate $\beta_1 \to 0$, then

$$K = \wp^{-1/2}, \ A = -\frac{3g_3}{4}, \ B = 0$$

$$\hat{X} = \partial_t + \frac{1}{2}\left(\frac{\dot{\wp}}{\wp}x + \beta_2 \frac{\dot{\wp}}{\wp} W\right)\partial_x - \left[\frac{\dot{\wp}}{\wp}u + 3\dot{\wp}x^2 + \frac{\beta_2}{2}\left(\frac{1}{\wp} + 12\dot{\wp}W\right)x + \frac{1}{2}\beta_2^2 W\left(\frac{1}{\wp} + 6\dot{\wp}W\right)\right]\partial_u$$

$$z = x\wp(t)^{-1/2} + \frac{1}{3}\beta_2 g_3^{-1}\wp(t)^{-1/2}\int_0^t \wp(s)\, ds$$

$$u(x,t) = \omega(z)\wp^{-1} - \left(\frac{1}{2}\frac{\dot{\wp}}{\wp}x + \beta_2 \frac{\dot{\wp}}{2\wp}W\right)^2$$

$$W(t) = \int_0^t \frac{\wp(s)}{\dot{\wp}(s)}\, ds$$

$$\frac{\partial^4\omega}{\partial z^4} + \omega\frac{\partial^2\omega}{\partial z^2} + \left(\frac{\partial\omega}{\partial z}\right)^2 - \frac{3}{4}g_3\frac{\partial\omega}{\partial z} - \frac{3}{2}g_3\omega = \frac{9}{8}g_3^2 z^2 \to \text{P4} \tag{135}$$

Thus, the direct method and the method of conditional symmetries give the same result for the Boussinesq equation.

4.5 General comments

More generally the direct method (as originally formulated) gives the same results as conditional symmetries if ξ/τ is independent of u.

(E. Pucci and Saccomandi, C. Nucci and P. Clarkson)

If $\frac{\xi}{\tau} = f(x,t,u) \Rightarrow$ conditional symmetries give more solutions, but implicit ones $z = z(x,t,u)$ (we assume $\frac{\partial f}{\partial u} \neq 0$).

The Clarkson-Kruskal method can also be generalized.

Résumé on dimensional reduction

Several systematic methods around

1. Lie point symmetries: algebra integrable to group $\Rightarrow$ many other applications
   One equation $\Rightarrow$ invariant solutions.
   A system $\Rightarrow$ invariant and partially invariant solutions.
2. Conditional symmetries (nonclassical method).
   Determining equations are nonlinear. The conditions $\xi_i u^\alpha_{x_i} - \phi^\alpha = 0$
   Sometimes more reductions, sometimes the same as given by Lie symmetries (e.g. for the KdV equation).
   No useful group transformations.

3. Direct method
   $u(x_i) = U(x_i, \omega(z)), z = z(x_i)$
   Same results as conditional symmetries if $\frac{\xi_i}{\xi_1}$ independent of u (fiber preserving).

Caution: not every exact analytical solution of a PDE comes from a reduction to an ODE!

E.g. multisolitons ($n \geq 2$)

Also a PDE may be reduced to a coupled system of ODEs (Estevez)

Tool kit for obtaining exact solutions

1. Lie point symmetries, which provide both invariant solutions and partially invariant solutions.
2. Direct method and conditional symmetries
   Closely related; useful to know both; difficulties in calculations: complementary. Use Lie symmetries to simplify.
3. Other "side conditions"
4. The machinery of "integrability": "soliton theory"
5. Painlevé expansions, in particularly truncated ones, for "partially integrable equations".

Open question: can one say a priori when conditional symmetries exist?

We mention that a relation exists between conditional symmetries and Bäcklund transformations.

Example:

$$u_{xt} = -2\sinh u \tag{136}$$
$$X = \xi\partial_x + \tau\partial_t + \phi\partial_u .$$

Put $\tau \neq 0, \xi = 0, \tau \to 1$. Then we have $X = \partial_t + \phi\partial_u$ and the condition is $u_t - \phi = 0$. The determining equation is

$$\phi\phi_x\phi_{uu} - \phi\phi_u\phi_{ux} - \phi_{tx}\phi_u + \phi_{tu}\phi_x - 4\cosh(2u)\phi\phi_u - 2\sinh(2u)(\phi_u^2 + \phi_{ut} + \phi\phi_{uu}) = 0.$$

This is difficult to solve in general, however, a particular solution is

$$\phi = f_t + 2a\sinh(f + u)$$

where f is any solution of the initial equation (136).

The Bäcklund transformation is precisely of this form

$$u_t = f_t + 2a\sinh(f + u).$$

The other half of the Bäcklund transformation is

$$u_x = -f_x + \frac{2}{a}\sinh(f - u).$$

5 Concluding comments

This lecture series, presented at the 1999 CIME school in Cetraro contained two more lectures. One was on nonlinear ordinary differential equations with superposition formulas and their relation to Bäcklund transformations. The lecture was a brief summary of results contained in a series of articles, a list of which is attached. The final sixth lecture was devoted to symmetry methods for solving difference equations. The subject could be summed up as "Continuous symmetries of discrete equations". For recent references, containing references to earlier work, see the list attached.

5.1 References on nonlinear superposition formulas

1. R. L. Anderson, *A nonlinear superposition principle admitted by coupled Riccati equations of the projective type.* Lett. Math. Phys. 4, 1-7, (1980).
2. R. L. Anderson, J. Harnad and P. Winternitz, *Group theoretical approach to superposition rules for systems of Riccati equations.* Lett. Math. Phys. 5, 143-148, (1981).
3. R. L. Anderson, J. Harnad and P. Winternitz, Systems of ordinary differential equations with nonlinear superposition principles. Physica D 4, 164-187, (1982).
4. J. Harnad, P. Winternitz and R. L. Anderson, Superposition principles for matrix Riccati equations. J. Math. Phys. 24, 1062-1072, (1983).
5. D. W. Rand and P. Winternitz, Nonlinear superposition principles: A new numerical method for solving matrix Riccati equations. Comp. Phys. Commun. 33, 305-328, (1984).
6. S. Shnider and P. Winternitz, Nonlinear equation with superposition principles and the thory of transitive primitive Lie algebras. Lett. Math. Phys. 8, 69-78, (1984).
7. S. Shnider and P. Winternitz, Classification of systems of nonlinear ordinary differential equations with superposition principles. J. Math. Phys. 25, 3155–3165, (1984).
8. M. Sorine and P. Winternitz, Superposition laws for nonlinear equations arising in optimal control theory. IEEE Transactions, AC-30, 266-272, (1985).
9. M. A. del Olmo, M. A. Rodriguez and P. Winternitz, Simple subgroups of simple Lie groups and nonlinear differential equations with superposition principles. J. Math. Phys. 27, 14-23, (1986).
10. T. C. Bountis, V. Papargeorgiou and P. Winternitz, On the integrability of systems of nonlinear ODEs with superposition principles. J. Math. Phys. 27, 1215-1224, (1986).
11. J. Beckers, V. Hussin and P. Winternitz, Complex parabolic subgroups of $G(2)$ and nonlinear differential equations. Lett. Math. Phys. 11, 81-86, (1986).

12. M. A. del Olmo, M. A. Rodriguez and P. Winternitz, Superposition formulas for rectangular matrix Riccati equations. J. Math. Phys. 28, 530-535, (1987).
13. L. Gagnon, V. Hussin and P. Winternitz, Nonlinear equations with superposition formulas and the exceptional group $G(2)$. III. The superposition formulas. J. Math. Phys. 29, 2145-2155, (1988).
14. J. Beckers, L. Gagnon, V. Hussin and P. Winternitz, Superposition formulas for nonlinear superequations. J. Math. Phys. 31, 2528-2534, (1990).
15. L. Michel and P. Winternitz, Families of transitive primitive maximal simple Lie subalgebras of diff(n). In L. Vinet editor, Advances in Mathematical Sciences-CRM's 25 years, CRM Proceedings and Lecture Notes, 451-479, AMS, Providence, R. I.(1997).
16. B. Grammaticos, A. Ramani and P. Winternitz, Discretizing families of linearizable equations. Phys. Lett. A 245, 382-388, (1998).
17. A. Turbiner and P. Winternitz, Solutions of nonlinear ordinary differential and difference equations with superposition formulas. Lett. Math. Phys. 50, 189-201, (1999).
18. M. Havliček, S. Pošta and P. Winternitz, Nonlinear superposition formulas based on imprimitive group action. J. Math. Phys. 40, 3104-3122, (1999).

5.2 References on continuous symmetries of difference equations

1. D. Levi and P. Winternitz, *Continuous symmetries of discrete equations.* Phys. Lett. A 152, 335-338, (1991).
2. V. Dorodnitsyn, R. Kozlov and P. Winternitz, *Lie group classification of second order difference equations.* J. Math.Phys. 41, 11-24, (1999).
3. D. Levi, S. Tremblay and P. Winternitz, *Lie point symmetries of difference equations and lattices.* J.Phys.A 33, 8507-8524, (2000).
4. D. Levi and P. Winternitz, *Lie point symmetries and commuting flows for equations on lattices.* J.Phys. A 35, 2249-2262, (2002).

Acknowledgments

I would like to thank Professor Antonio Maria Greco for inviting me to present this series of lectures. I am specially indebted to him for taking upon himself the burden of reproducing the text from my hand written notes while I was temporarily incapacitated by an accident. Research grants from NSERC of Canada and FCAR du Québec are gratefully acknowledged.

References

1. M. J. Ablowitz and P. A. Clarkson, *Solitons, nonlinear evolution equations and inverse scattering.* Cambridge Univ. Press, Cambridge, (1991).

2. M. J. Ablowitz, A. Ramani, and H. Segur, *A connection between nonlinear evolution equations and ordinary differential equations of P-type.* J. Math. Phys. 21, 715-721, (1980).
3. M. J. Ablowitz, A. Ramani, and H. Segur, *A connection between nonlinear evolution equations and ordinary differential equations of P-type.* J. Math. Phys. 21, 1006-1015, (1980).
4. D. Arrigo, P. Broadbridge, and J. M. Hill, *Nonclassical symmetry solutions and the methods of Bluman–Cole and Clarkson–Kruskal.* J. Math. Phys. 34, 4692-4703, (1993).
5. J. Beckers, J. Patera, M. Perroud, and P. Winternitz, *Subgroups of the Euclidean group and symmetry breaking in nonrelativistic quantum mechanics.* J. Math. Phys.18, 72-83, (1977).
6. D. Bérubé and M. de Montigny, *A Mathematica program for the calculation of Lie point symmetries of systems of differential equations.* Preprint CRM-1822, Montréal (1992).
7. G. W. Bluman and J. D. Cole, *The general similarity of the heat equation.* J. Math. Mech.18 1025-1042 (1969).
8. G. W. Bluman and S. Kumei, *Symmetries and Differential Equations.* Appl. Math. Sci., vol. 81, Springer-Verlag, Berlin (1989).
9. J. Boussinesq, *Théorie de l'intumescence appelée onde solitaire ou de translation se propageant dans un canal rectangulaire.* Comptes Rendus, 72, 755-759, (1871).
10. J. Boussinesq, *Théorie des ondes et des remous qui se propagent le long d'un canal rectangulaire horizontal, en communiquant au liquide contenu dans ce canal des vitesses sensiblemant parielles de la surface au fond.* J. Math. Pures Appl. Ser. 2.17, 55, (1872).
11. C. P. Boyer, R. T. Sharp, and P. Winternitz, *Symmetry breaking interactions for the time dependent Schrödinger equation.* J. Math. Phys.17, 1439-1451, (1976).
12. B. Buchberger, *A survey on the method of Gröbner bases for solving problems in connection with systems of multivariate polynomials* Symbolic and Algebraic Computation by Computers eds. N. Inada and T. Soma World Scientific, Singapore. 69-83, (1985).
13. B. Buchberger, *Applications of Gröbner bases in nonlinear computational geometry,* Mathematical Aspects of Scientific Software. ed. J. Rice Springer Verlag, New York, 59-87, (1988).
14. G. Burdet, J. Patera, M. Perrin, and P. Winternitz, *The optical group and its subgroups.* J. Math. Phys. , 19, 1758-1780, (1978).
15. B. Champagne, W. Hereman, and P. Winternitz, *The computer calculation of Lie point symmetries of large systems of differential equations.* Comm. Math. Phys. , 66, 319-340, (1991).
16. B. Champagne and P. Winternitz, *A Macsyma program for calculating the symmetry group of a system of differential equations.* Preprint CRM-1278, Montréal (1985).
17. B. Champagne and P. Winternitz, *On the infinite-dimensional symmetry group of the Davey–Stewartson equation.* J. Math. Phys. , 29, 1-8, (1988).
18. P. A. Clarkson, *New exact solutions for the Boussinesq equation.* Europ. J. Appl. Math., 1, 279-300, (1990).
19. P. A. Clarkson, *Dimensional reductions and exact solutions of a generalized nonlinear Schrödinger equation.* Nonlinearity, 5, 453-472, (1992).

20. P. A. Clarkson, *Nonclassical symmetry reductions of nonlinear partial differential equations.* Math. Comp. Model. 18, 45-68, (1993).
21. P. A. Clarkson, *Nonclassical symmetry reductions of the Boussinesq equation.* Chaos, Solitons and Fractals 5, 2261-2301, (1995).
22. P. A. Clarkson and S. Hood, *Nonclassical symmetry reductions and exact solutions of the Zabalotskaya–Khokhlov equation.* Europ. J. Appl. Math. 3, 381-414, (1992).
23. P. A. Clarkson and M. D. Kruskal, *New similarity solutions of the Boussinesq equation.* J. Math. Phys. 30, 2201-2213, (1989).
24. P. A. Clarkson, D. K Ludlow and T. J. Priestley, *The classical, direct and nonclassical methods for symmetry reductions of nonlinear partial differential equations.* Meth. Appl. Anal. 4, 173-195, (1997).
25. P. A. Clarkson and E. L. Mansfield, *Symmetry Reductions and Exact Solutions of a class of Nonlinear Heat Equations.* Physica D, 70, 250-288, (1994).
26. P. A. Clarkson and E. L. Mansfield, *Algorithms for the nonclassical method of symmetry reductions.* SIAM J. Appl. Math. 54, 1693-1719, (1994).
27. P. A. Clarkson and E. L. Mansfield, *On a shallow water wave equation.* Nonlinearity 7, 975-1000, (1994).
28. P. A. Clarkson and E. L. Mansfield, *Symmetry reductions and exact solutions of shallow water wave equations.* Acta Appl. Math. 39, 245-276, (1995).
29. P. A. Clarkson, E. L. Mansfield, and A. E. Milne, *Symmetries and Exact Solutions of a $2+1$-dimensional Sine-Gordon System.* Phil. Trans. R. Soc. Lond. A 354, 1807-1835, (1996).
30. P. A. Clarkson, E. L. Mansfield, and T. J. Priestley, *Symmetries of a class of nonlinear third-order partial differential equations.* Math. Comp. Model. 25, 195-212, (1997).
31. P. A. Clarkson and T. J. Priestley, *Symmetries of a class of nonlinear fourth-order partial differential equations.* J. Nonlinear Math. Phys. (1999) to appear.
32. P. A. Clarkson and T. J. Priestley, *Shallow water wave systems.* Studies in Applied Mathematics (1999) to appear.
33. P. A. Clarkson and P. Winternitz, *Nonclassical symmetry reductions for the Kadomtsev–Petviashvili equation.* Physica D, 49, 257-272, (1991).
34. R. Conte, Chapter 1, this volume.
35. J. F. Cornwell, *Group Theory in Physics, vol. II,* Academic Press, New York (1984).
36. D. David, N. Kamran, D. Levi, and P. Winternitz, *Subalgebras of loop algebras and symmetries of the Kadomtsev-Petviashvili equation.* Comm. Math. Phys. 55, 2111-2113, (1985).
37. D. David, N. Kamran, D. Levi, and P. Winternitz, *Symmetry reduction for the Kadomtsev–Petviashvili equation using a loop algebra.* J. Math. Phys. 27, 1225-1237, (1986).
38. V. A. Dorodnitsyn, *Transformation groups in net spaces*, Itogi Nauki i Tekhniki **34** (1989), 149–191; J. Soviet Math. **55** (1991), 1490–1517.
39. E. B. Dynkin, *Semisimple subalgebras of semisimple Lie algebras*, Mat. Sbornik **30** (1952), 349–462; Amer. Math. Soc. Transl. **6** (1957), 111-244.
40. E. B. Dynkin, *Maximal subgroups of the classical groups,* Trudy Moskov. Mat. Obšč **1** (1952), 39–166; Amer. Math. Soc. Transl. **6** (1957), 245–378.
41. P. G. Estévez, *Nonclassical symmetries and the singular manifold method for the Fitzhugh–Nagumo equation.* Comm. Math. Phys. A, 171, 259-261, (1992).

42. P. G. Estévez and P. R. Gordoa, *Nonclassical symmetries and the singular manifold method: theory and six examples*Stud. Appl. Math. 95 (1995), 73-113.
43. R. Fitzhugh, *Impulses and physiological states in theoretical models of nerve membrane*Biophysical J.1 (1961), 445-466.
44. R. Floreanini and L. Vinet, *Lie symmetries of finite-difference equations* J. Math. Phys. 36 (1995), 7024-7042.
45. A. S. Fokas and M. J. Ablowitz, *On a unified approach to transformations and elementary solutions of Painlevé equations*J. Math. Phys. 23 (1983), 2033-2042.
46. D. A. Frank-Kamenetskii, *Diffusion and Heat Exchange in Chemical Kinetics*Princeton University Press, Princeton1955
47. W. I. Fushchich, *Conditional symmetry of the equations of Mathematical Physics*Ukrain. Math. J.43 (1991), 1456–1470.
48. W. I. Fushchich, *Ansatz '95*Nonlinear Math. Phys.2 (1995), 216–235.
49. W. I. Fushchich, W. M. Shtelen, and N. I. Serov, *Symmetry Analysis and Exact Solutions of the Equations of Mathematical Physics*Kluwer, Dordrecht1993
50. G. Gaeta, *Nonlinear Symmetries and Nonlinear Equations*Kluwer, Dordrecht1994
51. L. Gagnon and P. Winternitz, *Lie symmetries of a generalised nonlinear Schrödinger equation: I–The symmetry group and its subgroups* Comm. Math. Phys. 21 (1988), 1493–1511.
52. L. Gagnon and P. Winternitz, *Lie symmetries of a generalised nonlinear Schrödinger equation: II–Exact solutions*Comm. Math. Phys. 22 (1989), 469–497.
53. L. Gagnon and P. Winternitz, *Lie symmetries of a generalised nonlinear Schrödinger equation: II–Exact solutions* Comm. Math. Phys. 22 (1989), 469–497.
54. L. Gagnon and P. Winternitz, *Exact solutions of the spherical quintic nonlinear Schrödinger equation* Comm. Math. Phys. 134A (1989), 276–281.
55. V. A. Galaktionov, *On new exact blow-up solutions for nonlinear heat conduction equations with source and applications*Diff. and Int. Eqns.3 (1990), 863–874.
56. V. A. Galaktionov, *Quasilinear heat equations with first-order sign-invariants and new explicit solutions*Nonlinear Anal., Theory, Meth., and Appl.23 (1994), 1595 1621.
57. C. S. Gardner, J. M. Greene, M. D. Kruskal, and R. M. Miura, *Method for solving the KdV equation*Comm. Math. Phys. 19 (1967), 1095–1097.
58. P. Goddard and D. Olive, *Kac Moody and Virasoro algebras in relation to quantum physics* Int. J. Mod. Phys. A1 (1986), 303–414.
59. D. Gómez-Ullate, S. Lafortune and P. Winternitz, *Symmetries of discrete dynamical systems involving two species*Preprint CRM-2567, Montréal1998
60. E. Goursat, *??*Ann. Sci. Ec. Norm. Sup (3)618809??
61. B. Grammaticos, F. W. Nijhoff, and A. Ramani, this volume.
62. A. M. Grundland and L. Lalague, *Lie subgroups of the symmetry group of the equations describing a nonstationary and isotropic flow: invariant and partially invariant solutions* Can. J. Phys.72 (1994), 362–374.
63. A. M. Grundland and L. Lalague, *Lie subgroups of symmetry groups of fluid dynamics and magnetohydrodynamics equations* Can. J. Phys.73 (1995), 463–477.

64. A. M. Grundland and L. Lalague, *Invariant and partially invariant solutions of the equations describing a nonstationary and isentropic flow for an ideal and incompressible fluid in* (3 + 1) *dimensions* J. Phys. A. Math. gen29 (1996), 1723–1739.
65. A. G. Hansen, *Similarity Analyses of Boundary Value Problems in Engineering*Prentice-Hall, Englewood Cliffs1964
66. W. Hereman, *Application of a Macsyma program for the Painlevé test to the Fitzhugh–Nagumo equation*Partially Integrable Evolution Equations in Physicseds. R. Conte and N. BoccaraNATO ASI Series C, vol. 310Kluwer, Dordrecht, (1990), 585–586.
67. W. Hereman, *Review of symbolic software for the computation of Lie symmetries of differential equations*Euromath. Bull2 (1993), 45–79.
68. W. Hereman, *Symbolic software for Lie symmetry analysis*CRC Handbook of Lie Group Analysis of Differential Equationsed. N. Ibragimovvol. 3CRC Press, Boca Raton, (1996), 367–522.
69. R. Hernández Heredero, D. Levi and P. Winternitz, *Point symmetries and generalized symmetries of nonlinear difference equations*Preprint CRM-2568, Montréal1998
70. R. Hirota, *Direct methods in soliton theory* Solitonseds. R. K. Bullough and P. J. CaudreyTopics in Current Physics, vol. 17Springer-Verlag, Berlin, 1980, 157–176.
71. R. Hirota and M. Ito, *Resonance of solitons in one dimension*Comm. Math. Phys. 52 (1983), 744–748.
72. S. Hood, *New exact solutions of Burgers's equation—an extension to the direct method of Clarkson and Kruskal*J. Math. Phys. 36 (1995), 1971–1990.
73. V. Hussin, P. Winternitz, and H. Zassenhaus, *Maximal abelian subalgebras of complex orthogonal Lie algebras* Comm. Math. Phys. 141 (1990), 183–220.
74. V. Hussin, P. Winternitz, and H. Zassenhaus *Maximal abelian subalgebras of pseudo-orthogonal Lie algebras* Comm. Math. Phys. 173(1992),125–163.
75. N. H. Ibragimov, *Transformation Groups Applied to Mathematical Physics*Reidel, Dordrecht1985
76. E. L. Ince, *Ordinary Differential Equations* Longmans, Green and co., London and New York1926Reprinted Dover, New York, 1956
77. N. Jacobson, *Lie algebras*Dover, New York1979
78. M. Jimbo and T. Miwa, *Solitons and infinite dimensional Lie algebras* Publ. RIMS, Kyoto19 (1983), 943–1001.
79. N. Joshi and P. J. Vassiliou, *The existence of Lie symmetries for first-order analytic discrete dynamical systems* J. Math. Anal. Appl.195 (1995), 872–887.
80. E. G. Kalnins and P. Winternitz, *Maximal abelian sybalgebras of complex Euclidean Lie algebras* Comm. Math. Phys. 72 (1994), 389–404.
81. T. Kawahara and M. Tanaka, *Interactions of traveling fronts – an exact solution of a nonlinear diffusion equation*Comm. Math. Phys. 97A(1983), 311–314.
82. B. P. Komrakov, *Reductive subalgebras of semisimple real Lie algebras*, Doklady AN SSSR **308** (1989), 521–525; Soviet Math. Dokl. **40** (1990), 329–333.
83. B. P. Komrakov, *Maximal subalgebras of real Lie algebras and a problem of Sophus Lie*, Doklady AN SSSR **311** (1990), 538–532; Soviet Math. Dokl. **41** (1990), 269–273.
84. B. Kostant, *On the conjugacy of real Cartan subalgebras* Proc. Nat. Academy Sci. USA41 (1955), 967–970.

85. D. Levi and M. A. Rodríguez, *Symmetry group of partial differential equations and of differential-difference equations: the Toda lattice versus the Korteweg-de Vries equation* Comm. Math. Phys. 25 (1992), L975–L979.
86. D. Levi, L. Vinet and P. Winternitz, *Lie group formalism for difference equations*Comm. Math. Phys. 30 (1991), 633.
87. D. Levi, L. Vinet and P. Winternitz (eds.), *Symmetries and integrability of difference equations*CRM proceedings and lecture notes, Vol 9AMS, Providence, R.I.1996
88. D. Levi and P. Winternitz, *The cylindrical Kadomtsev-Petviashvili equation : its Kac-Moody-Virasoro algebra and relation to KP equation* Comm. Math. Phys. 129A (1988), 165–167.
89. D. Levi and P. Winternitz, *Continuous symmetries of discrete equations*Comm. Math. Phys. 152A1991335
90. D. Levi and P. Winternitz, *Symmetries of discrete dynamical systems*J. Math. Phys. 37199655515576
91. D. Levi and P. Winternitz, *Nonclassical symmetry reduction: example of the Boussinesq equation*Comm. Math. Phys. 22198929152924
92. D. Levi and P. Winternitz (eds.), *Painlevé transcendents, their asymptotics and physical applications*Plenum, New York1992
93. D. Levi and P. Winternitz, *Symmetries and conditional symmetries of differential-difference equations* J. Math. Phys. 34199337133730
94. D. Levi and R. Yamilov, *Conditions for the existence of higher symmetries of evolutionnary equations on the lattice* J. Math. Phys. 38199766486674
95. S.-Y. Lou, *A note on the new similarity reductions of the Boussinesq equation* Comm. Math. Phys. 151A1990133135
96. D. K. Ludlow, *Nonclassical similarity reductions of the Navier–Stokes equations*Ph.D. thesis (Department of Mathematics, University of Exeter, UK)1994
97. S. Maeda, *The similarity method for difference equations* IMA J. Appl. Math.381987129134
98. E. L. Mansfield`diffgrob2`:*A symbolic algebra package for analysing systems of PDE using Maple*1993 `ftp ftp.ukc.ac.uk`login: anonymous, password: your email address, directory: `pub/Liz/Maple`, files: `diffgrob2_src.tar.Z`, `diffgrob2_man.tex.Z`
99. E. L. Mansfield `diffgrob2`: *A symbolic algebra package for analysing systems of PDE using Maple*preprint, Department of Mathematics, University of Exeter, U.K.1993
100. E. L. Mansfield, *The differential algebra package* `diffgrob2`Maple Technical Newsletter319963337
101. E. L. Mansfield, *Applications of the differential algebra package* `diffgrob2` *to classical symmetries of PDEs*J. Symb. Comp.231997517533
102. E. L. Mansfield, *Symmetries and exact solutions for a* $2+1$*-dimensional shallow water wave equation*Math. Comp. Simul.4319973955
103. E. L. Mansfield and E. Fackerell, *Differential Gröbner Bases*preprint no. 92/108, Macquarie Univerisity, Sydney, Australia1992
104. E. L. Mansfield, G. J. Reid and P. A. Clarkson, *Nonclassical reductions of a coupled nonlinear Schrödinger system*Comp. Phys. Comm.1999to appear
105. L. Martina, G. Soliani, and P. Winternitz, *Partially invariant solutions of a class of nonlinear Schrödinger equations* Comm. Math. Phys. 25199244254435

106. L. Martina and P. Winternitz, *Analysis and applications of the symmetry group of the multidimensional three-wave resonant interaction problem* Comm. Math. Phys. 1961989231277
107. L. Martina and P. Winternitz, *Partially invariant solutions of nonlinear Klein-Gordon and Laplace equations* J. Math. Phys. 33199227182727
108. W. Miller Jr., J. Patera, and P. Winternitz *Subgroups of Lie groups and separation of variables* J. Math. Phys. 221981251260
109. J. D. Murray, *Mathematical Biology*Springer-Verlag, New York1989
110. M. Musette, Chapter 3, this volume.
111. J. S. Nagumo, S. Arimoto, and S. Yoshizawa, *An active pulse transmission line simulating nerve axon*Proc. IRE50196220612070
112. T. Nishitani and M. Tajiri, *On similarity solutions of the Boussinesq equation*Comm. Math. Phys. 89A1982379380
113. M. C. Nucci and P. A. Clarkson, *The nonclassical method is more general than the direct method for symmetry reductions: an example of the Fitzhugh-Nagumo equation*Comm. Math. Phys. 164A19924956
114. M. A. del Olmo, M. A. Rodríguez, P. Winternitz, and H. Zassenhaus, *Maximal abelian subalgebras of pseudounitary Lie algebras* Comm. Math. Phys. 135199079151
115. P. J. Olver, *Applications of Lie Groups to Differential Equations*2nd edition, Graduate Texts Math., vol. 107Springer-Verlag, New York1993
116. P. J. Olver, *Direct reduction and differential constraints* Proc. R. Soc. Lond. A4441994509523
117. P. J. Olver and P. Rosenau, *The construction of special solutions to partial differential equations*Comm. Math. Phys. 114A1986107112
118. P. J. Olver and P. Rosenau, *Group-invariant solutions of differential equations*SIAM J. Appl. Math.471987263275
119. P. J. Olver and E. M. Vorob'ev, *Nonclassical and conditional symmetries*CRC Handbook of Lie Groups Analysis of Differential Equations. III.Theoretical Developments an Computational Methodsed. N. H. IbragimovCRC Press, Boca Raton1996291328
120. J. Ondich, *A differential constraints approach to partial invariance* Europ. J. Appl. Math.61995631637
121. A. Yu. Orlov and P. Winternitz, *Algebra of pseudodifferential operators and symmetries of equations in the Kadomtsev-Petviashvili hierarchy* J. Math. Phys. 38199746444674
122. L. V. Ovsiannikov, *Group Analysis of Differential Equations*Academic, New York1982
123. E. V. Pankrat'ev, *Computations in differential and difference modules*Acta Appl. Math.161989167189
124. J. Patera, R. T. Sharp, P. Winternitz, and H. Zassenhaus, *Continuous subgroups of the fundamental groups of physics. III. The de Sitter groups* J. Math. Phys. 181977)22592288
125. J. Patera, P. Winternitz, and H. Zassenhaus, *Continuous subgroups of the fundamental groups of physics. III. General method and the Poincaré group* J. Math. Phys. 16197515971614
126. J. Patera, P. Winternitz, and H. Zassenhaus, *Continuous subgroups of the fundamental groups of physics. III. The similitude group* J. Math. Phys. 16197516151624

127. J. Patera, P. Winternitz, and H. Zassenhaus, *Quantum numbers for particles in de Sitter space* J. Math. Phys. 171976717728
128. J. Patera, P. Winternitz, and H. Zassenhaus, *Maximal abelian subalgebras of real and complex symplectic Lie algebras* J. Math. Phys. 24198319731985
129. A. H. Pincombe and N. F. Smyth, *Initial boundary value-problems for the Korteweg–de Vries equation*Proc. Roy. Soc. Lond. A4331991479498
130. E. Pucci, *Similarity reductions of partial differential equations*Comm. Math. Phys. 25199226312640
131. G. R. W. Quispel, H. W. Capel and R. Sahadevan *Continuous symmetries of differential-difference equations* Comm. Math. Phys. 1701992379383
132. G. R. W. Quispel, F. W. Nijhoff, and H. W. Capel, *Linearization of the Boussinesq equation and the modified Boussinesq equation*Comm. Math. Phys. 91A1982143145
133. D. Rand, *Pascal Programs for Identification of Lie Algebras, I: RADICAL, A Program to Calculate the Radical & Nilradical of Parameter-Free and Parameter-Dependent Lie Algebras* Comm. Math. Phys. 411986105125
134. D. Rand, *Pascal Programs for Identification of Lie Algebras, III: Levi Decomposition and Canonical Basis* Comm. Math. Phys. 461987311322
135. D. Rand and P. Winternitz, *ODEPAINLEVE A Macsyma Package for Painlevé Analysis of Ordinary Differential Equations* Comm. Math. Phys. 421986359383
136. D. Rand, P. Winternitz, and H. Zassenhaus, *Pascal Programs for Identification of Lie Algebras, II: SPLIT, A Program to Decompose Parameter-Free and Parameter-Dependent Lie Algebras into Direct Sums* Comm. Math. Phys. 461987297309
137. D. Rand, P. Winternitz, and H. Zassenhaus, *On the Identification of a Lie Algebra Given by its Structure Constants, I: Direct Decompositions, Levi Decompositions and Nilradicals* Comm. Math. Phys. 1091988197246
138. G. J. Reid, *A triangularization algorithm which determines the Lie symmetry algebra of any system of PDEs*Comm. Math. Phys. 231990L853L859
139. G. J. Reid, *Algorithms for reducing a system of PDEs to standard form, determining the dimension of its solution space and calculating its Taylor series solution*Europ. J. Appl. Math.21991293318
140. G. J. Reid, *Finding abstract Lie symmetry algebras of differential equations without integrating determining equations*Europ. J. Appl. Math.2 (1991), 319–340.
141. G. J. Reid and A. Wittkopf, *Finding abstract Lie symmetry algebras of differential equations without integrating determining equations*A Differential Algebra Package for Maple 1993 `ftp 137.82.36.21` login: anonymous, password: your email address, directory: `pub/standardform`
142. G. J. Reid, A. D. Wittkopf and A. Boulton *Reduction of systems of nonlinear partial differential equations to simplified involutive forms*Europ. J. Appl. Math.7 (1996), 635–666.
143. C. Rogers and W. F. Ames, *Nonlinear Boundary Value Problems in Science and Engineering*Academic Press, New York1989
144. P. Rosenau and J. L. Schwarzmeier, *On similarity solutions of Boussinesq type equations*Comm. Math. Phys. 115A (1986), 75–77.
145. J. Rubin and P. Winternitz, *Point symmetries and conditionally integrable nonlinear evolution equations* J. Math. Phys. 31 (1990), 2085–2090.

146. F. Schwarz, *A REDUCE package for determining Lie symmetries of ordinary and partial differential equations* Comm. Math. Phys. 27 (1982), 179–186.
147. F. Schwarz, *An algorithm for determining the size of symmetry groups*Computing49 (1992), 95–115.
148. A. C. Scott, *The application of Bäcklund transforms to physical problems*Bäcklund TransformationsR. M. Miura, ed.Lect. Notes in Math., vol. 515, Springer, Berlin1975, 80–105.
149. H. Segur and A. Finkel, *An analytical model of periodic waves in shallow water*Stud. Appl. Math. 73 (1985), 183–220.
150. N. F. Smyth, *The effect of conductivity on hotspots*J. Aust. Math Soc., Ser. B33 (1992), 403–413.
151. H. Stephani, *Differential Equations, Their Solutions Using Symmetries*Cambridge Univ. Press, Cambridge1989
152. D. A. Suprunenko and R. I. Tyshkevich, *Commutative matrices*Academic Press, New York1968
153. Z. Thomova and P. Winternitz, *Maximal Abelian Subgroups of the Isometry and Conformal Groups of Euclidean and Minkowski Spaces* Comm. Math. Phys. 31 (1998), 1831–1858.
154. Z. Thomova and P. Winternitz, *Maximal Abelian subalgebras of pseudo-Euclidean Lie algebras*Lin. Alg. Annl.1999(to appear)
155. M. Toda, *Studies of a nonlinear lattice*Comm. Math. Phys. .8 (1975), 1–125.
156. V. L. Topunov, *Reducing systems of linear-differential systems to passive form*Acta Appl. Math.16 (1989), 191–206.
157. F. Ursell, *The long-wave paradox in the theory of gravity waves*Proc. Camb. Phil. Soc.49 (1953), 685–694.
158. E. M. Vorob'ev, *Symmetries of compatibility conditions for systems of differential equations*Acta Appl. Math.24 (1991), 1–24.
159. J. Weiss, *The Painlevé property for partial differential equations.* II. *Bäcklund transformation, Lax pairs, and the Schwarzian derivative*J. Math. Phys. 24 (1983), 1405–1413.
160. J. Weiss, M. Tabor, and G. Carnevale, *The Painlevé property for partial differential equations*J. Math. Phys. 24 (1983), 522–526.
161. E. E. Whittaker and G.M. Watson, *Modern Analysis*4th edition, Cambridge Univ. Press, Cambridge 1927
162. P. Winternitz, *Kac–Moody–Virasoro symmetries of integrable nonlinear partial differential equations*Symmetries and Nonlinear Phenomenaeds. D. Levi and P. WinternitzWorld Scientific, Singapore 1988, 358–375.
163. P. Winternitz, *Group theory and exact solutions of partially integrable differential systems*Partially Integrable Evolution Equations in Physicseds. R. Conte and N. BoccaraKluwer, Dordrecht1990515567
164. P. Winternitz, *Lie groups and solutions of nonlinear partial differential equations*Integrable Systems, Quantum Groups and Quantum Field Theorieseds. L. A. Ibort and M. A. RodríguezKluwer, Dordrecht 1993, 429–495.
165. P. Winternitz and J.-P. Gazeau, *Allowed transformations and symmetry classes of variable coefficient Korteweg-de Vries equations* Comm. Math. Phys. 167A (1992),246–250.
166. N. N. Yanenko, *Theory of consistency and methods of integrating systems of nonlinear partial differential equations* Proceedings of the Fourth All-Union Mathematics Congress Leningrad 1964, 247–259.

167. N. J. Zabusky, *A synergetic approach to problems of nonlinear dispersive wave propagation and interaction*Nonlinear Partial Differential Equationsed. W. F. AmesAcademic Press, New York 1967, 233–258.
168. V. E. Zakharov, *On stocastization of one-dimensional chains of nonlinear oscillations*Sov. Phys. JETP38 (1974), 108–110.
169. H. Zassenhaus, *Lie groups, Lie algebras and representation theory*Les Presses de l'Université de Montréal, Montréal1981
170. Zhuhan Jiang, *Lie symmetries and their local determinacy for a class of differential-difference equations* Comm. Math. Phys. 240A (1998), 137–143.
171. S. Zidowitz, *Conditional symmetries and the direct reduction of partial differential equations*Modern Group Analysis: Advanced Analytical and Computational Methods in Mathematical Physicseds. N. H. Ibragimov, M. Torrisi and A. ValentiKluwer, Dordrecht 1993, 387–393.

List of Participants

1. Abenda Simonetta, University of Bologna, Italy
   Abenda@ciram.ing.unibo.it
2. Abrahamson Mikael, Lund University, Sweden
   mikab@maths.ith.se
3. Adami Riccardo, University of Roma "La Sapienza" Italy
   adami@mat.uniroma1.it
4. Anders Igor, Paris, France
   igor_anders@yahoo.com
5. Brugarino Tommaso, University of Palermo, Italy
   brugarin@unipa.it
6. Class Waksjö, Linköping University, Sweden
   class.waksjo@math.lin.se
7. Conte Robert, CEA Saclay, France (lecturer)
   conte@spec.saclay.cea.fr
8. Delgado Marina, University of Madrid, Spain
   mdelgado@alum.math.uc3m.es
9. Greco Antonio, University of Palermo, Italy (editor)
   greco@gremat.math.unipa.it
10. Kamp Van der Peter, University of Amsterdam, the Netherlands
    peter@few.vu.nl
11. Kockelkoren Julien, CEA Saclay, France
    kockel@spec.saclay.cea.fr
12. Lafortune Stéphane, University of Montréal, Canada
    lafortus@crm.unomtreal.cn
13. Loris Ignace, University of Brussels, Belgium
    igloris@vub.ac.be
14. Magri Franco, University of Milano, Italy (lecturer)
    magri@vmimat.mat.unimi.it
15. Martinez Torres David, University of Madrid, Spain
    dmartinez@arrakis.es
16. Morgante Anna, CEA Saclay, France
    morgante@llb.saclay.cea.fr
17. Musette Micheline, University of Brussels, Belgium (lecturer)
    mmusette@vub.ac.be

18. Peruzza Rosanna, University of Palermo, Italy
peruzza@dipmat.math.unipa.it
19. Rinaldi Emanuela, University of Messina, Italy
rinaldi@dipmat.unime.it
20. Satsuma Junkichi, University of Tokyo, Japain (lecturer)
satsuma@poisson.ms.u-tokyo.ac.jp
21. Sciacca Vincenzo, University of Palermo, Italy
sciacca@dipmat.math.unipa.it
22. Sergyeyev Artur, University of Kiev, Ukrainian
arthurser@imath.kiev.ua
23. Seta Luciano, University of Palermo, Italy
seta@dipmat.math.unipa.it
24. Tokihiro Tetsuji, University of Tokyo, Japain
toki@poisson.ms.u-tokyo.ac.jp
25. Verhoeven Caroline, University Brussels, Belgium
cverhoev@vub.ac.be
26. Winternitz Pavel, University of Montréal, Canada (lecturer)
wintern@crm.umontral.ca
27. Zalij Alexander, Kiev University, Ukrainian
Zaliy@imath.kiev.ua
28. Zanoni Alberto, University of Pisa, Italy
zanoni@posso.dm.unipi.it

Index

N-soliton solution 4, 137, 145, 156, 164
ψ-series 14, 49
singular part transformation 18, 19

AKNS equation 32, 61
AKNS system 59, 61

Bäcklund
  parameter 139, 150, 157, 160
Bäcklund parameter 3, 5, 26
Bäcklund transformation 3, 18, 49, 137, 139, 144, 147, 150, 152–154
  auto– 3, 59, 138–140, 154, 155, 157, 159–161
  bilinear 49, 59, 157
  hetero– 3–5, 10, 26, 31, 35, 36, 44, 46, 59, 156, 158
Bianchi
  diagram 151, 160
  permutability theorem 151, 160, 165
birational transformation 72, 140–142, 157
Boussinesq equation 20, 35, 54, 143, 149
  modified 59
branch 14
Broer-Kaup equation 20, 59, 61

characteristic 69, 138
classical
  method 69
closed form 2
complex Ginzburg-Landau equation 65

d'Alembert equation 4, 10, 21, 30, 31
Darboux covariance 144, 145
Darboux transformation 138, 140, 144, 152, 154, 155
  binary 39, 145, 164
  classical 144
differential Groebner basis 50
differential part 6
diophantine conditions 17
direct
  method 69

elliptic function 17, 140
expansion variable 11, 147, 148

family 142, 148, 149
family of movable singularities 14
Fordy-Gibbons equation 45
Fuchs indices 15

general analytic solution 52, 66

hierarchy 19, 157
Hirota-Satsuma equation 20, 37, 41, 54, 158
homographic
  group 12, 140, 148
  invariants 12

indices 142
indicial equation 15
integrability 5, 137, 138, 142, 149
isospectral 5
Ito equation 7

Kadomtsev-Petviashvili equation 143
Kaup-Kupershmidt equation 20, 43–45, 54, 61, 145, 156, 158, 159, 161
Korteweg-de Vries

hierarchy 7
Korteweg-de Vries equation 12, 18, 33, 59, 61, 151
fifth order 156
modified 57, 61, 148, 154
KPP equation 15, 60, 69
Kundu-Eckhaus equation 68
Kuramoto-Sivashinsky equation 2, 5, 8, 16, 18, 49

Lax pair 5, 18, 138, 140, 143
Lax representation 5
order 6
Riccati representation 6, 149
scalar representation 6
string representation 6
zero-curvature representation 6
Liouville equation 4, 10, 30, 56, 57, 61
Lorenz model 18

microdifferential operator 6
Miura transformation 35, 149, 158
Moutard transformation 146, 165

no-logarithm condition 15
nonclassical
method 69
nonlinear Schrödinger equation 7, 59, 61, 67
nonlinear superposition formula 137, 151, 153, 156, 157, 160, 161, 164

one-soliton solution 4, 152, 163

Painlevé
analysis 10, 142, 147, 148
equation 176
equation P2 72, 73
equation P4 73
equation P6 74
equation (P1) 7
equation (P6) 75
equations 19, 140
Gambier classification 140
property 10, 139
test 10, 11, 138, 140, 142

Rössler system 18
reduction 62, 65, 66, 69, 156
resonances 15, 142
resummation 20

Sawada-Kotera equation 20, 43, 44, 46, 54, 61, 145, 149, 156, 158, 160
Schlesinger transformation 19
Schur polynomials 190
Schwarzian 12
sine-Gordon equation 3, 4, 6, 8, 10, 18, 20, 55, 61, 139, 148, 154
singular manifold
equation 30, 33, 35, 36, 52, 54, 57, 59, 152
method 19, 20, 27, 69, 138
variable 11, 138, 142
singular part operator 16, 138, 147–149
solution
exact 2, 18
global 2
local 2
spectral parameter 5, 26, 145, 150

tau-function 8, 19, 138, 149, 160, 164, 165
truncation
equations 27
method 19, 20, 27, 138, 142, 147, 148, 150, 153
variable 27
two-soliton solution 153, 156, 161, 163–166
Tzitzéica equation 20, 38, 61, 146, 149, 165

vacuum 4, 145, 152, 153, 156, 163–166

Lecture Notes in Physics

For information about Vols. 1–587
please contact your bookseller or Springer-Verlag
LNP Online archive: springerlink.com

Vol.588: Y. Watanabe, S. Heun, G. Salviati, N. Yamamoto (Eds.), Nanoscale Spectroscopy and Its Applications to Semiconductor Research.

Vol.589: A. W. Guthmann, M. Georganopoulos, A. Marcowith, K. Manolakou (Eds.), Relativistic Flows in Astrophysics.

Vol.590: D. Benest, C. Froeschlé (Eds.), Singularities in Gravitational Systems. Applications to Chaotic Transport in the Solar System.

Vol.591: M. Beyer (Ed.), CP Violation in Particle, Nuclear and Astrophysics.

Vol.592: S. Cotsakis, L. Papantonopoulos (Eds.), Cosmological Crossroads. An Advanced Course in Mathematical, Physical and String Cosmology.

Vol.593: D. Shi, B. Aktaş, L. Pust, F. Mikhailov (Eds.), Nanostructured Magnetic Materials and Their Applications.

Vol.594: S. Odenbach (Ed.),Ferrofluids. Magnetical Controllable Fluids and Their Applications.

Vol.595: C. Berthier, L. P. Lévy, G. Martinez (Eds.), High Magnetic Fields. Applications in Condensed Matter Physics and Spectroscopy.

Vol.596: F. Scheck, H. Upmeier, W. Werner (Eds.), Noncommutative Geometry and the Standard Model of Elememtary Particle Physics.

Vol.597: P. Garbaczewski, R. Olkiewicz (Eds.), Dynamics of Dissipation.

Vol.598: K. Weiler (Ed.), Supernovae and Gamma-Ray Bursters.

Vol.599: J.P. Rozelot (Ed.), The Sun's Surface and Subsurface. Investigating Shape and Irradiance.

Vol.600: K. Mecke, D. Stoyan (Eds.), Morphology of Condensed Matter. Physcis and Geometry of Spatial Complex Systems.

Vol.601: F. Mezei, C. Pappas, T. Gutberlet (Eds.), Neutron Spin Echo Spectroscopy. Basics, Trends and Applications.

Vol.602: T. Dauxois, S. Ruffo, E. Arimondo (Eds.), Dynamics and Thermodynamics of Systems with Long Range Interactions.

Vol.603: C. Noce, A. Vecchione, M. Cuoco, A. Romano (Eds.), Ruthenate and Rutheno-Cuprate Materials. Superconductivity, Magnetism and Quantum Phase.

Vol.604: J. Frauendiener, H. Friedrich (Eds.), The Conformal Structure of Space-Time: Geometry, Analysis, Numerics.

Vol.605: G. Ciccotti, M. Mareschal, P. Nielaba (Eds.), Bridging Time Scales: Molecular Simulations for the Next Decade.

Vol.606: J.-U. Sommer, G. Reiter (Eds.), Polymer Crystallization. Obervations, Concepts and Interpretations.

Vol.607: R. Guzzi (Ed.), Exploring the Atmosphere by Remote Sensing Techniques.

Vol.608: F. Courbin, D. Minniti (Eds.), Gravitational Lensing:An Astrophysical Tool.

Vol.609: T. Henning (Ed.), Astromineralogy.

Vol.610: M. Ristig, K. Gernoth (Eds.), Particle Scattering, X-Ray Diffraction, and Microstructure of Solids and Liquids.

Vol.611: A. Buchleitner, K. Hornberger (Eds.), Coherent Evolution in Noisy Environments.

Vol.612: L. Klein, (Ed.), Energy Conversion and Particle Acceleration in the Solar Corona.

Vol.613: K. Porsezian, V.C. Kuriakose, (Eds.), Optical Solitons. Theoretical and Experimental Challenges.

Vol.614: E. Falgarone, T. Passot (Eds.), Turbulence and Magnetic Fields in Astrophysics.

Vol.615: J. Büchner, C.T. Dum, M. Scholer (Eds.), Space Plasma Simulation.

Vol.616: J. Trampetic, J. Wess (Eds.), Particle Physics in the New Millenium.

Vol.617: L. Fernández-Jambrina, L. M. González-Romero (Eds.), Current Trends in Relativistic Astrophysics, Theoretical, Numerical, Observational

Vol.618: M.D. Esposti, S. Graffi (Eds.), The Mathematical Aspects of Quantum Maps

Vol.619: H.M. Antia, A. Bhatnagar, P. Ulmschneider (Eds.), Lectures on Solar Physics

Vol.620: C. Fiolhais, F. Nogueira, M. Marques (Eds.), A Primer in Density Functional Theory

Vol.621: G. Rangarajan, M. Ding (Eds.), Processes with Long-Range Correlations

Vol.622: F. Benatti, R. Floreanini (Eds.), Irreversible Quantum Dynamics

Vol.623: M. Falcke, D. Malchow (Eds.), Understanding Calcium Dynamics, Experiments and Theory

Vol.624: T. Pöschel (Ed.), Granular Gas Dynamics

Vol.625: R. Pastor-Satorras, M. Rubi, A. Diaz-Guilera (Eds.), Statistical Mechanics of Complex Networks

Vol.626: G. Contopoulos, N. Voglis (Eds.), Galaxies and Chaos

Vol.627: S.G. Karshenboim, V.B. Smirnov (Eds.), Precision Physics of Simple Atomic Systems

Vol.628: R. Narayanan, D. Schwabe (Eds.), Interfacial Fluid Dynamics and Transport Processes

Vol.629: F. Sylos Labini, A. Gabrielli, L. Pietronero, Statistical Physics for Cosmic Structures

Vol.630: T. Brandes, S. Kettemann (Eds.), Anderson Localization and Its Ramifications

Vol.631: D. J. W. Giulini, C. Kiefer, C. Lämmerzahl (Eds.), Quantum Gravity, From Theory to Experimental Search

Vol.632: A. M. Greco (Ed.), Direct and Inverse Methods in Nonlinear Evolution Equations

Printing: Strauss GmbH, Mörlenbach
Binding: Schäffer, Grünstadt

GPSR Compliance

The European Union's (EU) General Product Safety Regulation (GPSR) is a set of rules that requires consumer products to be safe and our obligations to ensure this.

If you have any concerns about our products, you can contact us on ProductSafety@springernature.com

In case Publisher is established outside the EU, the EU authorized representative is:

Springer Nature Customer Service Center GmbH
Europaplatz 3
69115 Heidelberg, Germany

Batch number: 10584469

Printed by Printforce, the Netherlands